Amino Acids and the Asymmetry of Life

Advances in Astrobiology and Biogeophysics

springer.com

This series aims to report new developments in research and teaching in the interdisciplinary fields of astrobiology and biogeophysics. This encompasses all aspects of research into the origins of life – from the creation of matter to the emergence of complex life forms – and the study of both structure and evolution of planetary ecosystems under a given set of astro- and geophysical parameters. The methods considered can be of theoretical, computational, experimental and observational nature. Preference will be given to proposals where the manuscript puts particular emphasis on the overall readability in view of the broad spectrum of scientific backgrounds involved in astrobiology and biogeophysics.
The type of material considered for publication includes:

- Topical monographs
- Lectures on a new field, or presenting a new angle on a classical field
- Suitably edited research reports
- Compilations of selected papers from meetings that are devoted to specific topics

The timeliness of a manuscript is more important than its form which may be unfinished or tentative. Publication in this new series is thus intended as a service to the international scientific community in that the publisher, Springer-Verlag, offers global promotion and distribution of documents which otherwise have a restricted readership. Once published and copyrighted, they can be documented in the scientific literature.

Uwe Meierhenrich

Amino Acids and the Asymmetry of Life

Caught in the Act of Formation

With a Foreword by Henri B. Kagan

Springer

Prof. Dr. Uwe Meierhenrich
Université de Nice-Sophia Antipolis
Faculté des Sciences
CNRS UMR 6001
Parc Valrose
06108 Nice
France
uwe.meierhenrich@unice.fr

ISBN: 978-3-540-76885-2 e-ISBN: 978-3-540-76886-9

Library of Congress Control Number: 2008930865

Cover design: eStudio Calamar S.L.

Printed on acid-free paper

9 8 7 6 5 4 3 2 1

springer.com

for Eva

Foreword

A book on amino acids in the context of the origin of life and life's homochirality is unusual. Professor Meierhenrich has written a very detailed scenario on the attempts of scientists to propose hypotheses taking into account the stereochemical information given by the analysis of organic molecules. One key tool refers to the famous statement of Louis Pasteur in his lectures: "Optical activity is a signature of life".

The author presents many facets of the chemical and physical processes involved on earth or in space and he enters into a fascinating area of ongoing scientific debate. At times the diversity of the topics involved might be demanding for undergraduates or newcomers to the field, but the book, while challenging, illuminates the way science is progressing among a wide diversity of techniques, scientific areas, and hypotheses.

When I started preparing my Ph.D. thesis at the Collège de France in Paris under the supervision of Dr. Jean Jacques, I was attracted by the possibility of deducing the shape of molecules, including their absolute configuration, from experimental data, either spectroscopic or chemical. Organic chemistry remained strongly connected to pharmaceutical chemistry, where many compounds of biological interest are asymmetric and optically active. The manipulation of molecular models as Dreiding models (there were no computers at that time) allowed for analysing the three-dimensional behaviour of many rigid or mobile structures. Conformational analysis of cyclohexane and polycyclic systems such as steroids have been recently rationalized by D. H. R Barton and O. Hassel (Nobel prize winners in 1969). The prediction of the steric course of reactions that create a new asymmetric center became possible and was based on simple concepts. After my Ph.D., I was involved in a fruitful collaboration with Prof. Alain Horeau, Head of the Organic Chemistry Laboratory at the Collège de France. He proposed a chemical method to establish the absolute configuration of an asymmetric center of secondary alcohols RCHR(R')OH as well as deuterated primary alcohols RCHD(R)OH. I was associated with the latest developments of the Horeau's method. Later, I started, always in association with A. Horeau, to approach the difficult problem of efficiently creating an asymmetric center in an achiral molecule (asymmetric synthesis). For that purpose, we used chiral auxiliaries temporarily connected to a substrate or a reagent. Especially successful was an asymmetric synthesis that provided almost enantiomerically pure aspartic acid.

When I became scientifically independent, I launched two completely different projects simultaneously in Orsay. One was the exploration of asymmetric catalysis using chiral transition metals. The other was to revisit the old question of the use of a chiral physical agent such as circularly polarized light (CPL), already suggested by Pasteur and Le Bel in 1860–1874. In asymmetric catalysis, we made important discoveries in the 1970s by devising chiral ligands (C_2-symmetric chiral diphosphines) for rhodium complexes. These new catalysts allowed preparation of α-amino acids by asymmetric hydrogenation with up to 80% enantiomeric excess (*e.e.*), values which were the highest at that time. Nowadays, some enantioselective catalysts are used in industry to prepare chiral compounds with 99% *e.e.*

Our second project was more in line with the topic of the present book. The presence of circularly polarized light, a chiral agent, has been suggested to be present when the prebiotic soup started to produce biomolecules. Pasteur, Le Bel, van't Hoff, Cotton, Kuhn, and many others considered this possibility seriously, but it was W. Kuhn, who, in 1929–1930, first performed a partial asymmetric photolysis of a racemic compound (*N,N*-dimethyl α-azidopropionamide). The recovered material was optically active, although of very small *e.e.* This case of kinetic resolution was not followed by other examples until 1974, when we clearly showed that kinetic resolution of camphor can be carried out with CPL and afforded recovered camphor with 20 % *e.e.* at high conversion. The key factor for a successful experiment is to select a compound with a good *g* factor ($g = \Delta\varepsilon/\varepsilon$). Later several groups established that CPL can also be used to perform similar experiments on amino acids. In 1971, we described the first example of an asymmetric synthesis by using an achiral substrate and CPL. A photolytic cyclization gave rise to hexahelicene, which was slightly enriched in one enantiomer. This was easy to measure with a polarimeter because hexahelicene has a huge specific rotation. With regard to the question of whether CPL is able to produce optically active compounds, we learned that the answer is "yes". But what is its prebiotic significance? This question still remains unanswered.

We were also involved in a third project that may be of interest for prebiotic chemistry: the amplification of a small enantiomeric excess by catalytic or stoichiometric methods. In fact, we showed in 1986 that it is possible to use an enantioimpure catalyst to create a product of higher enantiomeric excess than the one of the catalyst. This effect was named the "nonlinear effect" and it can be useful in autocatalytic reactions. In 2007, we established experimentally, as proposed by I. Ugi in 1977, that some racemic reagents can increase the *e.e.* of partially resolved samples thanks to a selective destruction of some racemic content in the samples.

The present book discusses many ways to initiate or develop optical activity from achiral or racemic materials. The outlined space exploration and the study of chiral molecules are certainly of great interest to better understand the origin of homochirality of biological systems. The underlying problem, however, may remain unsolved for a long period of time, if not forever. The scientific subject is intellectually appealing, and it has great potential for attracting young people towards chemistry as well as the physical and natural sciences.

Orsay, France
May 2008

Henri B. Kagan

"Now, if you'll only attend, Kitty, and not talk so much, I'll tell you all my ideas about Looking-glass House. First, there's the room you can see through the glass – that's just the same as our drawing room, only the things go the other way. I can see all of it when I get upon a chair – all but the bit behind the fireplace. Oh! I do so wish I could see that bit! I want so much to know whether they've a fire in the winter: you never can tell, you know, unless our fire smokes, and then smoke comes up in that room too – but that may be only pretence, just to make it look as if they had a fire. Well then, the books are something like our books, only the words go the wrong way; I know that, because I've held up one of our books to the glass, and then they hold up one in the other room."

"How would you like to live in Looking-glass House, Kitty? I wonder if they'd give you milk in there? Perhaps Looking-glass milk isn't good to drink – But oh, Kitty! now we come to the passage. You can just see a little peep of the passage in Looking-glass House, if you leave the door of our drawing-room wide open: and it's very like our passage as far as you can see, only you know it may be quite different on beyond. Oh, Kitty! how nice it would be if we could only get through into Looking-glass House! I'm sure it's got, oh! such beautiful things in it!

Let's pretend there's a way of getting through into it, somehow, Kitty. Let's pretend the glass has got all soft like gauze, so that we can get through. Why, it's turning into a sort of mist now, I declare! It'll be easy enough to get through"– She was up on the chimney-piece while she said this, though she hardly knew how she had got there. And certainly the glass was beginning to melt away, just like a bright silvery mist.

In another moment Alice was through the glass, and had jumped lightly down into the Looking-glass room. The very first thing she did was . . .

Lewis Carroll in
Through the Looking-Glass, and What Alice Found There

Preface

The story of this book begins with the question "*How did life originate and why were left-handed molecules selected for its architecture?*" It is widely known that in processes triggering the origin of life on Earth, the equal occurrence, the parity between left-handed amino acids and their right-handed mirror images, is violated. The balance is inevitably tipped to the left – provoking that life's proteins today exclusively implement the left form of amino acids.

Why to the left?

If we rerun the tape of life's origin, would we once more obtain today's life using the left form of amino acids? Can't we think of mirror image life made of right-amino acids? Did mirror image life exist on Earth or does it exist somewhere else? What would mirror image life look like and how would it interact with the environment and with us? Puzzling questions of this kind contributed to the awakening of public interest in symmetry, the hypothetical existence of mirror image life, and its properties. Symmetry phenomena like these made also me curious and I began to separate molecules from their mirror image forms.

After several attempts, we were lucky to be engaged in the construction of the "Chirality-Module" of the ambitious cometary mission ROSETTA, an instrument designed to land on a comet and search for molecules and their mirror image forms. The idea is to trace the origin of life's left-handedness back to outer space.

Indeed, when we produced a comet analogue in the laboratory, we found a wide variety of amino acid molecules and their mirror image forms therein. The results supported not only the concept of the extraterrestrial delivery of prebiotic molecules triggering the appearance of life on Earth, now well-known and highly cited, but it provided a source of motivation in devoting an entire book to all the fascinating phenomena related to the asymmetry of life and its origin.

It is probably true that most scientists are solving more applied problems in biology, chemistry, physics, and mathematics. But we should always be alert to the great problems, such as understanding the processes leading to the origin of the one-handedness of life. Given past experience, such great questions will be solved – not by chance but by the prepared mind (Penny 2006). This book aims to be a starting point for the preparation of that mind by becoming a valuable resource for courses in chemistry, biology, biochemistry, and physics.

As Professor in Physical and Bioanalytical Chemistry, I was fortunate to work with and learn from Wolfram H.-P. Thiemann at the University of Bremen. He introduced me to the intriguing field of chirality, to theories on the origin of biomolecular asymmetry, and to the ROSETTA-COSAC community.

The international collaboration in my research field on the origin of life's molecular asymmetry was originally initiated by Helmut Rosenbauer at the Max Planck Institute for Solar System Research in Katlenburg-Lindau, Germany, via the ambitious conception and preparation of the COSAC-experiment onboard the ROSETTA Lander. My thanks go to Helmut Rosenbauer for this scientific support and his colleagues Fred Goesmann, Martin Hilchenbach, and Reinhard Roll for their fascinating basic work on COSAC and for plenty of conferences in the midst of fruitful discussions.

It was also a great pleasure to carry out scientific research within the international COSAC-team including Jan Hendrik Bredehöft, Jean-Francis Brun, Antonio Casares, David Coscia, Pascale Ehrenfreund, Guy Israel, Laurent Janin, Elmar K. Jessberger, Takekiyo Matsuo (†), Guillermo M. Muñoz Caro, François Raulin, Robert Sternberg, Cyril Szopa, Hermann Wollnik, and Sandrine Zubrzycki, who provided their full cooperation and scientific support. I also appreciate related discussions with Franz R. Krueger, Kensei Kobayashi, and Jun-ichi Takahashi.

I acknowledge Alexandra J. MacDermott, University of Cambridge, UK and Clear Lake University in Houston, Texas, for innovative ideas and suggestions with respect to the detection of an enantiomeric excess in samples of non-terrestrial origin. I am glad for a number of instructive discussions on the origin of biomolecular asymmetry with Dilip Kondepudi, Wake Forest University, North Carolina, for valuable comments of Thomas Buhse, Morelos State University in Cuernavaca, Mexico, on chiral symmetry breaking, and for the very helpful and enlightening discussions on the theoretical complexity of magneto-chiral effects with Laurence Barron, University of Glasgow, and Geert Rikken, Grenoble High Magnetic Field Laboratory. Many thanks go to Wilfried A. König (†), University of Hamburg. John R. Cronin and Sandra Pizzarello, both of Arizona State University, and Volker Schurig, University of Tübingen, are acknowledged for helpful suggestions and discussions.

Intensive experimental collaborations were performed with J. Mayo Greenberg (†) at the Raymond and Beverly Sackler Laboratory for Astrophysics at the Leiden Observatory and his group composed of Guillermo M. Muñoz Caro, Willem A. Schutte, and Almudena Arcones Segovia. I particularly acknowledge the pleasant cooperation with Greenberg's Ph.D. student Guillermo M. Muñoz Caro, now at the Astrobiology Center in Madrid, and his carefully performed experiments on the simulation of interstellar ices with isotopically labelled reactants.

I particularly acknowledge the generous support from the Centre de Biophysique Moléculaire in Orléans. In October 2000, I became a member of the prominent laboratory of André Brack and Bernard Barbier and I would like to thank them as well as the whole CBM-team, including Ms. Marylene Bertrand-Urbaniak, François Boillot, Corinne Buré, Annie Chabin, Romain Jacquet, Mai-Julie Nguyen, and Frances

Westall for generous support in the laboratory and in language instruction. I explicitly thank Paul Vigny, former director of CBM, and Maryse Fauquembergue for the marvellous and complete organization of my French studies.

Advanced experiments with circularly polarized light were performed at the Synchrotron Centers LURE and SOLEIL in Paris at beamlines SA-61, SU-5 and DESIRS. I would like to acknowledge the staff of these research centers and especially the substantial and advanced studies of Laurent Nahon, Christian Alcaraz, and Bertrand Pilette on both the generation and detection of circularly polarized synchrotron radiation. Thanks also go to Louis d'Hendecourt and Michel Nuevo from the Institut d'Astrophysique Spatiale in Paris-Orsay and to Martin Schwell from the Université Paris VII Denis Diderot. I also appreciate to work with Søren V. Hoffmann at beamlines UV-1 and CD-1 at Århus University, Denmark, in the Center of Storage Ring Facilities.

Thanks also go to Ray Wolstencroft, Royal Observatory at the University of Edinburgh, Scotland, for discussions and suggestions for the detection of circularly polarized light reflected from the planet Mercury by means of the BepiColombo mission of the European Space Agency (ESA).

I wish to thank Max P. Bernstein for his contributions and Jason P. Dworkin (both at NASA Ames Research Center, Moffet Field, California) for discussions on enantioselective chromatographic techniques for samples of interstellar ice analogues.

Moreover, I acknowledge lively discussions on the multifaceted chirality-phenomena with all members of my research team at the University of Nice-Sophia Antipolis namely Nicolas Baldovini, Katharina Breme, Cécilia Castel, Céline Delasalle, Jean-Jacques Filippi, and Rodolphe Perriot.

The author thanks the Deutsche Forschungsgemeinschaft (DFG), Bonn, and the Agence Nationale de la Recherche (ANR), Paris.

And I would like to thank the well-disposed reader not only for following the ideas and concepts exposed in this book, but also for feedback, frank criticism, and suggestions.

Nice, Cap d'Ail, France,
June 2008

Uwe Meierhenrich

Contents

List of Symbols and Abbreviations

Symbols and abbreviations were used according to the list below. In some cases, such as α or Z, one symbol has more than one meaning. The proper meaning becomes clear from the context in which the symbol is used.

α	Separation factor and optical rotation angle
α	Prefix of an amino acid carrying the amino function at the carbon atom in α-position and amplitude for asymmetry
a	Stereodescriptor for axial chirality and semimajor axis of elliptically polarized light
$[\alpha]$	Specific rotation
a_0	Non-helical term in Moffit-Young equation
A	Absorption
aeg	Aminoethylglycine
AIB	α-Aminoisobutyric acid
a.m.u.	Atomic mass unit
a.u.	Atomic unit
AU	Astronomical unit
β	Prefix of an amino acid carrying the amino function at β-carbon and decay process of unstable atomic nuclei
b	Semimajor axis of elliptically polarized light
b_0	Helical term in Moffit-Young equation
B	Magnetic field vector
BZ	Belousov-Zhabotinskii
C_1	Non-axial point group
CD	Circular dichroism and cyclodextrin
CG	Comet 67P/Churyumov-Gerasimenko
CGS	Unit-system based on cm, g, and sec
c.i.i.	Cyclic immonium ion
CIP	Cahn-Ingold-Prelog
C_n	Rotational group
COSAC	Cometary sampling and composition experiment

D	Absolute configuration of an enantiomer related to D-glyceraldehyde
Δ	Prefix of a chiral octahedral system
da	Diamino acid
DAH	Diaminohexanoic acid
DAP	Diaminopentanoic acid
d.l.	Detection limit
DMA	Dimethylacetal
DMF	Dimethylformamide
D_n	Rotational group
ε	Molar extinction coefficient
E	Configuration of stereoisomers and electric field vector
e.e.	Enantiomeric excess
ECEE	Ethoxycarbonyl ethylester
en	Ethylenediamine
ΔE^{PNC}	Parity non-conserving energy difference
ESR	Electron spin resonance
FKN	Field-Körös-Noyes
$f_n(t)$	Normalized Gaussian white noise
FTT	Fischer-Tropsch type
g	Anisotropy factor
GC-MS	Gas chromatography-mass spectrometry
GCxGC	Multi-dimensional gas chromatography
h	Planck constant
H	Hamiltonian operator
HFIP	Hexafluoroisopropanol
HIV	Human immunodeficiency virus
HMT	Hexamethylenetetramine
I	Stokes parameter describing the total intensity of electromagnetic radiation
I_0	Light intensity
ID	Inner diameter
ISO	Interstellar Space Observatory
ITMS	Ion trap mass spectrometer
k	Rate constant and Boltzmann constant
K	Equilibrium constant
l	Prefix of a laevo rotatory crystal
l	Axis of coordinate system to determine the Stokes parameters
λ	Distance from the chemical equilibrium
λ_C	Critical wavelength
Λ	Prefix of a chiral octahedral system
L	Absolute configuration of an enantiomer related to L-glyceraldehyde
LB	Linear bifringence
LCPL	Left-circularly polarized light

LD	Linear dichroism
LURE	Laboratoire pour l'utilisation du rayonnement électromagnétique
μ	Electric dipole moment
μCE	Micro-capillary electrophoresis
m	Magnetic dipole moment
M	Prefix of a chiral helical structure
$[M]$	Specific molar rotation
MALDI	Matrix assisted laser desorption ionisation
MCD	Magnetic circular dichroism
MC	Magnetochiral dichroism
MOMA	Mars organic molecule analyser
MORD	Magnetic optical rotation dispersion
M_{θ}	Molar ellipticity
MW	Molecular weight
$\bar{\nu}_e$	Antineutrino
NCA	N-Carboxyanhydride
OMC	Orion molecular cloud
ORD	Optical rotation dispersion
π	Molecular orbital
φ	Torsion angle and phase angle
Φ	Quantum yield
Φ	Torsion angle
[Φ]	Molecular rotation
P	Prefix of a chiral helical structure
$P1$	Triclinic space group
$P2_1$	Chiral space group
$P2_12_12_1$	Chiral space group
$P2_13$	Achiral cubic space group
Pa3	Achiral cubic space group
Pca2_1	Achiral orthorhombic modification
PEM	Photo-elastic modulator
Phen	1,10-Phenanthroline
PNA	Peptide nucleic acid
PNC	Parity non-conserving
p-RNA	Pyranosyl-RNA
PVED	Parity-violating energy difference
Q	Stokes parameter
r	Axis of coordinate system to determine the Stokes parameters
R	Absolute enantiomer configuration according to CIP notation
$R\bar{3}c$	Rhombohedral space group
RCPL	Right-circularly polarized light
ROA	Raman optical activity
R_S	Chromatographic resolution
σ	Molecular orbital

s	Propagation direction of light
S	Absolute enantiomer configuration according to CIP notation
S_{0-3}	Stokes parameters
SIM	Single ion monitoring
S_N	Fractional circular polarization of a northern hemisphere
S/N	Signal-to-noise
S_S	Fractional circular polarization of a southern hemisphere
SU-5	Synchrotron beamline at LURE
θ	Ellipticity
$[\theta]$	Specific ellipticity
t	Retention time
t'	Reduced retention time
T	Tesla
TFA	Trifluoracetyl
TIC	Total ion current
t_M	Hold-up time
TNA	Threofuranosyl nucleic acid
TOF	Time-of-flight
U	Stokes parameter
UKIRT	United Kingdom Infrared Telescope
V	Stokes parameter
vCD	Vibrational circular dichroism
V^{Coul}	Coulomb interaction
V^{PNC}	Weak part of the binding energy
VU	Vester-Ulbricht
VUV	Vacuum ultraviolet
$W^{\pm}$	Charged massive boson
w_b	Base width
WCOT	Wall-coated open-tubular
ξ	Extend of reaction
z	Charge
Z	Proton number and configuration of stereoisomers
Z^0	Neutral vector boson

Chapter 1
Tracing Life's Origin: From Amino Acids to Space Mission ROSETTA

1.1 The Origin of Life

The origin of life is one of the unsolved fundamental mysteries in modern natural science. The endeavour to discover some of its secrets continues to challenge researchers from many different scientific domains such as astrophysics and planetary sciences, geology, mineralogy, geochemistry, organic, inorganic, analytical, and physical chemistry, biophysics, biochemistry, biology, and even mathematics. Not to forget the profound interest of philosophers, writers, theologians, and artists in a wide spectrum of aspects on the scientific basics of our origin. How could life originate? In spite of the lack of the whole process, some clear-cut steps in the evolutionary origin of life seem to be established and common sense, today.

We assume, in brief, that before biological evolution more than 3.5 billion years ago defined phases of chemical evolution started with the synthesis of simple prebiotic organic molecules, such as amino acids, either on Earth or in interstellar environments. Subsequent prebiotic reactions are assumed to require liquid water to form the first polymers. After the molecular selection of a self-replicating system and the so-called RNA-world scenario (Gilbert 1986; Joyce 2002) the DNA-protein-world led via the last universal common ancestor[1] to the starting point of biological evolution.

Despite this knowledge, we are far from a coherent understanding of the life-originating phases chronologically separated into cosmic, chemical, and biological evolution. It is, in particular, the incapability to link chemical and biological evolution that continues to puzzle scientists. How can a mix of non-living chemicals be transformed into something as complex as the living cell? Is there some intrinsic aspect of the structure of the involved molecules such as amino acids that will enable us to conclude whether their origin was biological or chemical? Were they active precursors of living structures or "junk" molecules not used up by evolutionary

[1] The last universal common ancestor (LUCA) is the first living organism from which all currently living organisms, including Archaeans, Eukaryotes, and Bacteria, descended (Woese 1998). Tracing the phylogenetic tree backwards by ribosomal RNA sequence comparisons derives its hypothetical nature.

U. Meierhenrich, *Amino Acids and the Asymmetry of Life*. Advances in Astrobiology and Biogeophysics, © Springer-Verlag Berlin Heidelberg 2008

processes? The answer is "yes", each amino acid bears a fingerprint inside itself telling us the story of its origin. This fingerprint is carried and manifested by the chirality of the amino acid.

1.2 The Asymmetry of Life

This book is aimed at catching this fingerprint and discovering its evolutionary origin. We will not review *all* the general aspects of the origin of life,[2] but we will focus on one particularly fascinating and crucial period therein, namely the emergence of chirality. More scientifically spoken it is the emergence of "biomolecular homochirality". I will present cutting edge research on the elucidation of the ultimate cause of the "homochirality" phenomena on Earth and the possible correlation with physicochemical parameters. This first chapter will serve as a general, unifying introduction to all these phenomena, letting us develop an overview of the manifold aspects of "biomolecular homochirality". The upcoming chapters will systematically go through all these phenomena in more detail.

So, what are we talking about when we say "biomolecular homochirality"? Life is not symmetric; the phenomenon of asymmetry can well be observed in living organisms commencing with microorganisms, to plants, animals, and even human beings. Just to give two examples from everyday life: soccer players prefer to use their right or left food to shoot, and people favour to write distinguishingly either with their right or left hand. In scientific terms, biochirality describes the phenomenon that organisms display laterality. Here, symmetry is obviously broken. If we imagine living beings in a mirror-world their mirror-world properties would be non-identical to their real world properties.[3]

But does this well-known macroscopic phenomenon also exist on the microscopic molecular level? "Yes"! A recent advertisement for "right-handed yogurt" has introduced the general public to the concept that many biomolecules exist in distinguishable "right-" and "left-handed" versions. Molecules or parts of molecules can be structured as their image or mirror image. Image and mirror image molecules are often not identical and can possess different biological properties. In the frame of this book we will hence reflect on mirrors and their effects on molecules, we will discuss properties of mirror image molecules and speculate on the existence of mirror-image worlds including our difficulty in accessing them.

Back to the real world's molecular level: the basic building blocks of life are mostly found in exclusively one out of two principally possible mirror image

---

[2] A selection of recent literature on general aspects on the origin of life contains Brack (1998), Smith and Szathmáry (1999), Ward and Brownlee (2000), Schopf (2002), Adams (2002), Davies (2003), Rauchfuß (2005), and Luisi (2006).

[3] This phenomenon is witnessed, for example, by the high rate of accidents of central European drivers on left-running highways on the British Islands.

forms, designated as D or L.[4] Natural amino acids, the components of proteins (enzymes), are found in the L-form such as L-alanine, L-valine, L-leucine, L-aspartic acid, and others. Sugar molecules are – in contrast to this – biosynthesised in the right-form for implementation into nucleic acids. D-ribose and D-deoxyribose are used in ribonucleic acid RNA and deoxyribonucleic acid DNA respectively, and D-glucose in glycogen, starch, and cellulose. The mirror image structures of amino acids (called D-enantiomers) are not tolerated for the molecular architecture of proteins; in a similar selective way, mirror image sugars (L-carbohydrates) do not contribute to the molecular construction of the nucleic acids RNA and DNA. In natural sciences, this widely distributed phenomenon is called biomolecular asymmetry.

In other words, living organisms are based on biopolymers that are strongly selective towards the chirality of their monomer subunits. Monomers of biopolymers are almost exclusively homochiral. Homochirality is a property of matter, which is made up of only one "*hand*" (that is what is literally meant by "*cheir*" in the ancient Greek language) out of the potentially two "*hands*" available within confined boundaries of abundance. The boundary thus defined by this homochirality versus symmetry could even be interpreted as the borderline between living organisms here and non-living matter there. As we will see later, the intrinsic symmetry aspect of a chiral molecular structure will be used by future space missions, for example the upcoming ExoMars mission, as an elegant way of finding life traces somewhere else in the universe.

Also flavour and fragrance ingredients serve as typical examples of molecular asymmetry. Taste, scent, and other physiological properties can differ between right- and left-handed molecules. One can easily use one's nose and establish that enantiomers can smell different. Right-handed carvone for example shows the odour of caraway, whereas its mirror image, the left-handed carvone, disposes an odour of spearmint. Right- and left-handed limonene (orange versus lemon odour) and right- and left-handed menthol are other examples of differences in organoleptic properties of enantiomers. This is not surprising, because – as we have learned above – proteins of the olfactory system are composed of homochiral L-amino acids linked together via peptide bonds. The highly asymmetric arrangement of the odorant receptor protein is fine-tuned to distinguish between odorants in general and right- and left-handed odorants in particular. In addition to odour-reception, our taste is concerned by this homochirality phenomenon in a way that *S*-asparagine tastes bitter whereas the mirror image molecule called *R*-asparagine tastes sweet (Quack 1993). The fact that two enantiomers can have such different functions illustrates that stereochemistry controls whether, and how, molecule and receptor recognize one another and "shake hands" (Siegel 2001).

A dramatic example of the importance of the homochirality phenomena is the chemical thalidomide, in Europe better known under the trade name contergan. After medicinal application of the mixture of the *R*- and the *S*-form, it became dramatically apparent that the *S*-form of this drug is able to cause severe teratogenic effects, whereas its mirror image, the *R*-contergan, showed tranquillising properties.

---

[4] The prefix D is derived from the Latin *dexter*, right; the prefix L is deduced from *laevus*, left. Chap. 2 will introduce stereochemical terms systematically.

Today, the useful medicinal properties of *R*-contergan are employed against leprosy, HIV, and several types of cancers.[5] In fact, differences between mirror image forms concern numerous medicinal drugs. *S*-penicilliamine serves as an antiarthritic drug whereas *R*-penicilliamine is known to be highly toxic (Quack 1989).

We can conclude from the above paragraphs that the use of homochiral monomers is considered vital for the construction of biopolymers and must therefore be considered a fundamental step in chemical evolution towards the origin of life. The absolute homochirality of "simple", non-complex organic molecules is a prerequisite for an efficient self-replication of the first living systems. As a consequence, numerous examples of evolving structures and systems, which develop asymmetrically in timescales that can be observed in the laboratory, have been studied in great detail. However, despite a manifold of research activities, the underlying question is not yet fully understood: how did life select the left-handed form of amino acids for the construction of its proteins? Why did living organisms not opt for the right-handed form of amino acids? Was this a random or a determinate process at life's origin? In spite of this well known violation of molecular symmetry it is astonishing that so far there is no one convincing theory yet explaining the cause of the driving force behind the long process of evolution of such a biosphere as we observe today on Earth – astonishing the more while searching through the literature, where a large number of papers of more theoretical or experimental nature are devoted to this very essential problem (Thiemann 1998).

Would we get something a bit different if we wind back the tape of life again? Will there again be a mix of plants, animals, and humans? Certainly, we are unlikely to get a creature using heterochiral biocatalytic (enzymatic) and genetic systems. But focusing on homochiral systems only, we may ask whether life would have again selected L-amino acids for the construction of enzymes or whether a mirror world on the molecular level would have been possible as well. This will be one of the great, interesting questions discussed in this book.

Our aim will be to spread out the different attempts in detail to uncover the secrecy behind the phenomenon, why we see living nature in this one-handed form only today. We will collect and present information on the origin of biomolecular asymmetry, a mysterious phenomenon still that obviously continues to receive tremendous attention in science, as it is manifested by a constant high number of letters published in the high-impact journals Nature and Science. Strong and ongoing scientific debates on the international top-level will be outlined involving contributions from various scientific disciplines.

Furthermore, as scientists we are interested in questions like: what are the next experiments? Are the chemical reactions that determined life's handedness reproducible in the lab? Are fundamental stages assumed to have occurred in atmospheric, hydrothermal, or even extraterrestrial and interstellar environments?

---

[5] The medicinal application of a specific thalidomide enantiomer is more complicated than described above. The use of the pure *R*-thalidomide form that shows the useful effects is hampered by its slow interconversion (racemization) in the human body to harmful *S*-thalidomide. Consequently, thalidomide is not given to pregnant women today, in order to avoid the teratogenic potential of its *S*-enantiomer.

Are concentrations, temperature, pressure, pH values, and other physico-chemical parameters known to reproduce this crucial event? This book aims to contribute to the better understanding specifically of these questions by giving the answers in the appropriate chapters.

1.3 How Life's Asymmetry Originated

Theories on the origin of life's handedness are to a specific degree of a hypothetical character, as the evolution that lead to its formation cannot be repeated in a 1:1 experimental timescale. But in the past few decades our idea of how biomolecular homochirality evolved has advanced from highly speculative musing to a true scientific theory with a firm base in hard facts. Precise environmental conditions able to generate this essential biomolecular parity violation are of particular interest.

Primarily, we are confronted with the alternative between "terrestrial" theories that describe the process resulting in biomolecular asymmetry as evolutionary reactions which took place on early Earth during specific steps of chemical and/or biological evolution and "extraterrestrial" models that use chiral agents from outside Earth to explain the origin of biology's homochirality. Second, homochirality related research distinguishes between random mechanisms (Sect. 1.3.1) on the one side and determinate mechanisms (Sect. 1.3.2) on the other side (Bonner 1991). Furthermore we might ask: was the choice between left and right internally imposed by the system or externally imposed by some peripheral driving force? The main theories that have been developed to explain an abiogenic origin of the biomolecular chiral purity in terms of the physico-chemical processes involved are summarized in Fig. 1.1. They are highly controversial.

1.3.1 Random Mechanisms

Random mechanisms interpret the appearance of biomolecular asymmetry by a mere chance process inside the system. This process is undirected towards the chiral configuration of the produced or affected chiral molecules. Experimental studies on random mechanisms were often realized in the laboratory.

Examples include the asymmetric precipitation from super-saturated solutions: an aqueous solution of the amino acid asparagine containing its two mirror-image forms called enantiomers in a 50/50 racemic ratio was crystallized at various temperatures resulting in an enantiomeric excess in the precipitate and the opposite excess in the remaining aqueous solution (Thiemann 1974; Thiemann and Darge 1974). Here, a small asymmetry in an elementary nucleation effect was amplified in macroscopic crystals where it could be observed experimentally (see also Thiemann and Wagener 1970). The obtained enantiomeric excess is non-predictable

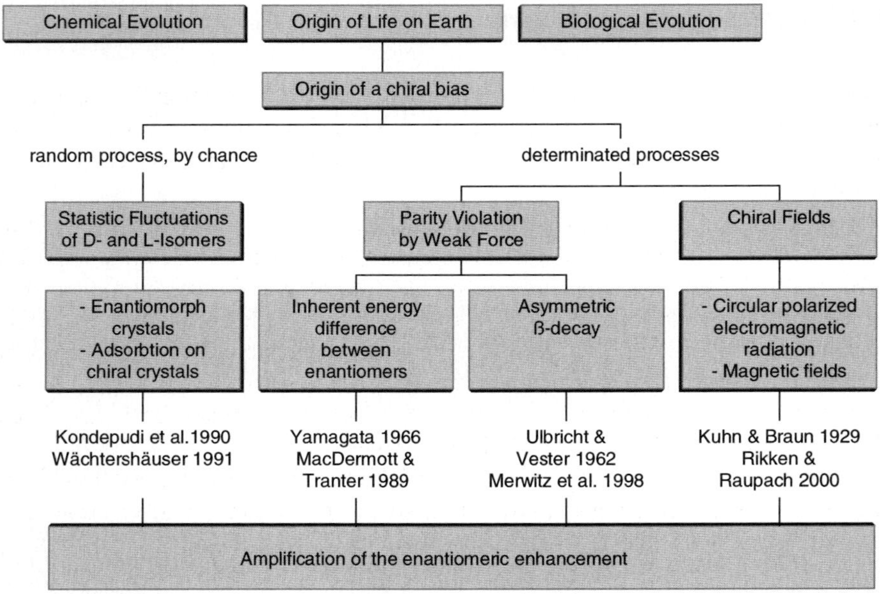

Fig. 1.1 Theories for an abiogenic origin of life's chiral bias. One distinguishes between random processes (*left branch*) and deterministic processes (*right branch*). Enantiomeric excesses produced by these models are considered generally to be small and insufficient for the evolutionary generation of the first biomolecules; experimental approaches for their amplification substantiated by mathematical models have been developed

and random. The asymmetry-creating crystallization was recently reproduced, adapted, and actualised for the amino acid tyrosine (Shinitzky et al. 2002), leaving higher concentrations of L-tyrosine in solution.

The spontaneous chiral symmetry breaking in a similar autocatalytic crystallization system caused a sensation in the scientific community (Kondepudi et al. 1990; Kondepudi and Asakura 2001) that will be systematically presented in Chap. 4. In these and similar scenarios the amplification of relatively small enantiomeric excesses by suitable mechanisms is of high importance. Stereospecific autocatalysis was proposed as such a process of amplifying capacity: the presence of one enantiomer encourages the generation of the same one but inhibits automatically the synthesis of its corresponding mirror image, leading ultimately to a state of homochirality (Frank 1953). Also, this process is to be classified as a random mechanism, since the initial small asymmetry tipping the balance of the whole system towards homochirality occurred just by chance. Promising mathematical formalisms of reaction sequences including an autocatalytic step (Kondepudi 1987), their experimental models (Buhse et al. 1993a,b), and successful autocatalytic amplification experiments observed in chemical reactions (Soai et al. 1995) were proposed and tested and will be outlined in the appropriate chapters (4 and 10) later in this book.

1.3.2 Determinate Mechanisms

Determinate mechanisms create predictable non-random enantiomeric enhancements. They involve the interaction of either internal or external physical driving forces – cogencies – with racemic or prochiral substances that are causing prevalence of one enantiomer.

Examples of such deterministic scenarios include the adsorption of organic molecules on the enantiomorph surface of quartz (Karagunis and Coumoulos 1938; Soai et al. 1999), or calcite (Hazen et al. 2001) as we will discuss later. Here, the resulting enantioenrichment in the organic molecules is determined by the crystal's asymmetry in a way that chiral information is transferred from the crystal's asymmetric surface towards initially racemic organic molecules via host-guest interactions. The asymmetric crystal's host surface serves as an external driving force determining the chirality of the organic guest system.

Nowadays, a directed vortex in a liquid is also to be considered an external driving force since this vortex was somehow surprisingly proven to induce an asymmetry into an arrangement of organic molecules. It was indeed the direction of stirring in terms of left or right revolution that determined the left or right helicity of a spiral association of compounds in a predictable manner (Ribó et al. 2001).

Other driving forces that determine the outcome of an enantiomeric excess in a non-random process include directed magnetic fields, the weak force as an internal driving force (Yamagata 1966), spin-polarized electrons (Musigmann et al. 1999), and circularly polarized light (Kuhn and Braun 1929; Kuhn and Knopf 1930) that is capable of inducing asymmetries into racemic organic molecules under particular interstellar conditions. Before describing this "extraterrestrial" scenario, I will briefly introduce the most important alternative physical driving forces.

1.3.2.1 Magneto-Optical Effects

Oriented magnetic fields have been suggested as physical driving forces, capable of inducing enantioenrichments into racemic or prochiral organic molecules. Here, the direction of the magnetic field vector applied to a chemical reaction would determine the stereochemical configuration ((+)- or (−)-enantiomers) of the products. In order to figure out such an intriguing possibility of magneto-optical effects, photoreactions were studied in strong directed magnetic fields in which the incoming photons disposed of a fixed orientation relative to the magnetic field. To give an example: hexahelicene, a chiral molecule, was produced photochemically from the *cis*-phenanthrylnaphthyl-ethene precursor within a homogeneous static magnetic field of 1.1 T, resulting in an enantiomeric excess of 0.07% hexahelicene. Interestingly, the configuration of the photoproduct hexahelicene was determined by the direction of the magnetic field vector. Liquid crystals were often used as hosts for such asymmetric photoreactions (Stöbel 1985; Teutsch and Thiemann 1986; Teutsch 1988; Thiemann and Teutsch 1990). For decades, enantiomeric excesses caused only by magneto-optical effects had been small ($<1\%$) and often hard to detect.

In one spectacular case, however, a tremendous enantiomeric excess of 98% was reported. Ph.D. student Guido Zadel performed an enantioselective reaction in a static magnetic field at the University of Bonn. Chiral arylethanols were synthesized starting from prochiral ketones. The resulting enantiomeric excess (20–98%) of arylethanols was published to be a linear function of the applied magnetic field ranging from 0.2 to 1.2 T (Zadel et al. 1994). Other laboratories that immediately investigated this spectacular finding were not able to reproduce Zadel's results (Feringa et al. 1994). Finally, the manuscript had to be withdrawn due to intentional "manipulation" of the data (Gölitz 1994).

A significant enantioselective magnetochiral effect was observed by Geert Rikken and Ernst Raupach (2000) that caused the introduction of enantiomeric excess into a prior racemic mixture. A photoreaction was used showing an enantioselective capacity that was dependent on the direction of the magnetic field vector. The handedness of the enantiomeric excess achieved in these experiments was induced by the relative orientation of the non-circularly polarized light towards the magnetic field. May such a magnetochiral anisotropy have been decisive for life's selection of its enantiomerically pure molecular building blocks? The answer will be given in Chap. 6, where we will see that the Earth's magnetic field is, referring to the authors' conclusion, too weak and periodically changing over geological time periods its direction, to have a suitable effect. In order to explain the origin of biomolecular dissymmetry Rikken and Raupach considered extraterrestrial magnetic fields for introducing an enantiomeric excess into organic racemates.

1.3.2.2 Biochirogenesis Based on the Weak Force

Originating in theoretical physics and elementary particle physics, a particularly intriguing explanation for a deterministic origin of biomolecular asymmetry continues to receive international attention: inside atomic nuclei, a tiny intrinsic asymmetry has been identified. This inner-nuclear asymmetry was proposed to influence chiral molecules such as amino acids from their own innermost interior serving as an internal deterministic driving force having favoured today's biomolecular homochirality of living organisms.

This explanation finds its origin in 1956, when Tsung Dao Lee and Chen Ning Yang discovered the violation of the parity law; they found out that the weak force is – among the four fundamental forces – an asymmetric one. The Nobel Prize in Physics honoured this discovery in 1957, as we will see in Chap. 5 referring to the herewith-triggered worldwide discussions among Wolfgang Pauli, Richard P. Feynman, Freeman J. Dyson, and others. The described inherent asymmetry of atomic nuclei was later assumed to cause "parity non-conserving energy differences" between enantiomers (Yamagata 1966), which themselves were proposed by a cluster of researchers mainly originating from Oxford University and London King's College to explain the optical asymmetry of the biosphere (Mason 1984; Tranter 1985b; MacDermott and Tranter 1989).

The first experimental evidence for the above hypothesis was claimed by Otto Merwitz (1976) and Bengt Nordén et al. (1985) investigating a radiolysis favouring the selective degradation of D-amino acids compared to their L-enantiomers. These studies used spin-polarized ("chiral") electrons emitted as β-radiation for the interaction with racemic amino acids. We will see later how β-radiation is indeed influenced by the weak nuclear force. In addition to this, carefully measured differences in the chiroptical properties of octahedral cobalt and iridium enantiomers were observed by Andrea Szabó-Nagy and Lajos Keszthelyi (1999). Also, these authors considered the recorded differences between enantiomers as being caused by parity non-conserving energy differences. These and other experimental trials to ultimately evaluate parity non-conserving energy differences will be exposed in Chap. 5.

As a consequence of this attractive and far-reaching theory based on the asymmetry of the weak nuclear force it has consequently been acknowledged that even atoms are principally optically active and that L- and D-molecules are truly diastereoisomers instead of enantiomers in the strict sense of the term (MacDermott 1993). This is a consequence that would indeed require considerable rewriting and reediting of the majority of stereochemistry textbooks.

Nevertheless, some authors question this theory because there is little experimental evidence for parity non-conserving energy differences (Bonner 1995a). This is not surprising in view of the extremely small magnitude of these calculated energy differences (in the order of 10^{-14} eV), that is considered probably too small to break the racemic state of the environment (Keszthelyi 1984; Goldanskii and Kuz'min 1988).

1.3.2.3 Photochirogenesis[6]

In contrast to the majority of nuclear physicists, photochemists may favour involving "chiral photons" in processes explaining the origin of biomolecular asymmetry. From the experiments of Werner Kuhn and E. Braun (1929) and W. Kuhn and E. Knopf (1930), it is a well known photochemical fact that the interaction of circularly polarized ultraviolet light with chiral but racemic organic molecules is able to introduce an enantiomeric excess into these racemates by disrupting the better absorbing enantiomer, resulting in an excess of the other. Among the above mentioned determinate mechanisms, particular processes involving circularly polarized light as the external chiral force have proved capable of producing significant and reproducible enantiomeric excesses from racemic or prochiral precursors. Nevertheless, these enantioselective photoreactions suffer in principle from two disadvantages in explaining satisfactorily the origin of biomolecular homochirality (Bonner 1995a). On the Earth's surface the circularly polarized light is extremely weak in a) intensity and b) handedness in the ultraviolet and visible region of the spectrum. Due to the discovery of both circularly polarized light in interstellar regions and prochiral

[6] In this context, the term "chirogenesis" is used to describe the induction of an enantiomeric enrichment into a physico-chemical system composed of prochiral or chiral molecules. An enantiomeric enrichment is generated; the system was composed of chiral (but racemic) molecules before, even if we talk about "chirogenesis".

and chiral organic molecules in meteorites, interstellar ices, and simulated comets, a renewed interest in extraterrestrial scenarios for a circularly polarized light-induced origin of molecular parity violation on Earth has recently arisen (Griesbeck and Meierhenrich 2002).

1.3.3 Chirogenesis in Outer Space

This section will briefly introduce the context in which interstellar chirogenesis is of interest. Chap. 6 will then summarize modern concepts and advances in the field of asymmetric photochemistry in interstellar space, and discuss our current understanding of the underlying mechanisms involved.

1.3.3.1 Interstellar Formation of Chiral Molecules

The chemical generation of simple prebiotic organic molecules from inorganic precursors has important implications for the understanding of the crucial first steps of Chemical Evolution. The topical discussion focuses on the question of whether these processes occurred in the atmosphere or hydrosphere of the early Earth or whether the required prebiotic organic molecules were spontaneously generated under interstellar conditions including their subsequent delivery to the early Earth via (micro-) meteorites and/or comets. In spite of the fact that neither any single process, nor any single source of energy is likely to account for all the organic compounds on the primitive Earth, this book will present arguments for an interstellar origin of prebiotic amino acid structures and critically review the main alternatives.

Until very recently, prebiotic amino acids were believed to have been mainly generated in the atmosphere of the early Earth, as successfully simulated by the Urey-Miller experiments (Miller 1953). More recently, two independent groups, one by Max Bernstein (Bernstein et al. 2002) of NASA Ames in the US and the other by our team from European universities and institutes (Muñoz Caro et al. 2002), put forward another scenario (Shock 2002). This involves chemical reactions on small ice grains that develop in the interstellar medium as it is illustrated in Fig. 1.2. In the laboratory, ultraviolet irradiation of ice mixtures containing well-known interstellar molecules (such as H_2O, CO_2, CO, CH_3OH, and NH_3) in conditions of vacuum and low temperature found in the interstellar medium generated amino acid structures including glycine, alanine, serine, valine, proline, and aspartic acid. After warm-up, hydrolysis, and derivatization, 16 amino acids as well as furans and pyrroles were identified. With the exception of glycine, all of these amino acids were chiral, but not homochiral. A more detailed presentation of these amino acid structures including comments on constitutional isomers, their chirality and the identification of a "new" family of amino acids namely diamino carboxylic acids will be given in Chaps. 7 and 8. We will discuss the mechanism of their formation and comment on the hydrolysis step required to release "free" chiral amino acids.

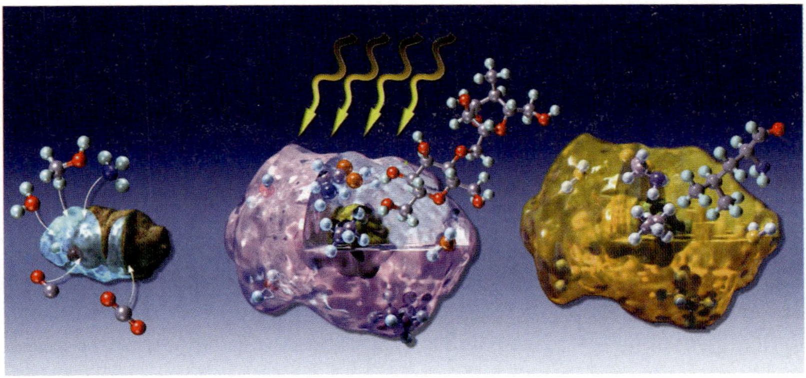

Fig. 1.2 Defined developing phases of interstellar dust particles in dense and diffuse interstellar clouds. First, in the dense interstellar medium tiny silicate grains of sub-micrometer diameter accrete ice layers that contain simple organic molecules such as H_2O, CO_2, CO, CH_3OH, and NH_3 (*left*). During growth process the ice layers are irradiated with energetic UV-photons in the diffuse interstellar medium (*center*) resulting in photoreactions that form radicals and polymers of yellowish color including new organic molecules like amino acids (*right*). These amino acids are chiral. The aggregation of such dust particles occurs as a result first of grain-grain collisions and then by collisions between aggregates leading to intermediate cometesimals and ultimately to comets. Illustration: Stéphane Le-Saint, Université de Nice-Sophia Antipolis

Today, one assumes that dust particles, such as those depicted in Fig. 1.2, serve as containers for chiral amino acids. These particles coagulate forming eventually larger bodies like comets. The chiral products of interstellar photochemical reactions could be preserved in these bodies, and in time be delivered to the Earth during the heavy bombardment geological period, which ended about 3.8 billion years ago. The comet components were heated during the atmospheric entry and impact with the planet's surface. The delivered chiral organic molecules might have played an important role in the appearance of primitive life and the emergence of biomolecular asymmetry on Earth. A prebiotic interstellar origin of amino acid structures is now accepted to be a plausible alternative to the Urey-Miller mechanism.[7]

Today, we assume that chiral organic molecules do form in interstellar environments. This formation has successfully been simulated in the laboratory in the case of amino acids. Furthermore, different meteorites have been analyzed and – supporting the above model – chiral organic molecules have been detected therein as well. A detailed description of meteoritic amino acids and diamino acids will be given in Chap. 8; here, the occurrence of chiral molecules in meteorites will be overviewed briefly.

[7] In personal discussions during a conference on Chemical Evolution and the Origin of Life in the Abdus Salam International Center for Theoretical Physics in Triest, Italy, 2003, the late Stanley Miller conceded that based on scientific argumentation today one should not assume that all prebiotic amino acids were formed either in the Early Earth's atmosphere or in interstellar environments. One should rather ask for the quantities of amino acids formed on Early Earth versus in interstellar regions. According to S. Miller, they were formed here and there.

1.3.3.2 Chiral Molecules in Meteorites

The detection of chiral molecules and the identification of eventual enantioenrichments in extraterrestrial samples such as meteorites have always been considered crucial for the understanding of the origin of biomolecular homochirality. Are chiral molecules present in meteorites and can one resolve and quantify single enantiomers therein?

Early reports on the analysis of carbonaceous chondrites – a particular family of meteorites showing high amounts of organic compounds – claimed the identification of chiral amino acids. However, these studies often examined terrestrial contaminations only (e.g. Oró et al. 1971; Engel and Nagy 1982). Nowadays, they have been superseded by experimental work employing better samples and more powerful analytical techniques (Cronin and Chang 1993). Most of this work has focused on the Murchison meteorite, which fell spectacularly in several pieces close to Murchison village in Australia in 1969. This meteorite belongs to the class of the carbonaceous chondrites. The high carbon content of carbonaceous chondrites is largely macromolecular but also contains a complex mixture of organic compounds. Some of these contents can be chiral such as carboxylic acids, dicarboxylic acids, amino acids, diamino acids, hydroxy acids, amines, alcohols, and others.

The breakthrough in the successful enantioselective detection of chiral amino acids in the Murchison meteorite was in 1997. It was performed simultaneously by independent research teams in the US and reported twice (Engel and Macko 1997; Cronin and Pizzarello 1997; see also Chyba 1997). Particularly for α-methyl amino acids such as isovaline enantiomeric excesses of up to 15% were reported. Enantiomeric excesses of classical (α-H) amino acids such as alanine, valine, leucine etc. were examined up to 3%. In all these enantioselective analyses of carbonaceous chondrites L-amino acids were found in excess towards their D-counterparts.[8] This finding is in intriguing coherence with the fact that biological organisms such as plants, animals, and human beings exclusively use L-amino acid enantiomers for the construction of their proteins.

As mentioned above, α-methyl amino acids were identified in the Murchison carbonaceous chondrite showing higher enantiomeric excesses compared to α-H amino acids. A higher resistance of α-methyl amino acids to racemization could explain this observation. Here, the observed enantiomeric excesses likely represent the original state of these α-methyl amino acids, whereas the more easily racemized α-H amino acids could have initially existed in non-racemic proportions too, but racemized during alteration of the carbonaceous chondrite parent body (Cronin and Pizzarello 1997).

Later on in this book we will discuss any asymmetric influence of meteoritic L-amino acid excesses on organic chemical evolution and the origin of life itself.

[8] During a personal discussion at the 10th ISSOL conference in Oaxaca, Mexico in 2002, Sandra Pizzarello, professor at Arizona State University and one of the world's leading experts in meteorite analysis confirmed that the enantioselective analyses in her institute of chiral amino acids in carbonaceous chondrites never revealed an excess of the D-enantiomers. All samples – even those from different pieces of the inhomogeneous Murchison meteorite – showed exclusively excesses of the L-amino acids.

So far, however, no direct evidence verifies that both the generation of amino acids themselves and the formation of an enantiomeric excess therein have initially occurred in an interplanetary or interstellar environment. So, alternatively both processes might have taken place during the entry of achiral and reactive amino acid forming molecules on Earth only. Some experimental results may not exclude this latter assumption because significantly higher amounts of amino acids were detected in samples of interstellar ice analogues and/or parent bodies of meteorites after sample warm-up and acid-mediated hydrolysis in the laboratory (see Chaps. 7 and 8). During the warm-up step – simulating the arrival on Earth – reactive molecules that had been generated under hard solar irradiation in space might have produced complex organic precursors that form free amino acids after contact with liquid water – the required hydrolysis step.

On the other hand, the occurrence of an excess of L-enantiomers in six α-methyl-α-amino acids, more stable against racemization reactions than their α-H homologues, detected by Sandra Pizzarello and John Cronin (2000) in both the Murchison and Murray meteorites might be a hint that robust non-racemic chiral molecules were originated in the interplanetary or interstellar medium. A mechanism that describes the early delivery of organic compounds incorporated in comets or meteorites to early Earth was published by J. Mayo Greenberg et al. (1994) and Pascale Ehrenfreund (1999). Based on these observations, non-racemic chiral molecules might have been delivered to early Earth where crucial processes of chemical evolution started in the required dissymmetric environment.

The essential question, whether considerable enantiomer excesses – such as was reported in 1997 by Engel & Macko and Cronin & Pizzarello – were formed due to the interaction of interstellar circularly polarized light or any other mechanism prior to the entry of the extraterrestrial material on Earth, or during the entry process itself as a remote possibility, too, could only be answered finally by in situ measurements on meteoritic parent bodies, cometary nuclei, or interstellar ice analogues at interstellar temperature. These experiments have not yet been performed. But interestingly, we are very close to realizing them in the frame of the ROSETTA Mission. Enantioselective analyses of chiral cometary species are foreseen in the frame of the ROSETTA-COSAC project and will be presented and illustrated in Chap. 9. Here is an appetising introduction:

1.3.3.3 Chiral Molecules in Comets

It is generally assumed that organic molecules must have existed in solar nebular grains before comets formed. These molecules were incorporated into the comets' nuclei during their formation (Huebner and Boice 1992; Greenberg et al. 1994; Muñoz Caro et al. 2002). In addition to this, there is increasing evidence that these organic molecules showed enantiomeric excesses since they had been subjected to chiral interstellar fields (Bonner 1995a; Meierhenrich et al. 2005b). According to a further idea, comets were assumed to have delivered some of their volatile inventory (Chyba 1990) and essential "biogenic" compounds (Oró 1961; Chyba and Sagan 1992) in a non-racemic ratio to the terrestrial (and lunar (Oró et al. 1970)) surface during the final impact stages of the early inner Solar System (Oberbeck and Aggarwal 1992). This delivery of cometary asymmetric organic compounds was suggested to have triggered not only the appearance of life on Earth but also the selection of left-handed amino acids for incorporation into early organisms. This is a rather spectacular hypothesis, but could it really be true?

Which role did comets play in spreading out non-racemic components over large interplanetary distances? Could comets indeed be used as vehicles bringing asymmetric molecular building blocks of life to Earth? And would there be – against any pessimistic point of view (Siegel 1998) – an experimental option to elucidate this kind of active participation of comets in chemical evolution processes e.g. by measuring the occurrence of enantiomeric excesses on a comet? Based on the above, comets are probably the most exciting place to search for enantioenrichments in our Solar System (MacDermott 1997). Comets represent unprocessed relics of the pre-solar nebula.

With the ambitious aim of investigating the above hypothesis the European Space Agency ESA is, for the very first time, going to perform an experiment with cometary matter, in which enantiomeric excesses of a number of cometary organic molecules will be determined in-situ. More than 20 years ago, in 1986, the European GIOTTO mission took the first photo of a cometary nucleus. After this highly successful mission, ESA focussed on cometary research and started to design, plan, and construct the spacecraft ROSETTA, an official cornerstone-mission, which was launched with an Ariane 5 rocket in March 2004. Since then, the ROSETTA probe has been in space and already performed several fly-by manoeuvres at Earth and Mars in order to accelerate towards the target comet that is called 67P/Churyumov-Gerasimenko. ROSETTA will reach the orbit of comet "Chury" in 2014 and subsequently detach its robotical Lander named Philae, which is designed to land softly on the comet's surface (Ulamec et al. 1997; Bibring et al. 2007). An artist's image of the Lander Philae on the comet's nucleus is given in Fig. 1.3.

In the context of this book, the "Chirality-Module" of the ROSETTA mission is of particular interest. This module is an integral part of the Cometary Sampling and Composition Experiment (COSAC) which is a sophisticated gas chromatograph coupled to a mass spectrometer (GC-MS) system capable of separating, identifying, and quantifying enantiomers (Rosenbauer et al. 1999; Goesmann et al. 2007b). COSAC's multicolumn gas chromatograph GC is equipped with both a thermoconductivity TCD and a mass spectrometric MS detector. It is constructed to analyse cometary matter in situ, i.e., after landing on the surface of the cometary nucleus from where it will telecommunicate the recorded data to Earth. COSAC's GC/TCD-MS equipment was mainly designed and fabricated at the Max Planck Institute for Solar System Research in Katlenburg-Lindau, Germany.

In supplement to COSAC's achiral stationary phases whose selection, development, and combination are described by Szopa et al. (1999, 2000, 2007), three capillary columns that are doted with a chiral liquid film on its inner surface were designed to separate a variety of chiral organic molecules into their enantiomers in order to determine their enantiomeric excess (Meierhenrich et al. 1999, 2001b). Cometary non-complex alcohols, amines, diols, hydrocarbons, carboxylic acids, and

Fig. 1.3 The ROSETTA Orbiter swoops low over the ROSETTA Lander Philae soon after touchdown on the nucleus of comet 67P/Churyumov-Gerasimenko. This landing is scheduled for November 2014. Graphical simulation, photo courtesy of EADS Astrium

amino acids are intended to be resolved into their enantiomers in order to quantify individual enantiomers and thus to verify the above hypothesis.

After introducing the required stereochemical terms, nomenclature, and enantioselective techniques in Chap. 2 and up-to-date theories on the origin of biomolecular asymmetry in Chaps. 3–6, this book will expose the formation of chiral amino acid structures in interstellar space (Chap. 7). We will discuss the analysis of amino acids and diamino acids in extraterrestrial samples such as meteorites in Chap. 8, before we outline in Chap. 9 the basic concept of the COSAC instrument onboard the ROSETTA Lander. I will describe ROSETTA's "Chirality-Module" and summarize preliminary experiments in order to investigate the operation of the chiral capillary columns with material that is expected to show similar physico-chemical properties compared with original cometary matter. I will report on our first investigations of chiral stationary phases like Chirasil-Val coated with chiral valine molecules and cyclodextrin coated phases applicable for gas chromatographic resolution of enantiomers of underivatized alcohols and diols.

Rather far-reaching conclusions are expected from ROSETTA's "Chirality-Module". In the case of a successful mission, we would be able to better specify the molecular precursors of life based on chiral carbon chemistry in its prebiotic forms. ROSETTA's "Chirality-Module" will moreover try to shed some light on all the theories on the origin of biomolecular asymmetry. Doing so, it might contribute to the elucidation of the ultimate cause of the spectacular biochirality phenomena on Earth.

Chapter 2
Stereochemistry for the Study of the Origin of Life

Before exploring up-do-date theories and cutting-edge research activities on the origin of life's molecular asymmetry, this Chap. 2 will provide a brief, comprehensible, and illustrated introduction on common stereochemical terms like chirality, stereogenic center, enantiomer, racemic mixture, optical activity, and homochirality. Stereochemical descriptors like R/S, D/L, (+)/(−), d/l, Δ/Λ, and P/M will be explained.

After being familiar with these basic terms, we will focus on techniques often applied in bioanalytical laboratories to gain experimental information on the handedness of molecules. Advantages and limitations of chiroptical techniques such as optical rotation dispersion (ORD) including the Cotton effect, and in particular the often used circular dichroism (CD) spectroscopy will be outlined. Briefly, modern vibrational circular dichroism (vCD) techniques and – based on recent advances in the field and nowadays commercially available instruments – the Raman optical activity (ROA) will be introduced.

We will establish the phenomenon that not only molecules and macroscopic structures can be chiral but also electromagnetic waves in form of circularly polarized electromagnetic radiation. We will learn how to characterize "chiral light" with the help of the Stokes parameters I, Q, U, and V. Finally, the chromatographic resolution of enantiomers on chiral stationary phases will be summarized, as their use is common practice in numerous analytical laboratories today. Molecular structures of the most important chiral selectors will be given. Chapter 2 will thus provide a sound basis in order to meet and fully understand the stereochemical vocabulary, the experiments' design, and the measurement strategies, which we will encounter in the upcoming chapters.

2.1 Amino Acids and Chemical Reactions: A Guided Tour

Amino acids serve as ideal molecules to introduce stereochemical terms: The classical epitome of a chiral (handed) molecule is the amino acid α-alanine. α-Alanine possesses of the sum formula $C_3H_7NO_2$ and can be depicted by two distinct molecular structures, given in Fig. 2.1. These two structures illustrate the identical

U. Meierhenrich, *Amino Acids and the Asymmetry of Life*. Advances in Astrobiology and Biogeophysics,

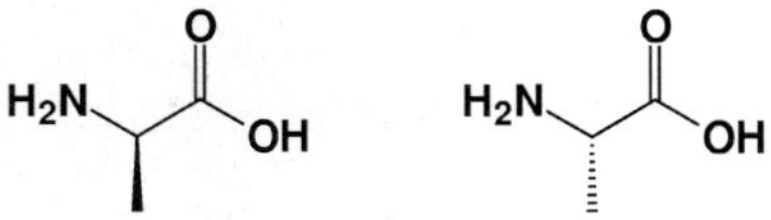

Fig. 2.1 Two three-dimensional non-superimposable molecular structures of the amino acid α-alanine. The structures are unsymmetrical and chiral; each of them is called an enantiomer

constitution of all chemical bonds connecting the functional groups of α-alanine. The central carbon atom, called α-carbon, holds the amino group and the carboxylic function as well as a methyl group. Consider that the structure is given in three dimensions: The methyl group at the bottom of the left structure points towards the viewer as it is illustrated by the thick line, whereas the methyl group in the right structure points away from the viewer as indicated by the dashed line. As a result of α-alanine's three-dimensionality, these structures are non-superimposable and, therefore, not identical. In fact, these molecular structures are unsymmetrical and mirror images of each other.

Structures of organic molecules (here: α-alanine) that are not identical, but mirror images of each other are called *enantiomers* (from Greek εχθρός, "enemy". The old term is *optical antipode*). Enantiomers belong to the group of isomers called stereoisomers, so do *E* and *Z* configurations at double bonds. In contrast to constitutional isomers, stereoisomers are identical in the connectivity of the atoms, but differ in the overall shape of the molecules. The prefix stereo- (Greek στερεό, "solid") refers to the fact that stereoisomerism involves structures that must be regarded as three-dimensional.

Structures are *chiral* (Greek χειρ, cheir, "hand") if they cannot be superimposed onto their mirror image or – as Vladimir Prelog put it in his 1975 Nobel Prize lecture "... if it cannot be brought into congruence with its mirror image by translation or rotation" (Cronin and Reisse 2005). The two distinct forms of chiral molecules that can exist are said to have opposite *absolute configuration*. Achiral structures – on the other hand – are superimposable on their mirror images. The amino acids glycine and β-alanine are superimposable on their mirror images and give examples of achiral amino acids (Fig. 2.2).

By definition, the α-carbon atom of an amino acid is located next to the carboxylic group. In an α-amino acid the amino group is connected at the α-carbon atom. In a β-amino acid such as β-alanine, the amino group is linked to the β-carbon atom.

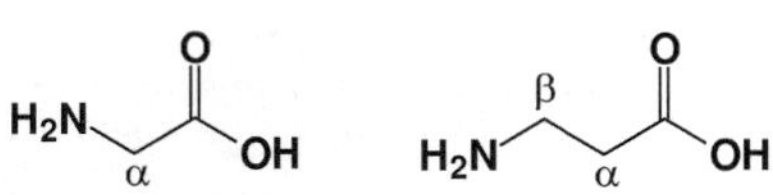

Fig. 2.2 Molecular structures of the α-amino acid glycine (*left*) and the β-amino acid β-alanine (*right*). Both molecules are superimposable with their mirror images and therefore achiral. Keep in mind: Amino acids are not necessarily chiral

The essential difference between chiral and achiral molecules is that chiral molecules do not possess a plane of symmetry;[1] achiral molecules, on the other hand, do have a plane of symmetry. In Fig. 2.2, the paper plane is this plane of symmetry. Any structure that has no plane of symmetry can exist as enantiomers. Any structure – on the other hand – with a plane of symmetry cannot exist as two enantiomers.

Using the definition of symmetry, the terminus "chiral" is by no means limited to molecular structures only. The same rules apply to everyday objects such as hands, gloves, scissors, pencil sharpeners, cars (with the non-symmetric position of the steering wheel[2]), dice (allowing to distinguish between original and imitation), snails, and shells, which have no plane of symmetry and are therefore chiral. Shoes provide another example. Your ears are chiral as well. Screws are chiral. They are manufactured so that they move forward by turning them clockwise with a screwdriver. Sometimes left-handed screws are prepared for special purposes. For example, pedals are fixed with right-handed screws on the one side and with left-handed screws on the other side of the bicycle in order to prevent them from loosening. The Moebius strip, the folding of arms, the way a person applauds, and the crossing of legs are asymmetric and chiral.

Rotating cyclones and tornados are chiral. They have a strong tendency to spin counterclockwise in the northern half of the globe, going clockwise in the southern half. The Coriolis effect based on earth's rotation causes this "symmetry-violating" spinning. As for the direction of bathtub vortices, the situation is not that clearly understood. Here, the question arose whether the Coriolis effect is strong enough to influence the right- versus left-handedness of the vortex. If the Coriolis effect had a dominating influence, this spiralling tendency of water down drains would be strongest at the poles, decreasing gradually as the tub moved toward the equator, where the tendency would vanish (Gardner 2005). Probably, however, many other factors are responsible for the vortex's right or left rotation such as the circulation the water acquires when the tub is filled and slight irregularities in the surface of the tub (see Chap. 4). Anyway, rotating vortices are chiral.

In many sports we can observe the counterclockwise travel around the field: In car racing, horse racing, bicycle racing (Bremer 6-Tage Rennen), dog racing, skating rinks, Olympic games, and so on. Baseball players dash counterclockwise around the bases. Carousels and carnival rides as well. Even customers in supermarkets are guided into a couterclockwise shopping tour. As Scot Morris, a correspondent, put it (in Gardner 2005) "it seems that everything goes counterclockwise except clocks".

---

[1] In specific cases a molecule is achiral if it shows a center of symmetry. (3*S*,6*R*)-3,6-dimethylpiperazine-2,5-dione serves as example for a cyclic molecule with a center of symmetry, but without a plane of symmetry, that is achiral. In an alternative classification based on the group theory, chiral molecules belong to the non-axial point group C_1 or the purely rotational groups C_n and D_n.

[2] In continental Europe cars have steering wheels on the left for right-of-the-road driving. This convention is almost universal throughout the world. Main exceptions are the British Isles, Malta, India, Australia, New Zealand but also Japan. Sweden, where the tourist accident rate was especially high, was the last nation on the continent to switch from left to right driving. The changeover that took place in 1967 was accompanied with a high costly preparation (see also Gardner 2005).

But even they are counterclockwise if you consider their motion from the clock's point of view.

A chair – on the other side – shows a plane of symmetry and is therefore achiral. An empty room shows three planes of symmetry and is achiral. Throw your glove into it, and the room will become chiral, too. Wonderful examples and impressive, colourful illustrations of chiral macroscopic structures were given by Brunner (1999). The lettering on book spines is chiral and differs between Europe and the Anglo-Saxon room. University libraries are faced with this irregularity particularly when half of the books are from Continental Europe and the other half from England and the United States. Clothing tends to follow the bilateral symmetry as well: Cravats (ties) in the United States show diagonal stripes in an inverse direction to European ones.

Occasionally, chemical reactions develop beautiful macroscopic chiral structures. They can be visualized – to give an intriguing example – by the Belousov-Zhabotinskii reaction (Fig. 2.3). Here, spiral structures form and move due to spatial

Fig. 2.3 Formation and growth of helical structures (spiral-waves) formed during the Belousov-Zhabotinskii reaction. In the vicinity of a right-handed helix, a left-handed helix is formed, moving synchronously. The photo was taken 40 min after mixing aqueous solutions of 0.33 mol L^{-1} malonic acid, 0.33 mol L^{-1} $NaBrO_3$, 0.2 mol L^{-1} H_2SO_4, and putting them onto a petri dish covered with a 1.4 mm gel produced by 5 mL filtered 14% (*w*/*w*) $Na_2O \times SiO_2$ and 2 mL 25 mmol L^{-1} ferroin solution. Gel formation was induced by adding 1.2 mL of 1.4 mol L^{-1} H_2SO_4. The diameter of the petri dish is 85 mm. Photo: Christoph Gottschling, Dept. Phys. Chemistry, University of Bremen

oscillations in a chemical system far away from chemical equilibrium but consistent with the laws of thermodynamics. It is clearly visible that 50% of the chiral waves evolve as right-hand spirals and 50% evolve as left-hand spirals. Often, in the near vicinity of a right-hand spiral a left-hand spiral evolves simultaneously and vice versa.[3]

Liesegang rings serve as a further beautiful example of the formation of macroscopic chiral helical structures developing during chemical reactions (Fig. 2.4). Here also, "ordered" structures evolve from a highly disordered mixture of molecules in solution: Visible helices develop spontaneously in this system far from chemical equilibrium involving reactions of higher order kinetics. 50% of the evolving spirals are right-helices and 50% are left-helices. It is worth mentioning that similar thermodynamic and kinetic conditions involved in the BZ-reaction and Liesegang ring structures, creating "order" out of "disorder" are discussed to be required for distinct steps of chemical and biological evolution as well. Other molecular systems to be studied in the laboratory that lead to observable chiral macroscopic phenomena include the left and right rotation of Bénard-cells, chiral crystals, chiral liquid crystals, and chiral membranes.

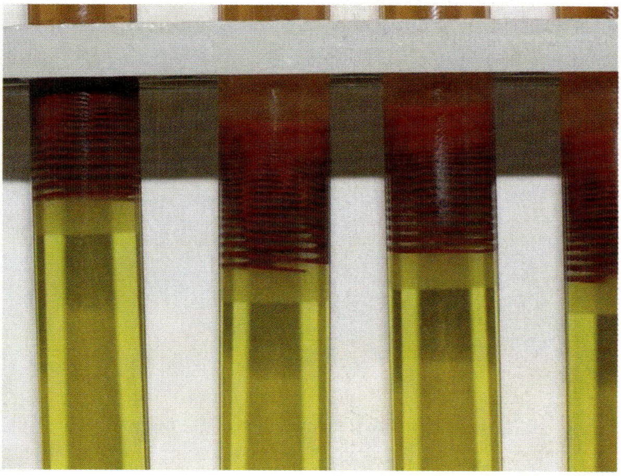

Fig. 2.4 Formation and growth of helical Liesegang rings in a viscous medium. In order to perform this "swinging" chemical reaction with oscillating properties in the lab, 4 g gelatin were added to 20 g of an aqueous potassium dichromate solution (0.2% (*w/w*)) by heating slightly. 10 mL of the warm yellow solution were given into a test tube. After cooling, 10 mL of 10% (*w/w*) silver nitrate solution were added cautiously onto the gel. Because of the light-sensitivity of silver salts, the test tube was mantled with aluminum foil. After 24 h, macroscopic chiral helical structures develop due to silver ion diffusion into the gel and precipitation of brownish silver chromate. By introducing the silver nitrate solution into the gel with the help of a Pasteur pipette, a capillary, or a dissolving medicinal drug capsule, onion layer-like three-dimensional structures form

[3] Chemical kinetics of the Belousov-Zhabotinskii (BZ) reaction are described by the Field-Körös-Noyes (FKN) model and can be followed using *Mathematica*® software (Kondepudi and Prigogine 1998). The BZ-reaction can easily be reproduced in the chemical laboratory at secondary school or high school level.

Coming back to chiral structures on the microscopic molecular level we note that bromochlorofluoromethane is chiral because it does not have a plane of symmetry. In fact, it cannot have a plane of symmetry, because it contains a tetrahedral coordinated carbon atom carrying four different groups (substituents): Br, Cl, F, and H. Such a carbon atom is known as a *stereogenic center*.[4]

For quite a while and beside amino acids, bromochlorofluoromethane was the "uncontested star" of chiral molecules. Among other experiments, one was able to determine its absolute configuration by a specific technique called Raman optical activity (Costante et al. 1997). Nowadays, particular interest is given to a chirally deuterated version of neopentane that represents the archetype of all molecules that are chiral as a result of a dissymmetric mass distribution (Barron 2007). The absolute configuration of enantioenriched *R*-[$^2H_1$, $^2H_2$, $^2H_3$]-neopentane (Fig. 2.5) was determined by Raman optical activity as well (Haesler et al. 2007).[5] Also here, the central carbon atom is called a stereogenic center.

If the two enantiomers of bromochlorofluoromethane (or [$^2H_1$,$^2H_2$,$^2H_3$]-neopentane) are formed in equal quantities, the pair of enantiomers is called a *racemic mixture* (compare Latin racema, "grape") or more loosely a *racemate*. The abbreviation is *rac*-bromochlorofluoromethane. This principle is very important. Never forget that if the starting materials of a reaction are achiral and the products are chiral, they will be formed as a racemic mixture of enantiomers. Amino acids produced by the Strecker synthesis must be racemic, because the starting compounds (aldehyde, ammonia, and KCN) are achiral. However, if we isolate alanine from a natural protein-rich source by hydrolysis, we find solely one enantiomer (Fig. 2.1 structure on the right). Samples of chiral compounds that contain only one enantiomer are called *enantiomerically pure*. We know from X-ray crystal structure

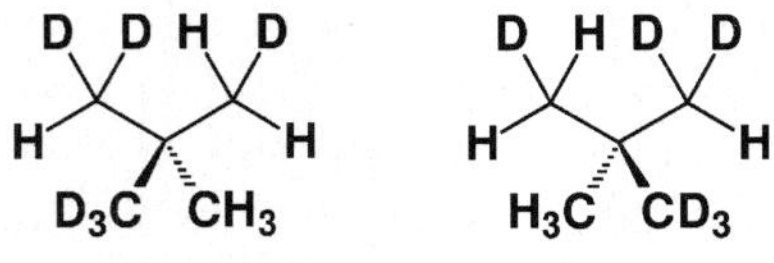

Fig. 2.5 Molecular structures of [$^2H_1$, $^2H_2$, $^2H_3$]-neopentane enantiomers. The molecules are mirror images, non-superimposable, and therefore chiral. The *R*-enantiomer is depicted on the left, the *S*-enantiomer on the right. The specific rotation of these molecules is not yet known

[4] If, on the other hand, a molecule possesses a stereogenic center, it is not necessarily chiral. The mirror plane of symmetry may pass through a (pseudo-) stereogenic center which has four different substituents among them one substituent in *R*-configuration and the same substituent in *S*-configuration. 2,3,4-Trihydroxyglutaric acid serves as example for an achiral molecule although it has a stereogenic center. Here, the stereogenic center is called pseudochiral center. The substituents are called *enantiotopic* groups (see for example Denmark 2006).

[5] For the ROA measurements Haesler et al. (2007) synthesized enantiopure *R*-[$^2H_1$, $^2H_2$, $^2H_3$]-neopentane. In order to eliminate deterministic offset problems the authors did require neither the physical optical antipode *S*-[$^2H_1$, $^2H_2$, $^2H_3$]-neopentane nor the racemic neopentane mixture. Instead, *S*-[$^2H_1$, $^2H_2$, $^2H_3$]-neopentane was "created optically" by the use of optical half-wave retarders (Hug 2003). This "virtual enantiomer" has the optical properties of the "strict" enantiomer, i.e., the geometrical enantiomer composed of antimatter (see Chap. 5).

analysis that biologically produced alanine contains only this enantiomer. Life uses *homochiral* monomers for the construction of proteins (enzymes) and DNA. In fact, nature does rarely use the deviant "wrong" enantiomer, which we will learn in more detail in Chap. 3.

Before continuing, one issue that often gives rise to confusion should be addressed. All molecules that do not have a plane of symmetry are chiral. If the molecules are of the same enantiomer, such as naturally isolated α-alanine, they are enantiomerically pure or homochiral. But all α-alanine is chiral. Even racemic α-alanine that is produced by the Strecker-mechanism is chiral.

Students have to be aware of the fact that – on the other hand – processes as such are never chiral. Terms as chiral synthesis, chiral catalysis, chiral recognition, chiral chromatography, or the recent book titles "Chiral Separations" (Humana Press), and "Chiral Photochemistry" (Marcel Dekker) should be avoided. Chromatography on a chiral stationary phase should better be called "enantioselective". An asymmetric synthesis or a chemical catalysis might be "enantiospecific".

If a mixture of chiral molecules is neither racemic nor homochiral the quantity of one enantiomer over the other is given according to Eq. 2.1 by the *enantiomeric excess* (*e.e.*), in which c designates the concentration of the indicated enantiomer. The enantiomeric excess is identical to the optical or enantiomeric purity or (sometimes) optical yield (Rau 2004).

$$e.e. = \frac{c_S - c_R}{c_S + c_R} \qquad (2.1)$$

Enantiomers show identical physico-chemical properties. Scalar (i.e. nonvectorial) values of enantiomers are identical. In sharp contrast to diastereoisomers, they have identical vapour pressures, boiling points, melting points, solubilities, diffusion constants, ionic mobilities, refractive indices, and reaction rates including excited state lifetimes. Enantiomers show identical activation energies, heat of formation, entropies, and thus Gibbs energies, identical bond lengths, bond strengths, and angles between the bonds. They provide identical absorption spectra in the UV or in the IR,[6] they show identical NMR spectra, and their mass spectra show identical fragmentation patterns. The important exception is the distinguishable interaction of enantiomers with linearly-polarized electromagnetic radiation. Chiral molecules can be (but must not)[7] *optically active*. The experimental way to determine this basic chiroptical property will be outlined in the upcoming paragraph.

---

[6] Photochromic compounds change their absorption spectra by photoirradiation. Photochromism phenomena of chiral molecules are more and more studied e.g. in the context of the photochemical control of liquid crystalline properties. For a recent review on chirality in photochromism see Yokoyama and Saito (2004).

[7] Some chiral molecules possess an optical activity too small to be measured. They are not optically active. Examples are 1-lauryl-2,3-dipalmitylglycerid and trialkylmethanes such as enantiomerically pure 4-methylnonane.

2.2 Optical Activity

If linearly-polarized light passes a cuvette with a chiral molecule in solution, the plane of the linearly-polarized light can rotate. Direction and value of the rotation depend on the chiral molecule itself, its configuration, and parameters like wavelength (λ), concentration (c), path length (d), solvent, and temperature (T). Racemic mixtures let the light pass unrotated, since each enantiomer compensates the optical activity of the other.

The study of the rotation of linearly-polarized light caused by chiral molecules is called polarimetry. Polarimeters are composed of a monochromatic light source, a linear-polarizing filter that converts unpolarized light into linearly-polarized light, a cuvette including the chiral analyte under examination, and a detector indicating value and direction of the rotated light. By convention, rotation of the plane of linearly-polarized light to the left (counterclockwise) as viewed towards the light source gives a negative value for the rotation angle α and the medium is said to be *laevo rotatory*. Rotation to the right (clockwise) corresponds to a positive value for α and the medium is said to be *dextro rotatory* (Barron 2004).

The Biot-Savard law, given in Eq. 2.2, calculates the specific rotation $[\alpha]$, also called the *specific optical rotatory power*, of a chiral compound from the observed rotation angle α (in degrees), concentration c (in g per $100\,\text{cm}^3$), and path length d (in dm). Unfortunately, CGS units instead of SI units are still in use, here. The specific rotation $[\alpha]$ is usually quoted without units.

For data comparison, the specific rotation $[\alpha]$ of a chiral analyte is usually measured at $T = 20°\text{C}$ with light emitted from a sodium lamp at 589.3 nm, called the sodium D-line. Under such conditions, values for specific rotations are commonly quoted as $[\alpha]_{\text{D}}^{20}$.

$$[\alpha]_{\text{D}}^{20} = \frac{100\,\alpha}{c \cdot d} \tag{2.2}$$

The Biot-Savard law (Eq. 2.2) corresponds to the generally known Beer-Lambert law describing the absorption of light passing a liquid solution in the form of $\varepsilon = Ac^{-1}d^{-1}$ with ε molar extinction coefficient and A absorption. The absorption can also be expressed as $A = \log_{10}(I_0/I)$ with I_0 the intensity of the incident and I the intensity of the transmitted light.

Is there any application-oriented interest to determine the specific rotation of a chiral compound? Yes, indeed, its determination is often applied to investigate the enantiomeric purity of an optically active product. The knowledge of the specific rotation is, however, not necessarily sufficient to determine the absolute configuration of a chiral molecule. To present an industrial-scale example, sugar beets, delivered by the farmer to sugar factories at the end of a year, are immediately extracted and subjected to polarimetric measurement in order to determine the optical activity indicating the sugar yield. The higher the optical activity, the higher the amount of sugar and the more the farmer will be recompensed.

The specific rotation $[\alpha]$ is proportional to the specific molar rotation $[M]$ designating the optical rotation as function of the molecular weight (MW) as given in Eq. 2.3.

$$[M]_{\mathrm{D}}^{20} = \frac{MW \times [\alpha]_{\mathrm{D}}^{20}}{100} \tag{2.3}$$

In Chap. 9 we will discuss how polarimetric measurements of the optical activity are used to study the origin of life's molecular asymmetry by determining minute amounts of chiral organic molecules in extraterrestrial objects such as Mars and cometary matter. Before that, we will have to find a consensus of how to classify and to name chiral molecules systematically.

2.3 Stereochemical Nomenclature

Due to historical reasons, several stereochemical descriptors are in use to classify chiral molecules, each of them independent from the others. These are the Cahn-Ingold-Prelog (CIP) system with the S/R-notation, the (+) or (−) direction of rotation of linearly-polarized light, as well as the D/L, d/l, P/M, and Δ/Λ-nomenclatures. All these classifications arise from different physical or structural observations and thus, the lets say given S-configuration of a chiral molecule does not include sufficient information to deduce whether this molecule has (+)- or (−)-optical activity or D- or L-configuration! In this Sect. 2.3, the stereochemical nomenclatures will be introduced briefly and in a manner adapted to the study of homochirality phenomena. For more detailed information, particularly on the enantiospecificity and stereoselectivity of a wide range of asymmetric chemical reactions, the interested reader is referred to textbooks on organic chemistry and stereochemistry, for example the recent and new style one by Clayden et al. (2006).

2.3.1 The Cahn-Ingold-Prelog Notation: R- and S-Enantiomers

In order to describe the configuration of a stereogenic centre, IUPAC recommends the use of Cahn-Ingold-Prelog (CIP) rules yielding to S/R-notations. According to CIP-rules, a priority number is to be assigned to each of the four substituents at a given stereogenic centre. Substituents with higher atomic numbers (Z) are assigned with higher priorities. In the frame of this book, which is focussing particularly on the chirality of amino acids, we will use the most simple chiral amino acid which is α-alanine as an introducing example: The stereogenic center of α-alanine, which is the central carbon atom, is linked to one nitrogen atom ($Z = 7$), a carbon atom ($Z = 6$) carrying oxygen ($Z = 8$), another carbon atom ($Z = 6$) attaching hydrogen atoms ($Z = 1$), and one hydrogen atom ($Z = 1$). Priorities of the substituents are consequently decreasing from the amino- to the carboxyl- over the methyl- to the hydrogen-substituent (Fig. 2.6).

The alanine molecule has now to be arranged in a way that the substituent with the lowest priority, in our case hydrogen ($Z = 1$), is pointing away from the viewer, as indicated by the dashed line in Fig. 2.6. Now, one describes a circle from the substituent with highest priority to the substituent with lowest priority. If one rotates

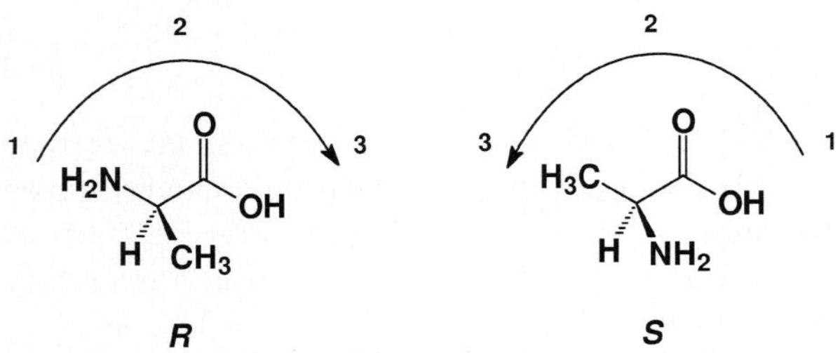

Fig. 2.6 Chemical structures of the enantiomers of $R-\alpha$-alanine (*left*) and $S-\alpha$-alanine (*right*), depicted with CIP-rules for labelling of the *R* and the *S*-configuration

in a clockwise manner, the stereogenic centre shows *R*-configuration (*R* for rectus), while an anticlockwise rotation will result in *S*-configuration (*S* for sinister). In the case of Figs. 2.1 and 2.6, the chemical structures given on the right depict the amino acid α-alanine in its *S*-configuration; the structures on the left show *R*-α-alanine. Following the same guidelines, the Cahn-Ingold-Prelog nomenclature can be applied for molecules with more than one stereogenic center as well.

2.3.2 (+)- and (–)-Enantiomers

Another stereochemical nomenclature for the classification of chiral molecules uses the optical activity of a chiral compound at a given wavelength to denominate (+)- and (−)-enantiomers. Enantiomers that rotate linearly polarized light at 589.3 nm to the right as seen towards the light-source show positive rotation. They are called (+)-enantiomers. Enantiomers that rotate linearly-polarized light at 589.3 nm to the left are called (−)-enantiomers.

However, the *R* or *S*-configuration of a given enantiomer does not necessarily determine the direction of optical rotation (+) or (−)! The relationship between absolute configuration and the sense of optical rotation is subtle and has exercised theoreticians for many years (Barron 2004). Chiroptical measurements showed that the enantiomer of *S*-α-alanine given in Figs. 2.1 and 2.6 on the right is (+)-*S*-α-alanine. As a consequence the enantiomer given on the left is (−)-*R*-α-alanine.

2.3.3 D- and L-Enantiomers

The most ancient classification of enantiomers refers to (+)-D-glyceraldehyde and (−)-L-glyceraldehyde as standards for the D/L-nomenclature (Fig. 2.7). Any enantiomerically pure compound that could be related by chemical transformation processes to (+)-D-glyceraldehyde is categorized as D-enantiomer, and any compound that corresponds to (−)-L-glyceraldehyde is categorized L-enantiomer. Today, D and L prefixes are only used for amino acids and sugars, since the process of

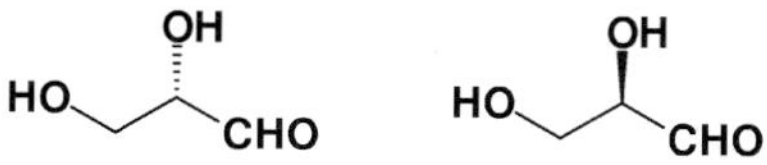

Fig. 2.7 Chemical structures of (−)-L-glyceraldehyde (*left*) and (+)-D-glyceraldehyde (*right*) used for labelling D- and L-enantiomers

denominating D or L by often complex and multiple chemical transformation processes is habitually slow and work-intensive.

2.3.4 *d*- and *l*-Enantiomorph Crystals

Chiral crystals are also called enantiomorph crystals. The best known examples of chiral crystals are quartz crystals. To distinguish between the two enantiomorph structures of enantiomorph crystals, the prefixes (+)- vs. (−)- can be used, which correspond to the prefixes *d*- vs. *l*-. Consequently, we distinguish between (+)-*d*-quartz and (−)-*l*-quartz. Other enantiomorph crystals which have played an important role in discussions on the origin of biomolecular asymmetry are sodium chlorate $NaClO_3$, the triclinic space group of pyrite FeS_2, as well as specific montmorillonite modifications. Crystals of calcite are, however, non-chiral. In general, the crystal's structure is determined by its space group. Among a total of 230 space group types that are in use by mineralogists for the systematic classification of three-dimensional structures of crystals and minerals, about 60 space group types give rise to chiral crystals.

2.3.5 *P*- and *M*-Descriptors

Some molecules are chiral even without having a stereogenic center. If two pairs of different substituents are coordinated along a chiral axis in perpendicular planes, but not attached to the same atom, we call this axial chirality. Stereodescriptors are *aR*, if the substituents' priority rotates in a clockwise manner, and *aS* for anti-clockwise rotation.

Helical chirality, a special case of axial chirality, is given when a molecular structure describes the form of a helix around an axis. If the molecule describes a helix in a clockwise manner away from the viewer, we call it a *P*-helix (*P* for plus), if it rotates in an anticlockwise manner away from the viewer, it is called a *M*-helix (*M* for minus). In DNA (*P*-helix), starch, and proteins (*P*-helix), helical chirality is of importance.

2.3.6 Δ- and Λ-Descriptors

Until now, we discussed tetrahedral, axial, and helical chirality. Additionally, octahedral systems such as chelate complexes occur regularly in the stereochemical literature and can be chiral as well. An example includes the octahedral

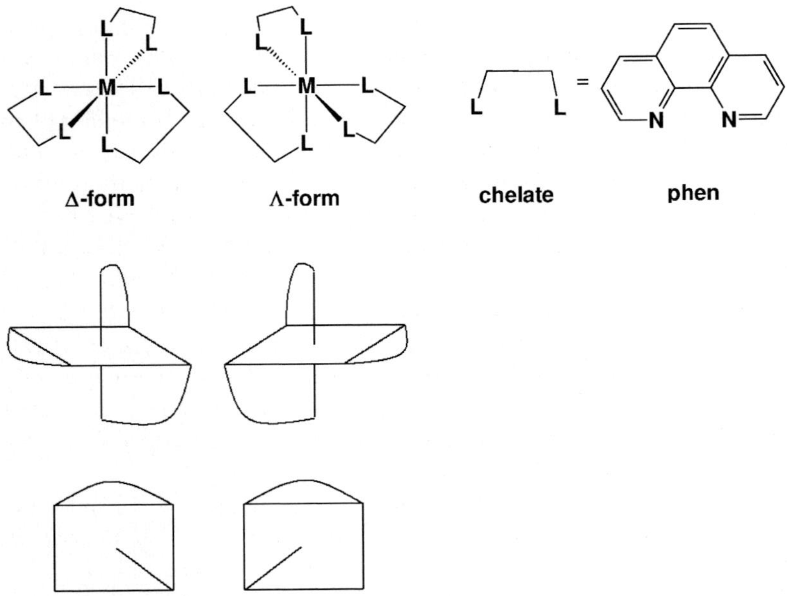

Fig. 2.8 Δ-form (*left*) and Λ-form (*right*) of an octahedral complex with three phen chelate ligands. For assigning the Δ- or Λ-form to a three-dimensional octahedral complex, the upper vertex is to be taken as a viewpoint. Its projection on the equatorial plane is represented as a straight line. After rotation that the bent line is on the top, the complex is the Δ-enantiomer if the straight line is on the right and the Λ-enantiomer if the straight line is on the left

$[Cr(phen)_3]^{2+}$ chelate complex in which the central metal ion M is Cr^{2+} surrounded by three 1,10-phenanthroline (phen) chelate ligands. Other central metal ions and/or chelating ligands such as ethylenediamine (en) are possible as well. The chirality of octahedral complexes composed of two or three chelate ligands is assigned by the Δ- (delta) and Λ- (lambda) symbol as illustrated in Fig. 2.8. In order to assign these stereochemical descriptors to an octahedral chelate system, an analogous three-dimensional projection is required. For the systematic stereochemical nomenclature of similar (but not analogous) octahedral systems with e.g. two chelate ligands the interested reader is referred to Herrero and Usón (1995).

2.4 Optical Rotation Dispersion and Cotton Effect

An optical method capable of differentiating between two enantiomers is referred to as a *chiroptical* technique. As indicated in Sect. 2.2, the optical rotation of a chiral molecule in solution depends on the wavelength of the incoming linearly-polarized light. Rotation angle α and specific rotation $[\alpha]$ change as a function of the wavelength, the measurement of which is called the optical rotation dispersion (ORD).

Often, the optical rotation changes only slightly by changing the wavelength, and ORD curves show no extrema (minimum or maximum) in the observed wavelength range in the visible spectrum. In this case, curves are called plain curves or normal dispersion curves. One distinguishes between positive and negative ORD plain curves. Positive ORD curves are obtained for positive values of the optical rotation. They are characterized by an increasing optical rotation towards short wavelengths in the ultraviolet range of the spectrum and show a negative slope. Negative ORD plain curves are for negative values of the optical rotation and show a positive inclination with more negative values towards shorter wavelengths. Achiral molecules and racemic mixtures are not optically active resulting in ORD curves with $[\alpha] = 0$ over the whole wavelength range. If one enantiomer of a given chiral molecule presents a positive ORD curve, the other enantiomer does necessarily present a negative ORD curve. Normal dispersion curves are obtained in a region of the spectrum without any absorption maxima of an analyte's chromophore, i.e., the functional group absorbing the visible or near ultraviolet light.

An abnormal dispersion curve is obtained if the wavelength of the incoming linearly-polarized light approaches the absorption maximum of a chromophore in the vicinity of a stereogenic center. Here – and this is a peculiar optical property of chiral molecules – the abnormal ORD curve describes an extremum, say a maximum, passes zero with a high slope, and enters in a minimum! The specific rotation changes its sign by a minute variation of the wavelength! This phenomenon in the abnormal ORD curve is called the *Cotton effect*. The Cotton effect of an optically active analyte is usually observed in solution and depends here on parameters like the dipole-character of the solvent, pH-value, temperature, etc.

The Cotton effect with its important change of sign for the optical rotation can be predicted even for an invisible or inaccessible region of the spectrum. A specific region of the ORD spectrum might be inaccessible due to solvent absorption and oxygen absorption below 200 nm and also ozone O_3 produced by the lamp of the polarimeter absorbs below 250 nm. Moreover, other wavelength-dependent optical effects may hide a region of the ORD spectrum. For the calculation and prediction of the Cotton effect, one uses the Drude equation 2.4 which denotes the specific rotation $[\alpha]$ given in experimental units of degrees per decimeter versus wavelength (λ) in nanometer. Previously to Drude, it had already been shown by Biot that the angle of rotation α is inversely proportional to the square of the wavelength λ of the light shining through a chiral medium. A positive Cotton effect calculated by the Drude equation is illustrated in Fig. 2.9.

In the Drude equation, A is a compound specific constant and λ_C denotes the critical wavelength at which the dispersion curve passes zero. If λ becomes λ_C the denominator becomes zero giving an infinite value for $[\alpha]$. The critical wavelength λ_C coincides with the maximum in the absorption spectrum of the chiral structure. Moreover it should be noted that modern molecular theories of optical rotation all provide equations of this form for transparent regions (Barron 2004).

$$[\alpha] = \frac{A}{\lambda^2 - \lambda_C^2} \qquad (2.4)$$

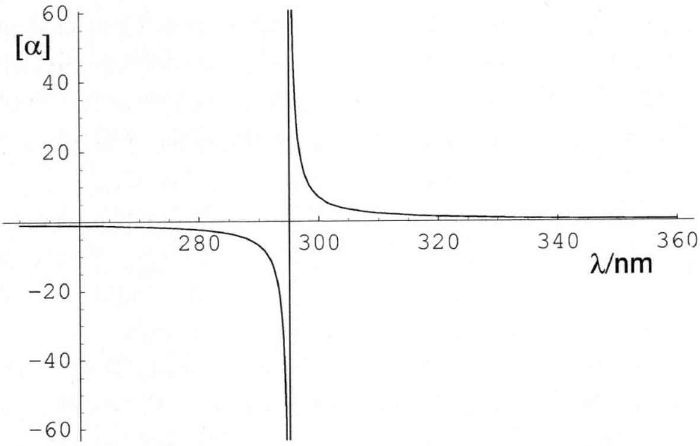

Fig. 2.9 Positive Cotton effect curve in the optical rotation dispersion (ORD) calculated by the Drude equation. Above 300 nm a positive ORD curve is given. By passing the critical wavelength λ_C =295 nm, the optical rotation changes sign and shows negative values! Below 290 nm the dispersion is given as negative ORD plain curve. In the selected and calculated example, the compound specific constant A is arbitrarily chosen as $A = 20\,500$ deg nm dm^{-1}. A Cotton effect of a chiral analyte recorded by an ORD spectropolarimeter shows a typical tangential curve. For comparison with values for the specific rotation $[\alpha]_D$, one has to be aware of its $\lambda = 589.3$ nm. As a consequence of the Cotton effect, the (+) and (−)-categorization of enantiomers requires a fixed wavelength

In order to determine the critical wavelength λ_C and the compound specific constant A for a given chiral molecule, the Drude equation can be transformed into Eq. 2.5.

$$\lambda^2[\alpha] = A + \lambda_C^2[\alpha] \tag{2.5}$$

Plotted values for $\lambda^2[\alpha]$ versus $[\alpha]$ provide a linear relationship giving access to the center distance A and the inclination λ_C. Typical values for Cotton effects with $[\alpha] = 0$ degrees dm^{-1} are at $\lambda_C = 232$ nm for the peptide chymotrypsin or $\lambda_C = 295$ nm for 3-β-hydroxy-5-androstan-17-one.

In the literature on chiroptical properties, rotation dispersion curves are often expressed by giving the molecular rotation [Φ] as a function of wavelength. The molecular rotation [Φ] is directly proportional to the specific rotation $[\alpha]$ and defined in Eq. 2.6, where MW is the molecular weight of the optically active substance in g mol^{-1}.

$$[\Phi] = \frac{MW \times [\alpha]}{100} \tag{2.6}$$

What about measuring the optical rotation dispersion of more complex molecules than an amino acid such as proteins? For biomolecules like proteins (enzymes) in their polymer form, the optical rotation dispersion is not only determined by the stereogenic centers of the monomers in the primary structure, but also by its helical chirality of the secondary structure. In proteins, the helical contribution to the

ORD signal is often not negligible[8] and can therefore be used to gain information on the three-dimensional structure of folded proteins. In order to distinguish between the ORD contribution caused by the stereogenic centers of the monomers and the helical chirality, the Moffit-Young equation 2.7 was developed particularly for interpretation of ORD signals of proteins.

$$[M]_\lambda^T = \frac{a_0 \lambda_C^2}{(\lambda^2 - \lambda_C^2)} + \frac{b_0 \lambda_C^4}{(\lambda^2 - \lambda_C^2)^2} \tag{2.7}$$

In this virial equation $[M]$ designates the specific molar rotation. The a_0-term of the Moffit-Young equation corresponds to the Drude equation 2.4. Values for b_0 correlate with the helical character of the ORD signal; b_0 varies between 0 and 1 or 0 and 100% and tells one directly the degree of "helicity" of the polymer in question. For the experimental determination of a_0 and b_0, the Moffit-Young equation is to be transformed into Eq. 2.8.

$$(\lambda^2 - \lambda_C^2)\,[M]_\lambda^T = a_0 \lambda_C^2 + \frac{b_0 \lambda_C^4}{(\lambda^2 - \lambda_C^2)} \tag{2.8}$$

Plotting of the left side of this equation versus $(\lambda^2 - \lambda_C^2)^{-1}$ results in a linear function with a center distance $a_0 \lambda_C^2$ and an inclination of $b_0 \lambda_C^4$. Following this way, the helical term b_0 can be determined by ORD measurements for a wide variety of helical analytes.

2.5 Through the Eye of a Chromophore: CD Spectroscopy

Some gemstones show different colours if one observes them from different angles. They are dichroic; the effect is called dichroism. In contrast to the optical rotation dispersion, circular dichroism (CD) spectroscopy does not describe the specific rotation $[\alpha]$ of an optically active substance as a function of the wavelength. CD spectroscopy measures the differential absorption of circularly polarized light by mirror image isomers. An explanation:

By definition, a chiral molecule does not show a plane of symmetry. Therefore, the excitation of electrons in a chiral molecule into an electronically excited state will not occur in a symmetric matter. The pushing of electron density from a starting state to a higher energy final state triggered by electromagnetic radiation proceeds in an asymmetric manner by some kind of helical electron movement. This helical

[8] As a matter of fact, the optical rotation dispersion in polyglycine is entirely determined by helical chirality, since glycine-monomers are achiral. Polyglycine can occur in left- and right-helices (*M*- and *P*-helices). In contrast to this, proteins (enzymes) are right-helices (*P*-helices), determined by the L-configuration of the α-amino acid monomers. So is DNA a right-helix (*P*-helix), determined by its chiral monomers the D-ribofuranose units. As an exception, Z-DNA, a type of DNA structure that is characterized by a regularity in its sequence adopts a left-helix (*M*-helix).

excitation of electrons in chiral molecules is affected by chiral photons in a way that right-circularly polarized light (RCPL) is not absorbed by chiral molecules in the same manner as left-circularly polarized light (LCPL). The interaction is different. The above explication is to illustrate and visualize CPL absorption and is, of course, not based on physical reality.

CD spectroscopy determines the difference of the molar extinction coefficient $\Delta\varepsilon$ of a given substance between LCPL and RCPL as written in Eq. 2.9. This differential absorption depends on the wavelength, a dependence that is measured by CD spectropolarimeters and given in a CD spectrum.

$$\Delta\varepsilon = \varepsilon_{\mathrm{LCPL}} - \varepsilon_{\mathrm{RCPL}} \tag{2.9}$$

Typically, CD spectra show an extremum (minimum or maximum) if the energy of the incoming circularly polarized light is closely matching the energy of absorption of a chromophore. The energy of this extremum coincides with the energy of the point of inflection described by the Cotton effect in abnormal ORD curves which itself ideally coincides with the maximum of an electronic absorption band (Barron 2004). Summarizing, a maximum in an ultraviolet and visible absorption spectrum results in an extremum in the CD spectrum and a Cotton effect with $[\alpha] = 0$ in ORD. In this context, it is sometimes stated that ORD and CD techniques are used to look at the stereochemistry of a chiral molecule "through the eyes of the chromophore".

A positive ORD plain curve shows a positive CD signal, a negative ORD plain curve shows a negative CD signal. In Fig. 2.10, the CD spectra of (−)-*R*-α-alanine and (+)-*S*-α-alanine enantiomers are given, in which each curve shows a maximum and a minimum and the two spectra are "opposite" to each other.

For CD measurements of chiral compounds spectropolarimeters are used, composed of

a. a light source, which is in most instruments a water-cooled xenon arc lamp for UV and visible CD measurements,
b. an optical device that produces monochromatic electromagnetic radiation that is alternatively left- and right-circularly polarized. This is performed by a series of mirrors and prisms creating linearly-polarized light, connected to a photoelastic modulator (PEM) or a Pockels cell, transforming linearly- into circularly polarized light, typically switching between right- and left-circular polarizations light with a frequency $\nu = 50\,\mathrm{kHz}$ by applying an AC voltage.
c. A cuvette for the analyte in solution, and
d. a photo-multiplier detection unit able to determine the energy-dependent differential absorption for the two circular polarizations.

For historical reasons, different instruments with different units are used in CD spectroscopy. In order to transform values for dichroic absorption $\Delta\varepsilon$ into the old but common unit of molar ellipticity M_θ in mol^{-1} dm^3 cm^{-1}, Eq. 2.10 can be applied. After absorption, the transmitted polarized light leaves a sample of a chiral molecule in solution as elliptically-polarized light. Formerly, such ellipticity was determined experimentally for recording circular dichroism spectra.

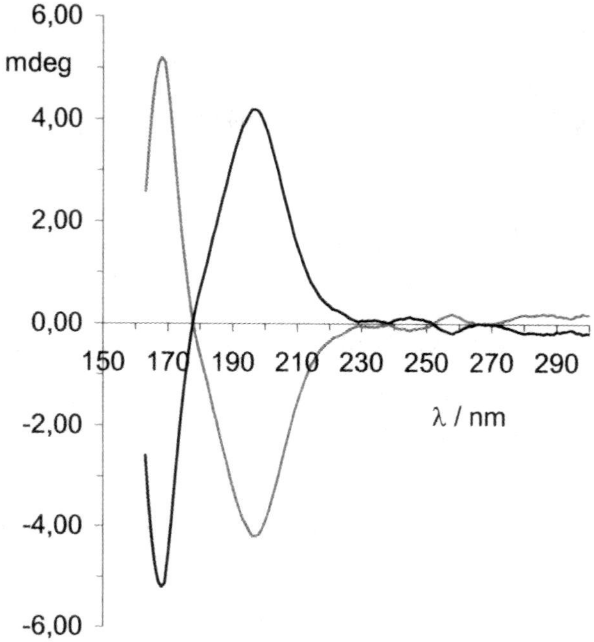

Fig. 2.10 Circular dichroism spectra of the amino acids (−)-*R*-α-alanine (*blue line*) and (+)-*S*-α-alanine (*black line*) recorded as zwitterions in hexafluoroisopropanol (HFIP) solution at the ISA synchrotron facility in Århus, Denmark. The spectra of the two enantiomers are depicted between 150 and 300 nm and show – as expected – opposite signs

$$M_\theta = 2.303\frac{4500}{\pi}\Delta\varepsilon \approx 3300\,\Delta\varepsilon \tag{2.10}$$

The molar ellipticity M_θ is related with the specific ellipticity $[\theta]$ in $g^{-1}\,dm^3\,cm^{-1}$ by Eq. 2.11 where *MW* is the molecular mass in g mol^{-1}.

$$[\theta] = \frac{100\,M_\theta}{MW} \tag{2.11}$$

The ellipticity θ itself is linked with the specific ellipticity by Eq. 2.12.

$$\theta = [\theta]c\,d \tag{2.12}$$

The differential absorption ΔA, often encountered in chemistry literature, is obtained by Eq. 2.13, the circular dichroism version of the Beer-Lambert law (Rodger and Nordén 1997).

$$\Delta A = \Delta\varepsilon\ c\ d \tag{2.13}$$

Note that even a molecule with defined *P*-helicity may show positive or negative CD signals since rearrangement of electron density with different handedness not necessarily following our *P*/*M*-notation might be involved.

Comparing ORD and CD techniques, the experimentator should note that CD spectroscopy is often advantageous and now generally preferred, since a better resolution of signals can be obtained, particularly if chromophores absorb at wavelengths close to each other. If there are several adjacent absorption bands, the net Cotton effect will be a superposition of the individual Cotton effect curves and the CD lineshape function drops to zero much more rapidly than the ORD lineshape function (Barron 2004). However, CD signals are generally weaker than ORD signals and the ORD technique allows determining values even in a certain distance from the Cotton effect.

In order to convert data on optical rotation $[m']_\lambda$ obtained from ORD measurements into the specific molar ellipticity $[\theta]_\lambda$ from CD records and vice versa, the Kronig-Kramers theorem represented in Eq. 2.14 can be used.

$$[m']_\lambda = \frac{2}{\pi} \int_0^\infty [\theta]_\lambda \left(\frac{\lambda_1^2}{\lambda^2 - \lambda_1^2} \right) d\lambda_1 \qquad (2.14)$$

Besides the above outlined experimental approach, one may attempt to theoretically predict the interaction of electromagnetic radiation with chiral organic molecules. Quantum mechanical calculations are required, which are based on the Hamiltonian operator $\hat{H}^{\text{int}}$ of the energy of interaction between light and matter. According to Eq. 2.15, this is the difference of the vector products between the electric field vector E and the operator (indicated with ^) for the electric dipole moment μ of the molecule and the vector product between the magnetic field vector B and the operator for the magnetic dipole moment m of the molecule. The software *Gaussian 03* offers a tool for quantum mechanical calculations of CD signals.

$$\hat{H}^{\text{int}} = -\hat{\mu} \cdot E - \hat{m} \cdot B \qquad (2.15)$$

Resulting from experimental data and quantum mechanical calculations, CD spectra of peptides and proteins in solution typically show electronic transitions of the occupied n-orbital to the unoccupied π-orbital (denoted with *) at about 220 nm with negative values. A negative (π^*,π)-electronic transition is at about 210 nm and a positive (π^*,π)-transition is often found at about 190 nm.

2.6 Miscellaneous Techniques

Furthermore, the elucidation of chiral structures in the context of the origin of biomolecular asymmetry is influenced and flanked by techniques such as magnetic optical rotation dispersion, magnetic and vibrational circular dichroism, linear dichroism, but also Raman optical activity. These techniques will be briefly introduced; interested readers will be referred to the literature.

Substances that are optically active in the absence of external influences are said to exhibit "natural" optical activity. Otherwise, all substances in magnetic fields are optically active, and electric fields can sometimes induce optical activity in special situations (Barron 2004). In a magnetic field, a helical character can be induced into a linear electron displacement in achiral molecules. Here, magnetic optical rotation dispersion (MORD) and magnetic circular dichroism (MCD) spectra are obtained even for achiral molecules. The observed values are biggest if the light is parallel to the magnetic field. Thiemann and Jarzak (1981) already described the use of and explained this technique in the context of research on the origin of biomolecular asymmetry. The theoretical fundament of which can be found in Barron (2004).

Conventionally, and as we have discussed above, optical activity phenomena have been related almost entirely with electronic transitions in chiral molecules. More recently, optical activity measurements have been extended to different vibrational states in chiral molecules making vibrational circular dichroism (vCD) spectroscopy more and more popular. As given in Eq. 2.16, vCD is the differential absorption (ΔA) of left- and right-circularly polarized *infrared* light by vibrating molecules (Berova et al. 2000).

$$vCD = A_{\mathrm{LCPL}} - A_{\mathrm{RCPL}} \tag{2.16}$$

Here, CD signals can be obtained in the infrared region of the electromagnetic spectrum. vCD techniques are particularly useful for the study of the conformational characteristics of proteins and nucleic acids but also smaller molecules like chiral pharmaceuticals. vCD spectroscopy can be used together with ab-initio calculations to determine the absolute configuration of chiral organic molecules.

Switching from the low-energetic range in vCD spectroscopy to the high-energetic spectral range, one can perform vacuum-ultraviolet (VUV) circular dichroism spectroscopy. VUV-CD spectroscopy attracted a certain interest in the study for the origin of life, which we will discuss in detail in Chap. 6. Synchrotron light can be used to measure CD spectra down to approx. 100 nm (Meierhenrich et al. 2005b).

Besides vCD and VUV-CD techniques, linear dichroism (LD) spectroscopy is applied to the study of an anisotropic arrangement of molecules that shows light absorption according to their orientation in space (see Rodger and Nordén 1997). Here, vertical linearly-polarized light is absorbed by vertically oriented molecules ($A_{|}$), such as oriented polyaromatic cyclic hydrocarbons, in a different way than horizontally linearly-polarized light ($A_{\perp}$). A non-oriented collection of molecules in solution is inactive in linear dichroism. A well-defined orientation of molecules for example in crystals, films, or on structured surfaces is required for LD spectroscopy. LD signals correspond to the differential absorption of oriented linearly polarized light as indicated in Eq. 2.17.

$$LD = A_{||} - A_{\perp} \tag{2.17}$$

A specific kind of spectroscopy called Raman spectroscopy can be used as supplementary tool to gain information on the chirality of organic molecules. This spectroscopy uses the Raman effect, which was observed by the Indian physicist and Nobel price winner Sir Chandrasekhara Venkata Raman. The Raman effect is based on photon scattering on the surface of matter. The spectrum of organic compounds, liquids, or solids irradiated with monochromatic light does not only show the Rayleigh scattering[9] at frequencies of ω_0, but also the so-called Raman-lines

[9] Rayleigh scattering is the scattering of electromagnetic radiation by particles that are much smaller than the wavelength of the electromagnetic radiation.

at frequencies of $\omega_0 \pm \omega$. In contrast to Rayleigh scattering, in the case of Raman scattering energy and momentum are exchanged between photons and scattering medium. Scattered light at the frequency of $\omega_0 - \omega$ corresponds to the Stokes-line, the frequency $\omega_0 + \omega$ corresponds to the Anti-Stokes-line.

It is worth mentioning that if chiral molecules are used as scattering medium, the Rayleigh and Raman scattered light is to a certain degree circularly polarized and the scattered intensity is slightly different in right- and left-circularly polarized light (Atkins and Barron 1969). Raman Optical Activity (ROA)-signals indicate the difference between the scattered intensities in right- and left-circularly polarized light as written by Eq. 2.18.

$$ROA = I^{\mathrm{R}} - I^{\mathrm{L}} \tag{2.18}$$

This observation led to the recent development of the ROA technique, applicable to a huge range of chiral samples of central importance for life, from small organic molecules over proteins, carbohydrates, nucleic acids to intact viruses (Barron et al. 2007). The significance of ROA is that it records – such as vCD techniques – vibrational optical activity and therefore provides more stereochemical information than standard chiroptical techniques of visible and ultraviolet CD, measuring electronic optical activity. For many molecules conformational details are visible by ROA, whereas visible and ultraviolet CD spectroscopy provide information on the stereochemical environment of the chromophore only. Molecules lacking a chromophore are inaccessible to visible and ultraviolet CD but can be studied with ROA. Very recently, a commercially available ROA instrument was introduced, using a visible laser beam at 532 nm from a frequency-doubled Nd/YAG laser (Chiral*RAMAN* by BioTools Inc.).

2.7 "Chiral Light" and the Stokes Parameters

Light is electromagnetic radiation which can be interpreted as wave of an electrical field vector E and a magnetic field vector B, each of them oscillating and perpendicular to the axis of translation and $E \perp B$. For unpolarized light, the electrical field vector has an infinite number of orientations. If one restricts its oscillations to one plane, for example by a linearly polarizing filter, we speak of linearly polarized light. Also here, $E \perp B$ is valid. This light is used in a polarimeter for the measurement of the optical rotation of a chiral molecule in solution.

Electromagnetic radiation can moreover consist of asymmetric light, called circularly polarized light. The photons then have angular momentum, called helicity, spinning either to the left or the right.

If one assumes that light can be described as an electromagnetic wave that oscillates in all planes, then circular polarized light only oscillates in one plane which rotates around the direction of propagation of the light. A more precise definition of circularly polarized light can be found in Barron (2004) saying that in a circularly polarized light beam, the tip of the electric field vector in a fixed plane perpendicular to the direction of propagation traces out a circle with time. Light is addressed right-circularly polarized light if the tip of the E vector rotates clockwise when viewed from the observer towards the light source. In circularly polarized light the electric field vector remains constant magnitude in time but proceeds in form of a helix in the propagation direction. Also here, at each point in space and time, the magnetic field is perpendicular to the electric field and to the direction of propagation.

Unpolarized light, linearly polarized light, and circularly polarized light including their wavelength λ are schematically illustrated in Fig. 2.11.

Circularly polarized electromagnetic radiation can be produced in a laboratory using a light source and a $\lambda/4$-plate. If tunable energy and a higher photon's energy of up to 12 eV are required, vacuum ultraviolet CPL can be produced artificially in a modern synchrotron.

More difficult is the *detection* of CPL at various energies. Here, polarimeters, such as the Onuki-type polarimeter can be applied. In Fig. 2.12 different polarizations of light, which were created at 7 eV at the synchrotron center LURE, Paris-Orsay, are given. The images clearly allow distinguishing between linearly, elliptically, and circularly polarized light.

But how can we experimentally specify the three parameters of electromagnetic radiation namely intensity, phase, and polarization? Four Stokes Parameters, $\boldsymbol{I}$, $\boldsymbol{Q}$, $\boldsymbol{U}$, and $\boldsymbol{V}$ (or in an alternative classification S_0, S_1, S_2, and S_3 as described by Barron (2004)) can do so and are to be determined. In a right-handed orthogonal coordinate system formed by **r** and **l** (see Fig. 2.13) the propagation direction is described by **s**. The total intensity $\boldsymbol{I}$ of electromagnetic radiation is given by Eq. 2.19.

$$\boldsymbol{I} \equiv \boldsymbol{I}_\mathrm{L} + \boldsymbol{I}_\mathrm{R} \tag{2.19}$$

For monochromatic light, $\boldsymbol{I}_\mathrm{L}$ and $\boldsymbol{I}_\mathrm{R}$ are uniquely specified if the degree of linear polarization (Eq. 2.20)

$$\boldsymbol{Q} \equiv \boldsymbol{I}_\mathrm{L} - \boldsymbol{I}_\mathrm{R} \tag{2.20}$$

is known. By $a = a' \sin\beta$ and $b = a' \cos\beta$ the equation $a/b \equiv \tan\beta$ is given, where a and b are the semimajor axes of the elliptically polarized light, a' is the hypotenuse, and γ is the angle between **l** and the major axis (Fig. 2.13). The definitions given in Eqs. 2.21 and 2.22

$$\boldsymbol{U} \equiv \boldsymbol{Q} \tan(2\gamma) \tag{2.21}$$

$$\boldsymbol{V} \equiv \boldsymbol{Q} \tan(2\beta) \sec(2\gamma) \tag{2.22}$$

are given for the Stokes parameters describing the elliptical polarization. In this notation, circular polarization is to be regarded as a particular case of elliptical polarization.

The four Stokes parameters are connected via Eq. 2.23

$$\boldsymbol{I}^2 = \boldsymbol{Q}^2 + \boldsymbol{U}^2 + \boldsymbol{V}^2. \tag{2.23}$$

For unpolarized light, Eq. 2.24 indicates

$$\boldsymbol{I}_\mathrm{L} = \boldsymbol{I}_\mathrm{R},\ \text{so}\ \boldsymbol{Q} = \boldsymbol{U} = \boldsymbol{V} = 0. \tag{2.24}$$

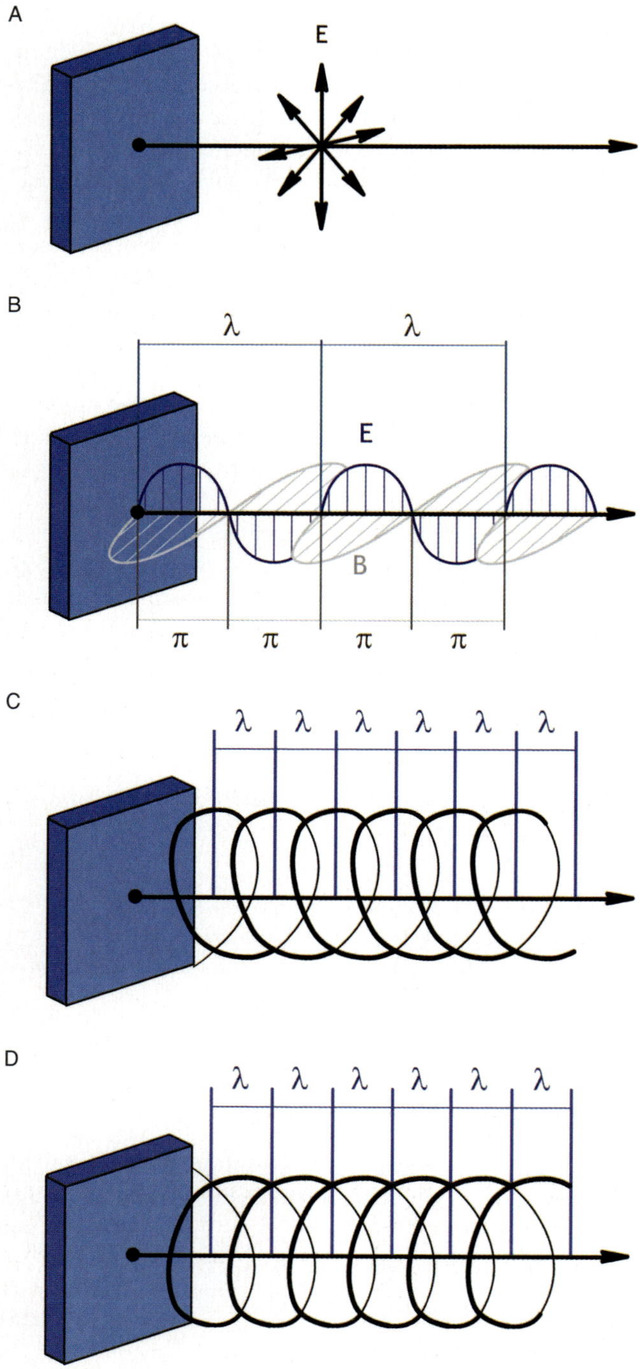

Fig. 2.11 The electrical field vector E of unpolarized light **A**, linearly polarized light **B**, simplified right-circularly polarized light **C**, and simplified left-circularly polarized light **D** are depicted. Illustration by Stéphane Le-Saint, UNSA

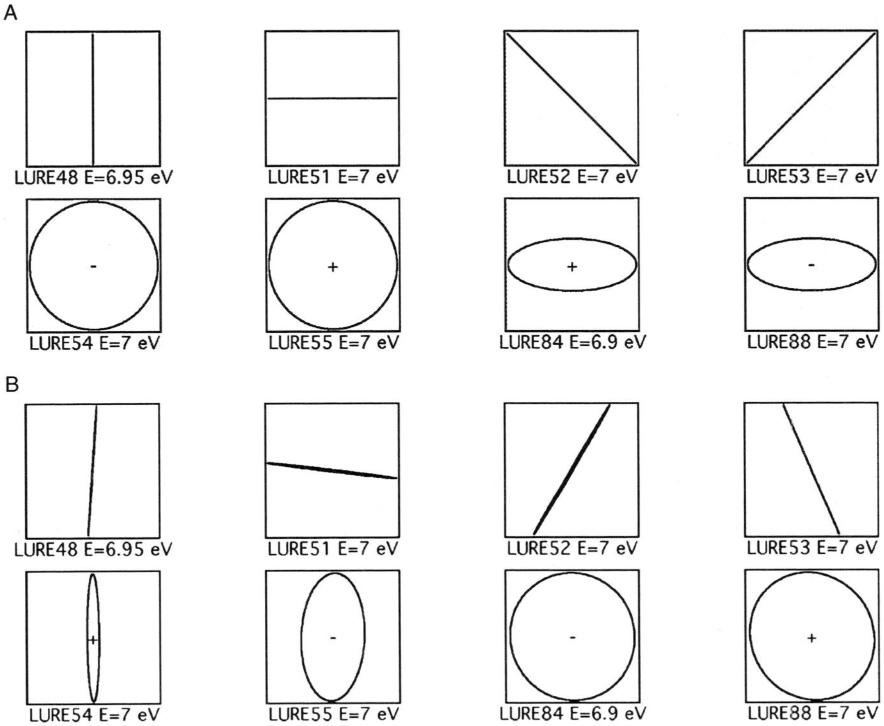

Fig. 2.12 Variations of the electrical field vector in the (x,y) frame at a fixed z versus time (observer is facing the beam light) as expected at the exit of the synchrotron undulator. The sign of Stokes parameter S_3 is displayed in the center of the corresponding figures (other examples and experimental conditions are given in Alcaraz et al. 1999)

The determination of the four Stokes Parameters allows the description of intensity, phase, and polarization of electromagnetic radiation.

2.8 Chromatographic Resolution of Enantiomers

Despite the wide range of applications of chiroptical techniques, the hitherto most powerful tool for qualitative and quantitative identification of chiral analytes of unknown nature is the chromatographic method. This method benefits from a number of great advantages: Chromatography – both gas-liquid as well as liquid-liquid – is able to separate and identify very small amounts of chemicals, in the order of a few nanograms and below; it allows the reliable identification of unknown compounds – especially if it is combined with a mass spectroscopic detection unit. In Analytical Chemistry, the separation of enantiomers is called “resolution” and is widely used to gain information on chiral systems. Today, most analytical laboratories are

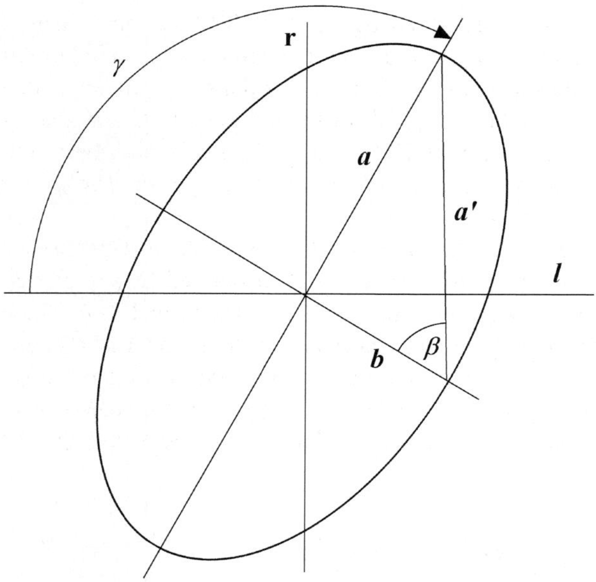

Fig. 2.13 Schematic view of elliptical polarized electromagnetic radiation including parameters deducing the four Stokes parameters. The orthogonal coordinate system is formed by **r** and **l**, the distances a, a', and b are given with the angles β and γ

equipped with the required enantioselective instrumentation for resolving mirror image isomers. We distinguish between indirect and direct resolution of enantiomers. For indirect resolution, the enantiomer analytes (*R*-analyte and *S*-analyte) are transformed by a chemical reaction with an enantiopure reactant (e.g. *R*-reactant) into diastereoisomers (*R*,*R*-derivative and *S*,*R*-derivative) which can be resolved with classical achiral stationary phases. We have seen that diastereoisomers have distinguishable physico-chemical properties such as boiling points, etc. Therefore they are susceptible to separation on achiral stationary phases. For direct resolution of mirror image analytes, a chiral stationary phase is required that contains an enantiopure selector (e.g. *R*-selector). Here, the stereogenic centers of the chiral analyte and the chiral stationary phase have to be brought together in such a way that there is an interaction between them. Thereby, separable diastereoisomeric associates are formed (*R*,*R*-associate and *S*,*R*-associate) from inseparable enantiomers. Chromatographic separation is based on a difference in partition coefficients of the analyte between a stationary phase and a mobile phase. In enantioselective gas chromatography, the stationary phase contains enantiomerically pure compounds such as the amino acid valine or cyclodextrin molecules.

It is in particular the development of wall-coated open-tubular (WCOT) capillary columns treated with chiral selectors that allows the highly efficient resolution of enantiomers. Different chiral selectors were developed and are still in use today. I will briefly present the most representative chiral selectors for enantioselective gas chromatography, because we will need them in Chap. 9 for the profound understanding of the "chirality-module" onboard the ROSETTA Lander which is now on the way to land on the surface of a comet where it will perform an important experiment on the origin of life's asymmetry. In practice a dazzling number of different chiral stationary phases has been applied for the separation of enantiomers (Gübitz and Schmid, 2004; Meierhenrich 2004). Here, Chirasil-Val phases, cyclodextrin phases, chemically bonded cyclodextrin phases, and acyclodextrin phases will be briefly described and illustrated.

Chirasil-Val phases: Quite often, the classical, thermally stable Chirasil-Val phases (Frank et al. 1978) are still in use. They are characterized by a high viscosity and low volatility of the organic polymeric siloxane (Schurig 1986). These "elastic" or "movable" dimethyl polysiloxanes possess covalently bonded diamide functions as one can see in Fig. 2.14, which are able to perform hydrogen bridge, dipole-dipole, as well as van der Waals bonds with the chiral analyte. One or two "arms" of the polymer are able to form an associative complex with the specific

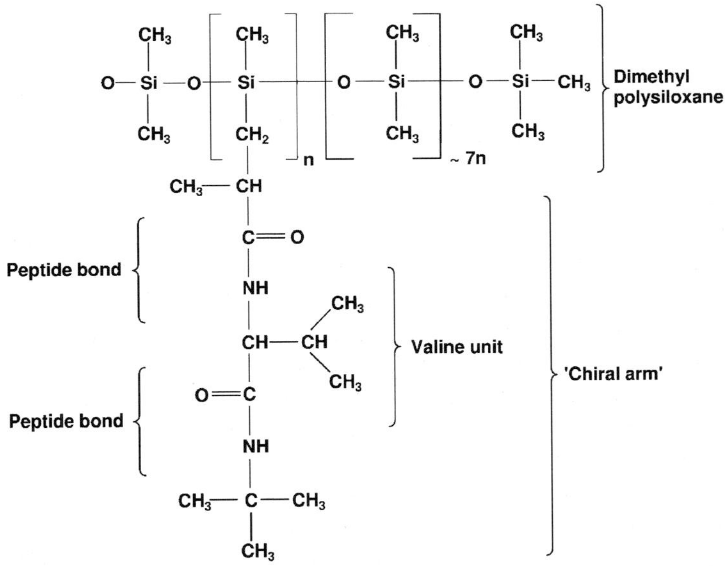

Fig. 2.14 Chemical structure of the Chirasil-Val polymer used as liquid film attached to the inner wall of a capillary column. The amino acid valine incorporated into the diamide chiral arm (side chain) of the dimethyl polysiloxane can be used in its L- or D-configuration resulting in Chirasil-L-Val and Chirasil-D-Val phases, respectively. Chirasil-L-Val and Chirasil-D-Val phases are commercially available[10]

[10] In enantioselective chromatography often both L-Val and D-Val phases are used in order to change the elution order of enantiomers. It is noteworthy that the L-Val phase is on the market with a layer thickness of 0.12 μm, whereas the commercially available D-Val phase shows a layer thickness of 0.08 μm only. As a consequence, for specific chiral analytes different separation factors and resolutions are observed.

enantiomer (Bayer 1983) in order to hold back one enantiomer stronger (and longer) than the other.

Chiral stationary phases with polymeric Chirasil-Val layers are often used for the resolution of amino acid enantiomers because amino acids interact perfectly with the chiral diamide side chain. Figure 2.15 gives an example of the amino acid leucine, base-line separated into its enantiomers. Consequently, Chirasil-Val phases are often applied in experiments for the study of various amino acid pools considered as "prebiotic".

Cyclodextrin phases: The classical and most commonly used chiral selectors are cyclodextrins (CDs), i.e., namely, cyclic oligosaccharides, which have been applied for 50 years for separating racemates. Since then, these selectors have made possible more than 1 000 chiral resolutions (Armstrong and Jin 1990; Armstrong et al. 1990; Schurig and Jakubetz 1997).

CDs are glycosidially bonded cyclic oligosaccharides of 6, 7 or 8 D-glucose-units called α-CD, β-CD or γ-CD, respectively.[11] The ring size of the CD molecule can be adapted for optimal resolution of chiral analytes. Typically, the space within an α-CD can accommodate a C_6 arene ring, a β-CD can admit molecules as big as naphthalene, while γ-CD can include anthracene or phenanthrene molecules. Furthermore, CDs can be chemically modified at the big entrance site in carbon position number 2 and 3 and alternatively at the small entrance site in position 6. Fig. 2.16 depicts the chemical structure of a β-cyclodextrin molecule.

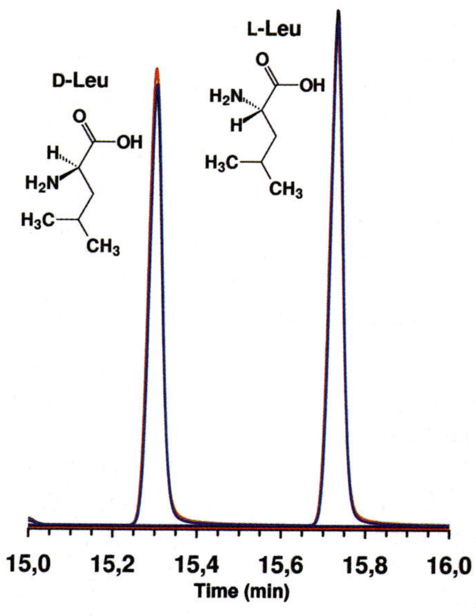

Fig. 2.15 Typical baseline resolution of the D- and L-leucine enantiomers in form of ethoxy carbonyl ethyl ester (ECEE) derivatives (Abe et al. 1996) on 25 m Chirasil-L-Val installed in an Agilent 6890/5973 GC-MSD system. Three overlaid chromatograms given in *blue*, *red*, and *black* color are shown. Splitless injection, oven temperature programmed to 3 min at 70°C, heated by 5°C/min to 155°C where it was kept constant for 5 min. Helium was used as a carrier gas with a constant flow of 1.5 mL/min

[11] Properties of δ-CD structures with 9 D-glucose-units have not yet been described in the literature. A retrosynthesis of these molecules would provide additional insight into chiral host-guest complexes and might challenge organic chemists.

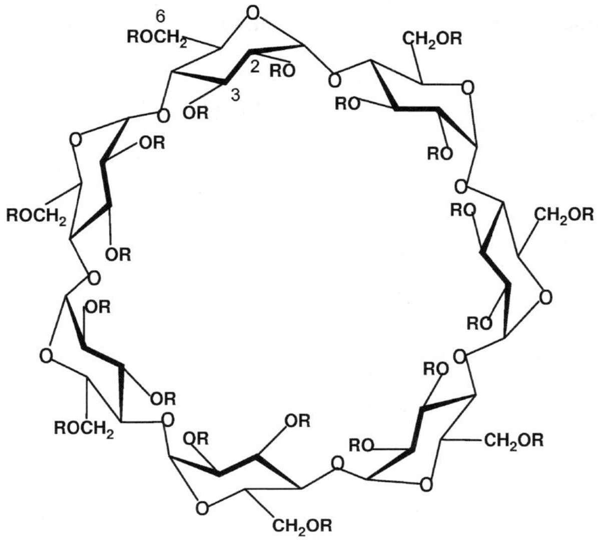

Fig. 2.16 Chemical structure of a β-cyclodextrin ring typically used for enantioselective gas chromatography. Here, cyclodextrins are dispersion-like embedded in the stationary phase. Highly enantioselective cyclodextrins can be created by changing the chemical function of the side chain. R = Methyl, permethylated cyclodextrin; R = *S*-2-hydroxypropyl cyclodextrin; commercial Astec Chiraldex B-PH cyclodextrin phase of Sigma-Aldrich. Alternatively, the Astec Chiraldex G-TA phase incorporates a 2,6-di-O-pentyl-3-trifluoroacetyl derivative of γ-cyclodextrin and the B-DA Chiraldex consists of a 2,6-di-O-pentyl-3-methoxy derivative of β-cyclodextrin. These columns are commercially available at the Sigma-Aldrich Co. who recently purchased Astec Inc.

Current research still tries to discover suitable substituents in order to modify CDs in these three positions (see e.g. Tisse et al. 2006). In an innovative manner, the research team of Jean-Philippe Bouillon at the University of Rouen actually seeks to create a hybrid between a Chirasil-Val phase and a cyclodextrin phase by attaching an amino acid to the CD molecule in position 6 in order to obtain highest enantiomeric resolution for a wide range of chiral analytes. Results are not published yet.

CDs are produced by partial degradation of natural starch. Consequently, they exist only in one-handed form. Synthetic chemistry has not yet been able to retrosynthesize the optical antipode for implementation into a stationary phase even if it would be very useful for specific kinds of experimental applications.

Free CDs are not suitable as chiral stationary phases in capillary gas chromatography: They are too polar and their melting points are too high. The introduction of specific hydrophobic groups leads to reduced melting points and better thermal stability. Stationary CD molecules are chiral and have been thought – as indicated above – to form associative complexes by inclusion (Schurig and Nowotny 1988) or other spatial arrangements (Lutz 1988) to interact with the chiral analyte.

In general, an optimum GC-separation of the enantiomers can be achieved when the stationary CD phase is able to include the chiral analytes into its cavities. By this inclusion, additional interactions between the CD molecule ("host") and the enantiomer ("guest") could become active that could not be formed by the conventional diamide substituted polymeric siloxane Chirasil-Val phases. The result of these additional interactions of the CD phases is – for a wide range of analytes such as amines, alcohols, and diols – a much better chiral recognition.

Chemically bonded cyclodextrin phases can be considered as the 3rd generation of chiral GC selectors. They contain permethylated β-cyclodextrin (β-CD) molecules, cyclic oligosaccharides composed of 7 D-glucose-units, chemically bonded to a polysiloxane film by octamethylene spacers which renders the column inert and robust against chemical and physical damage (Schurig et al. 1994; Jakubetz et al. 1997; Schurig and Jakubetz 1997). The chemical structure of the CD molecule anchored to the polysiloxane film is depicted in Fig. 2.17.

Linear oligodextrin phases: As outlined before, it has long been focussed on the cavities of cyclodextrin molecules contributing to the enantiomer separation of chiral analytes forming inclusion complexes dominated by non-polar interactions. The alkoxy groups located at the outside of the cyclodextrin torus were often ignored. These groups would respond to polar interactions with more polar analytes (Schurig et al. 1989). Surprisingly, linear oligodextrins, also entitled "acyclodextrins", were recently shown to have considerable potential as chiral selectors as well. In this case, host-guest interactions by inclusion can be excluded. Nevertheless, this 4th and most modern generation of chiral GC selectors provides good resolutions of chiral

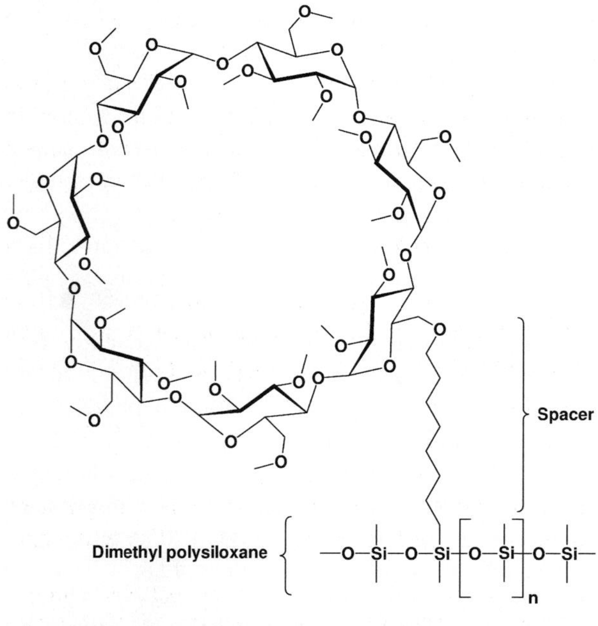

Fig. 2.17 Permethylated β-cyclodextrin chemically bonded to the polysiloxane phase. These polymers provide excellent separation factors for chiral hydrocarbon analytes. Trade name: Chirasil-Dex CB developed by Chrompack and now commercially available via Varian Inc.

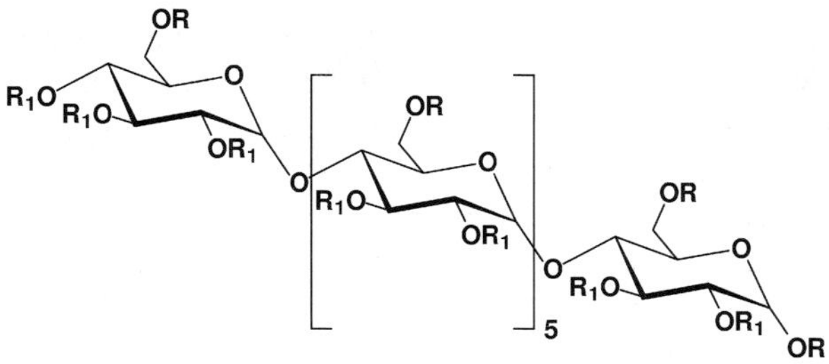

Fig. 2.18 Structural formula of a linear heptadextrin "acyclodextrin", a new selector in enantioselective gas chromatography for a variety of amino acids. R = *tert*-butyldimethylsilyl; R_1 = acetyl

analytes such as specifically derivatized amino acids (Sicoli et al. 2005). Its molecular structure is depicted in Fig. 2.18.

As a consequence of the enantiomer-resolving properties of linear oligodextrins, one should be careful in referring the resolution of enantiomers exclusively to inclusion complexes in cyclodextrin host molecules.

The merits of gas chromatography in the resolution of enantiomers have successfully stimulated research in liquid chromatography. Today, many chiral selectors have been developed for use in the high-performance (HPLC) mode including molecularly imprinted polymers (Gübitz and Schmidt 2004) but also in supercritical fluid chromatography (SFC), and thin layer chromatography (TLC). Other separation techniques in use are the various electromigration techniques such as capillary electrophoresis (CE), and capillary electrochromatography (CEC), a hybrid of HPLC and CE, which has recently been used rather successfully.

In order to determine the quality of an obtained resolution of chiral analytes, the separation factor α and the resolution R_S can be calculated. The chromatographic retention time t and the hold-up time t_M (also called time of passage) allows the calculation of the required reduced retention time t' (see Eq. 2.25).

$$t' = t - t_M \tag{2.25}$$

The ratio of the reduced retention times t'_R and t'_S corresponding to the R-enantiomer and the S-enantiomer results in the separation factor α as written in Eq. 2.26.[12] By convention, the separation factor α is equal or bigger than one ($\alpha \geq 1$).

$$\alpha = \frac{t'_R}{t'_S} \tag{2.26}$$

More precise information on the quality of enantiomer separation is given by the resolution R_S. The resolution is calculated according to Eq. 2.27 from the base

[12] For the calculation of the separation factor α reduced retention times t' are used. This is because the reduced retention time t' describes the time which the analyte stays during GC measurement (in between injection and detection) in the liquid polymer, i.e., the stationary phase. The hold-up time t_M corresponds to the time of the analyte in the gas phase (in the carrier gas).

width w_b of the chromatographic signal with w_{bR} for the *R*-enantiomer and w_{bS} for the *S*-enantiomer.

$$R = 2\frac{t_R - t_S}{w_{bR} + w_{bS}} \quad (2.27)$$

A resolution $R_S = 1$ corresponds to 85% of signals' resolution; 100% resolution of the signals corresponding to a base-line separation of the mirror image isomers is provided by a resolution of $R_S = 1.5$.

Having the information presented in Chap. 2 including both the chiroptical and chromatographic resolution of enantiomers in mind we are now equipped with the analytical background in order to enter the discussions in the following chapters presenting and illustrating up-to-date theories on the origin of life's molecular asymmetry.

Chapter 3
Minority Report: Life's Chiral Molecules of Opposite Handedness

Life is not symmetric. L-Mirror image isomers of amino acids play a crucial role in living organisms. They predominate in molecules produced via biological pathways such as protein sequences, whereas their optical antipodes, the D-enantiomers, are quantitatively of negligible occurrence in biological systems. In the case of life's genetic material, homochiral D-ribofuranose enantiomers are used for the molecular architecture of deoxyribonucleic acid (DNA) and ribonucleic acid (RNA) backbones. Here also, the L-configuration of the sugar enantiomers is largely underrepresented.

Based on the molecular asymmetry of today's life, one might assume that evolutionary processes at life's beginnings exclusively incorporated L-amino acids in an enantioselective manner for polycondensation and formation of peptides and proteins on the primitive Earth. Inappropriate D-amino acids and L-sugars might have been eliminated preliminarily due to physico-chemical processes of chemical evolution and excluded entirely from biological reactions. This assumption turns out not to be true.

Despite the dominance of L-amino acids and D-sugar molecules in the living world – the ratio of D- to L-glucose on Earth is estimated to be at least 10^{15} to 1 – a few deviant D-amino acids and L-sugars have found their way into the molecules of life where they occupy specific roles in various living higher organisms (MacDermott and Tranter 1989). In most of these roles homochirality is also maintained in their specific domain. New studies point to the assumption that some of these deviant amino acids and sugar molecules are not negligible, instead they are of considerable importance for upholding biochemical reactions for individual organisms.

In this chapter, a minority report on life's chiral molecules of opposite handedness will be given. Life's "desired" amino acids of the unusual D-configuration have been detected in plants (and food), bacteria, and also higher organisms such as frogs, snails, and spiders, but also rats, chickens, and even humans. Biochemical properties and the value of these D-amino acids in specific peptides will be given. We will then widen our view towards toxic and "undesired" amino acids in the living world, as some of these D-enantiomers are discussed to be involved in Alzheimer's disease, processes of eye-lens opacification, and also ageing phenomena. After the death of an organism, amino acids start to interconvert by racemization in a way that the ratio between D- and L-amino acids in archaeological and geochemical samples increases according to laws of chemical kinetics. This increase has been successfully measured and could be applied as a dating method called "the amino acid clock". The use of this method is complementary to the well-known ^{14}C-dating method and will be described briefly.

In view of the different theories on the origin of life, such a systematic collection of information on the occurrence of D-amino acids in living organisms is important. Some scientists argue that D-amino acids occurring in biological organisms today might be interpreted as molecular relicts of an ancient life form that was based on these "wrong" enantiomers. According to this theory, left (amino acid)-life and their mirror images right-life forms existed contemporaneously in early times and the selection of left-life occurred during biological evolution long after the first appearance of life on Earth. This model – called the biotic theory – will be discussed at the end of this chapter when we are aware of the occurrence of D-amino acids in living organisms.

3.1 Sympathy for the Devil: How D-Amino Acids Make Organisms Work

Often, D-amino acids are associated with destructive and toxic effects to living organisms. Recently, it became more and more apparent that some living organisms show "sympathy for the devil" by making use of the amino acids' "wrong" enantiomers. "Desired" D-amino acid enantiomers, i.e., D-amino acids produced "intentionally" by evolutionary processes, have been identified in plants (and in food), where considerable quantities seem to be omnipresent. Small peptides that use in their architecture at least one amino acid in its D-configuration have been found in vertebrates and invertebrates. In these cases, the amino acid's D-enantiomer can indeed be assumed to be "intentionally" implemented in the molecular construction of these peptides since the peptide looses its biological activity by inversion of the stereogenic center. Interestingly, "wanted" D-Amino acids have also been identified in the form of free amino acids in mammals, where their specific function, however, is less understood.

3.1.1 D-Amino Acids in Plants and Human Food

Recent studies have shown that certain D-amino acids are ubiquitously present in plants at higher levels than assumed earlier: D-amino acids were detected in leaves of ginkgo, maple, and sequoia trees. We consume D-amino acids regularly as they are molecular ingredients in fruits such as apple, pineapple, watermelon, papaya, mango, passion fruit, and coconut milk. In the most recent literature, D-amino acids are considered common constituents of plants and food.[1]

[1] D-Amino acids were identified in cactus, grasses, seeds, and seedlings of legumes such as maize, soybeans, runner beans, alfalfa, and also garden cress, and water cress (Brückner and Westhauser

In most of the cases, specific biochemical reaction pathways generate D-amino acids, which are not necessary to present in detail in the context of this book. But it is worth mentioning that the quantity of D-amino acids in foodstuff is often increased by food preparation: alkaline treatment as in the case of potatoes or Mexican tortillas increases the quantity of D-amino acids, acidic treatment as for example with gelatine, and moreover heat, pressure, and also microbial fermented foodstuffs such as cheese, yogurt, fish, beer, wine, and vinegar contain considerable quantities of D-amino acids. D-Amino acid enantiomers are furthermore expected to be created in the course of the Maillard reaction, which is a multistep reaction of reducing sugars with amino compounds leading finally to flavour compounds and melanoidins. The racemization of L-amino acids forming D-amino acids is assumed to occur at all reversible stages of the Maillard reaction. Consequently, D-amino acids in food are not *that* exotic. The quantity of individual D-amino acids in food can even be used to serve as an indicator for food processing, nutritional quality, bacterial contamination, age, and maturation, as well as authenticity.

Furthermore, in tobacco samples eight amino acids were found in their D-configuration usually in amounts of a few percent with the exception of D-α-alanine which was present in one smokeless tobacco sample at 50% (Kullman et al. 1999). These particularly high values were confirmed by Ali et al. (2006) who detected up to 34% D-α-alanine in different tobacco species. In this case as well, the Maillard reaction is assumed to be involved in the formation of D-amino acids.

3.1.2 D-Amino Acids in Peptides

Bacteria are known to make use of D-amino acids in several peptide antibiotics in which the "wrong" amino acids D-glutamate, D-phenylalanine, D-aspartate, and D-valine were found to be incorporated. It is important to note that the classical biosynthetic pathway in which DNA codes for the twenty L-amino acids of proteins cannot produce these exceptional D-amino acid-containing peptides. DNA does not code for D-amino acids. D-Amino acid containing peptides, on the contrary, are formed by specific biosynthetic routes that have nothing in common with usual protein synthesis. Bacteria implement D-α-alanine and D-glutamate not only in specific peptides, but also in the cell wall peptidoglycan layer (Nagata et al. 1998).

Moving from bacteria to eukaryotes, the peptide dermorphin serves as an example of a protein that makes use of a D-amino acid.[2] The small tetrapeptide achatin I

2003). In honey samples that usually contain 100 mg amino acids per 100 g honey, D-phenylalanine was found at up to 3%, and D-leucine up to 4%. In tea samples, the D-enantiomer of theanine, an amino acid commonly found in tea, was identified at up to 3%. Formosa Oolong tea even contained 13% D-theanine. Roasted cocoa beans, cocoa powder, chocolate, and cocoa shells include D-amino acids and in particular D-proline was found in amounts up to 37% (Pätzold and Brückner 2006). As expected, the quantities of D-amino acids increased on heating of the cocoa samples.

[2] Dermorphin was isolated from skin secretions of the South and Middle-American frog *Phyllomedusinae*. It is an opioid peptide of a total of seven amino acid residues including one D-α-alanine molecule. The biological activity of dermorphin was found to be 1 000 times higher than that of morphine. Substituting the D-α-alanine with an L-α-alanine residue provokes loosing the

is an example of a peptide that includes the D-enantiomer of phenylalanine.[3] A similar variation in the biological activity of a snail protein was reported for the peptide fullicin. Fullicin that includes a D-asparagine residue shows higher bioactivity compared to fullicin containing asparagine's L-enantiomer.

Moving to higher peptides, we can refer to ω-agotoxin that was isolated from the toxin of a spider and contained 48 amino acid residues. The amino acid serine at position 46 of ω-agotoxin was shown to be present in D-configuration.

At Himeji and Tokyo Institute of Technology, Yoko Nagata and her team followed an interesting hypothesis by investigating the occurrence of peptidyl D-amino acids among various organisms. She compared individual contents of D-amino acids for bacteria, archaea, and eukaryotes. At the beginning of the study, it seemed to her that D-amino acids widely distributed among bacteria have nearly been eliminated in eukaryotes during the evolution of these organisms. The few remaining examples of D-amino acids in eukaryotes were given above in this paragraph. If enantioselective analyses of different types of organisms confirmed this assumption, these results would indicate that the first organisms such as archaea may have used D-amino acids for their molecular architecture of biopolymers in an equivalent manner to L-amino acids and that the symmetry towards the exclusive use of L-amino acids in proteins today was violated only later, i.e., during evolution of eukaryotes. However, Nagata's team's results showed that bacteria and archaea, as well as eukaryotes, contain amounts of peptidyl D-amino acids in the soluble fraction which are not large but are comparable with each other (Nagata et al. 1998). Hence, these results could not be used to verify the attractive hypothesis on the evolutionary elimination of D-amino acids in eukaryotes.

How do we then interpret today's small but general occurrence of D-amino acids in the various domains of life such as bacteria, archaea, and eukaryotes? Were all these D-amino acids produced by successful mutations during biological evolution and racemization of L-amino acids meaning that the origin of life on Earth was exclusively based on the use of L-amino acids? Or are these D-amino acids molecular relicts of an ancient and parallel existing mirror-image life? This question remains difficult to answer (see Sect. 3.4), especially because ubiquitous L-amino acids of living organisms seem to have contaminated all niches of Earth erasing most molecular traces of any mirror-image life.

3.1.3 Do D-Amino Acids Affect Grey Matter?

Coming now to mammalians and human beings, we note that the identification of free D-amino acids has been reported for various mammalian tissues. Free

impressively high activity. The D-enantiomer in the peptide structure is thus essential for the biological activity of dermorphin.

[3] Achatin I was isolated in tissues of the African giant snail *Achatina fulica* and is described to enhance cardiac activity. Achatin II is the epimer peptide using L-phenylalanine. It was also isolated from the snail's tissue but shows, however, minor biological activity as compared to achatin I (Fujii 2002).

D-aspartic acid enantiomers were found in the brain of rats, chickens, and humans. Here, concentrations of D-aspartic acid seem to be related to developmental stages since concentrations increase and then rapidly decrease to trace amounts again in adult tissues. For example, in the prefrontal cortex of the human brain, as much as 60% of the total aspartic acid is present in its D-configuration at week 14 of gestation, but rapidly decreases to trace levels by the time of birth (Hashimoto et al. 1993). Nagata et al. (1994) identified a similar timely distribution pattern for the concentration of D-serine in the forebrain of mammals such as mouse, rat, and bull.[4]

D-serine, D-α-alanine, and D-proline were reported to be present in human plasma in D/L-ratios up to 0.24 (Nagata et al. 1992a) and also in mouse kidney in D/L-ratios up to 0.035 (Nagata et al. 1992b). In these experiments, particular attention had to be paid to the enantioselective analysis of serine. Unexpectedly, under the chosen hydrolysis conditions using 6 molar hydrochloric acid, the D-enantiomer decomposed faster than the L-enantiomer (Nagata and Tagashira 1998).[5] Experimental data on D/L-ratios of serine are therefore to be handled with care.

In contrast to the importance of frog dermorphin and snail achatin, the physiological role of free D-amino acids found in mammals is not yet known. Further investigations are strongly required to elucidate the source of D-amino acids as well as their physiological, pharmacological, and pathological significance.

3.2 The Dark Side of D-Amino Acids

Some D-amino acid enantiomers are associated with destructive and toxic effects for living organisms, in contrast to the above-presented "desired" deviant enantiomers. Consequently, biological organisms try to get rid of these enantiomers using various strategies. First of all, D-amino acids can be detected and eliminated by special proteins, called D-amino acid oxidases. Moreover, D-amino acids can be stored in deposits where they do not interfere with everyday metabolism (Rein 1992).

In some cases, the amount of D-amino acids constantly increases during ageing of the living organism. Eye-lenses serve as such an example where D-enantiomers of the amino acid D-aspartic acid were detected, which increase in concentration with age and denaturation induced by irradiation with β-particles. The mechanism of formation of D-aspartic acid probably involves stereoinversion of a specific aspartic acid residue (Asp-151) in the predominant lens protein called αA-crystallin. Additional post-translational modifications, including deaminations and oxidations were observed in crystalline lens protein samples of aged persons. The modifications

[4] Similar developmental changes were reported for other organs such as testis, adrenal and pineal glands, mainly focussing on the D-amino acids D-aspartic acid and D-serine.

[5] The authors proposed rather adventurously that the different behavior of the D- and L-serine enantiomers during hydrolysis might have caused the biomolecular asymmetry since the less decomposed L-serine might have had more opportunities to bind with peptide nucleic acid (PNA) than did D-serine (Nagata and Tagashira 1998).

appeared to be directly related to lens opacification, meaning that the eye-lens becomes less transparent (Momose et al. 1998; Fujii 2002). This by the way serves as a good example for Erwin Schrödinger's hypothesis that racemization is caused by an increasing incapability of the living organism to work *against* the force of increasing entropy. Here, the concept of "negentropy production" was considered as characteristic of the living matter.

These post-translational modifications were not observed in the lens protein samples of a young group. In these kinds of eye-medical studies, the eye-lens opacification was traced back to the amount of D-aspartic acid in the lens.

Moreover, other examples let us interpret the occurrence of D-amino acids in proteins as molecular markers for ageing (Fujii 2002): bones show relatively high concentrations of D-amino acids. In teeth, D-enantiomers of amino acids were identified as well. Profiles of D-aspartic acid in teeth revealed that aged living bodies deposit higher amounts of this deviant enantiomer, which was explained by an ongoing racemization of aspartyl residues in a protein over time (Fujii 2002). The older the organism, the higher is the concentration of D-amino acids to remove. There is thus speculation that molecules of deviating chirality may play a role in the processes of ageing itself (MacDermott and Tranter 1989).

Alzheimer's disease may also be linked to an increased amount of D-aspartic acid and D-serine enantiomers in the body. It is assumed that the deposition of the β-amyloid protein in the brain causes – or is involved in molecular processes in connection with – Alzheimer's disease. This protein is composed of 42 amino acid residues showing stereochemical inversion of the Asp-1, Asp-7, and Ser-26 amino acid residues in the case of Alzheimer's patients. It was thus suggested that the β-amyloid protein was aggregated by the racemization of these amino acid residues and accumulated in the brain (Fujii 2002 and references therein). Further studies confirmed that the racemization of an aspartic acid residue in the protein accelerated peptide aggregation and fibril formation observed for Alzheimer's patients.

3.3 "What Time was it?" Ask the Amino Acid Clock!

The racemization of biological L-amino acids towards D-amino acids really starts when counteractive processes, e.g. based on D-amino acid oxidases, become inactive after the death of a living organism which over time results in complete racemization. The ongoing increase of the D-amino acid quantity can thus be applied for archaeological and geochemical dating by a method called "the amino acid clock". This method has thermodynamic roots because the racemic mixture of chiral molecules is thermodynamically favoured over its enantioenriched or homochiral form, due to the mixture's increased entropy. Therefore, enantiopure amino acids approach spontaneously the state of a racemic mixture. The questions are: what does "spontaneously" mean in this context and how long does this racemization reaction take? At 0°C, α-alanine in aqueous solution racemizes in about one hundred thousand years. Aspartic acid racemizes faster, isoleucine more slowly. At room temperature, some thousand years are sufficient for α-alanine racemization (Rein 1992). If amino acids are stored in solid state without any humidity, the half-life time can be increased to some million years. The racemization of an amino acid in a peptide chain, however, is more complicated. As illustrated in Fig. 3.1, it is not important whether we start the racemization reaction from the enantiopure L-enantiomer or from the enantiopure D-enantiomer. The racemic mixture is favoured in both cases.

Fig. 3.1 Thermodynamically favoured racemization of chiral molecules. The "amino acid clock" is based on the racemization of L-amino acids accompanied by a gain of entropy towards a racemic D,L-mixture as depicted on the left side of this figure

When living organisms such as plants or animals die, the homochiral proteinaceous L-amino acids immediately start to racemize by changing their handedness at a very slow but rather uniform rate. The controlled recording of the racemization process hence provides a way of dating ancient objects by determining the enantiomeric excess of one or more amino acids. This way of dating is called the "amino acid clock". It is complementary to the ^{14}C-method of geochemical dating, since older objects can be dated. The half-life of ^{14}C is limited to only about 6 000 years making dating possible for approx. up to ten half-life times, i.e., up to 50 000 years.

Bone samples are important constituents of many archaeological sites and often the only biogenic material available. Hartwig Elster and Steve Weiner have investigated the amino acid racemization of fossil bones in the team of Emanuel Gil-Av at the Weizmann Institute of Science in Rehovot. Enantioselective analyses of aspartic acid in different fractions of 51 fossil bones of various ages showed that the collagen-rich fractions do manifest increasing D-aspartic acid contents with increasing age. No such correlation existed for the fractions rich in non-collagenous proteins (Elster et al. 1991). The accurate application of "the amino acid clock" is therefore limited to archaeological dating of collagen-rich samples of bones.

Amino acid racemization was also used in the case of human specimens of known age for the determination of their temperature history. The German emperor Lothar I, who died on 5 December 1137 in Bavaria, was buried 500 km north of the place of his death at his castle in Königslutter (20 km west of Braunschweig). Using 20–30 km per day as the estimated transportation speed for his corpse, Bada et al. (1989) calculated that an interval of several weeks is likely to have occurred between the time of his death in Bavaria and his subsequent burial at the castle. This long period would have obviously presented problems of post-mortem decay. Bada et al. (1989) have thus carried out aspartic acid racemization analyses of Lothar, and two of his relatives, his wife Richenza and brother-in-law Duke Heinrich der Stolze, both of whom died at Königslutter within a few years of Lothar's death and were buried in the castle along-side him. Enantioselective aspartic acid analyses of the obtained bones from the three individuals revealed that the corpse of Lothar I contained significantly higher amounts of D-aspartic acid than expected from the above calculations and was thus interpreted to be boiled in water for about 6 h before burial. Boiling was used to deflesh Lothar's corpse to prevent post-mortem decay during transit to his castle.

Despite its obviously wide range of applications, the "amino acid clock" has a considerable margin of error, because the rate of racemization of an amino acid is mostly not a simple linear function of time but also influenced by various parameters such as temperature, pH value, humidity, and position in a peptide linkage. Limitations of "the amino acid clock" became visible when Kunnas and Jauhiainen (1993) from the Joensuu University in Finland tried to estimate the age of a peat bog from the enantioselective determination of the D/L-ratios of free amino acids. Too many environmental factors influenced the D/L-ratios in natural conditions and the age of the selected peat samples was too young only by 9 000 years to allow the obtaining of reliable results on pure racemization without an additional source of interfering D-enantiomers, for example from the destruction of plant and microbial material.

Racemization reactions are not limited to amino acids. Sugar molecules such as glucose and ribose are also chiral and entropically favour their racemate. Since these molecules possess more than one stereogenic carbon atom, reactions that form a racemate are more complex. Here, the interconversion at one stereogenic center is called an epimerization instead of racemization since the molecule possesses more than one stereogenic center. Nevertheless, the epimerization of D-ribose can be accomplished within about one hundred years (Rein 1992) and is therewith too fast to be of interest for dating specific samples or other scientific applications in the context of archaeological research.

3.4 Biotic Origin: Were D-Life and L-Life Contemporaries?

3.4.1 Biotic and Selection Theories

So called biotic or selection theories were developed in the 1950s in order to explain the biomolecular asymmetry. According to these theories, two kinds of mirror-image organisms could have co-existed next to each other at a cellular or protocellular level in a very early phase of the biochemical or biological evolution. One type consisted of proteins (and nucleic acids) composed of L-amino acids (and D-sugars) and the other one of D-amino acid (and L-sugar) enantiomers. Both antipodal organisms co-existed for a certain time and the more favourable mirror image form, having a certain evolutionary advantage such as the development of "metabolic interdependence", could have been selected by interacting with other non-racemic substances (Wald 1957). This theory suggests that life arose in a racemic environment, and homochirality as such only developed later (Bonner 1995a). Wald suggested that the α-helix, formed during peptide growth, may function as the stereoselectivity agent, favouring the incorporation of one amino acid enantiomer over the other.

At the University of New Orleans, in the US, Stanley Goldberg and his team intensely tried to discover experimental evidence for Wald's theory. Goldberg et al. undertook an investigation in an effort to assess the tendency of an arbitrarily chosen system of a peptide assembly to display biased stereoselectivity. Derivatives of α-alanine, aspartic acid, and glycine were chosen for the formation of 34 different structures of di-, tri-, and tetrapeptides. These amino acids had been known for their probable presence in prebiotic environments. And indeed, each of the competitive peptide-forming processes was found to be stereoselective. Goldberg et al. (1987) concluded that these results supported the notion that reaction conditions may be found to allow or promote the stepwise assembly of chiral monomers into even small chains with significant levels of stereoregularity.

More evidence was compiled for the biotic theory by the fact that in contrast to "normal" proteins, D-amino acids were discovered in a number of specific biological samples such as bacterial cells (Nagata et al. 1998), vertebrate brains (Nagata et al. 1994), and others, which might be interpreted as relicts of mirror-image life and which might emphasize the biological selection process of specific antipodes during evolution.

3.4.2 How Polypeptides Form

Polypeptides and proteins are composed of amino acids. Their two most common structural motives are α-helix and β-sheet. The α-helix shows helical chirality with 3.6 residues per turn, a special case of axial chirality. As we have seen in Chap. 2, in proteins, the L-configuration of the monomer amino acids dictates the *P*-helical chirality of the helix.[6] This helicity arises from energetic constraints in the bonding of L-amino acids. A chain consisting of D-amino acids would produce spirals with an *M*-helical twist. This assumption was confirmed by chiroptical data of the β-amyloid peptide and its chemically synthesized all-D enantiomer. The circular dichroism spectra showed equal optical rotation in value but opposite optical rotation in sign (Cribbs et al. 1997), suggesting that the folded forms of the all-D and all-L peptides are mirror images when viewed in three dimensions (see also Petsko 1992; Milton et al. 1992). So, how did life's polypeptides and proteins form, and do they contain any structural motive?

The desired information on the formation of polypeptides and proteins can be gained by their synthesis under controlled conditions in the laboratory. The growth of the α-helical and β-sheet motifs under variation of the amino acid monomers' chirality is of particular importance. These studies have been systematically performed in André Brack's research team at the Centre de Biophysique Moléculaire in Orléans, France.

Here, a slight enantiomeric excess in a mixture of activated amino acid enantiomers was shown to be amplified by a stereoselective polymerization forming

---

[6] The structural protein collagen is an exception to the rule: each of the three amino acid chains of collagen is twisted into an *M*-helix with only three amino acids per turn (Borchers et al. 2004).

α-helical structures. Brack and Spach demonstrated that a right-handed α-helical peptidic seed composed of L-amino acids incorporates L-amino acid monomers 18 times faster than D-amino acids (Brack and Spach 1971). Similarly, a β-sheet conformation of synthetic polypeptides enriched in a dominant enantiomer was found to be more stable than a random coil of racemic amino acids (Brack and Spach 1980). Such macro-molecular routes of peptide growth towards α-helices or β-sheets show considerable potential in the amplification of enantiomeric enrichments in the reactants; they were theoretically modelled through the Majority Rule caused by the strong cooperativity effect claimed by Green and Selinger (1998) and Green et al. (1999). More recently, an enantiomeric enhancement through self-assemblies of amphiphilic activated amino acid analogues on the surface of water has been reported (Zepik et al. 2002). But do these items inevitably point towards the biotic or selection theories?

3.4.3 Counterarguments Against the Biotic Theory

A common criticism on the biotic or selection theories is based on experiments and a qualitative model of Goldanskii and Kuz'min (1988, 1991) and argues that pure homochiral protein macromolecules and pure polynucleotide chains are not self-replicable in a racemic medium. In short: life requires homochirality! Models of biogenesis do all demand a supply of sufficient enantioenrichments in precursor molecules. These enantioenrichments are required to trigger self-organization far from thermodynamic equilibrium (Krueger and Kissel 1989) in order to generate living structures. Manifold laboratory experiments on the generation of polymers that are composed of enantiopure monomers determined the effect of "enantiomeric cross inhibition" (Lundberg and Doty 1957; Idelson and Blout 1958; Joyce et al. 1984; Wächtershäuser 1991; Rein 1992). The statement "life requires homochirality" is supported by the "enantiomeric cross inhibition", an effect that is often encountered in the scientific literature and deserves an explanation.

3.4.3.1 Enantiomeric Cross Inhibition of Polypeptides

The effect of "enantiomeric cross inhibition" was discovered at the University of Boston by M. Idelson and E. R. Blout, who investigated the polymerization of an L-glutamate derivative[7] that forms polypeptides with up to 1 500 monomer units. Under a variety of experimental conditions kinetic data were recorded: solvent, pH values, and temperature were varied and molecular initiators were added in order to fully understand the polymerization reaction and to elucidate its mechanism. In the context of this book, it is of crucial importance that Idelson and Blout (1958) also added defined quantities of D-glutamate[7] to the reaction mixture of L-glutamate[7] in

[7] γ-benzyl-L-glutamate-N-carboxyanhydride (NCA) derivative

order to study the effect of the optical configuration of the reactants on the polymerization. Kinetic data were recorded and the molecular weights of the formed polymers were determined. The result of these studies is that the presence of small amounts of D-glutamate decreased the polymerization rate and lowered the molecular weight of the polypeptide formed as depicted in Fig. 3.2. This effect was later called the "enantiomeric cross inhibition": the cross enantiomer (here: D-glutamate) inhibits the chain formation of L-glutamate. In other words: D-glutamate incorporated at the end of an L-glutamate oligopeptide acts as chain terminator. Also the inverse effect was observed namely that a small quantity of L-glutamate[7] decreased the polymerization rate of a growing D-glutamate[7] polypeptide.

One year before, Lundberg and Doty (1957) had investigated the rate of polymerization of D,L-mixtures of the glutamate derivative showing a pronounced decrease in the propagation rate for the racemic mixture relative to that of the pure L-enantiomer. While the growing chain was non-helical, the lowering was less than a factor of two and was thereby accounted for in terms of a preference of the chain to add the isomer corresponding to its terminal residue. At a later stage of the reaction process, when the growing chain had reached the helical configuration, the lowering of the rate was greater than a factor of two and this indicated a preference exerted by residues lying behind the terminal one. For the L-glutamate derivative the helical

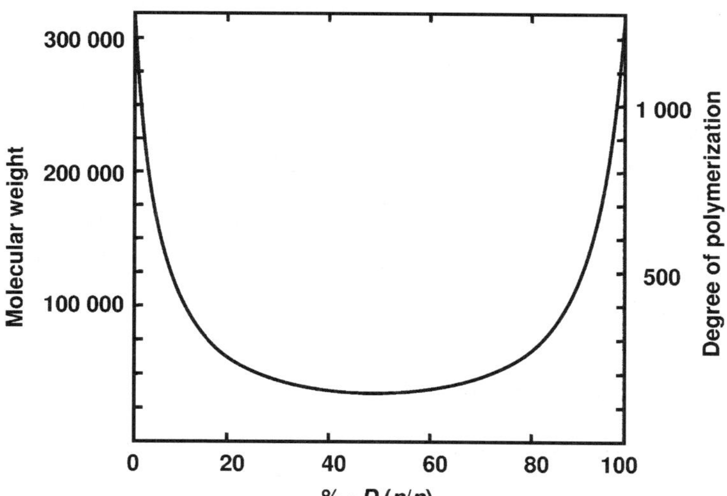

Fig. 3.2 Effect of optical isomer incorporation on the degree of polymerization and molecular weight. The more the initial enantiomer composition deviates from a racemic 50% D- and 50% L-mixture (see *x*-axis), the higher the degree of polymerization of the created polymer (see *y*-axis on the right) and the higher the molecular weight of this polymer (see *y*-axis on the left). As expected, the pure D-glutamate derivative has the same degree of polymerization as the pure L-glutamate derivative. However, when as little as 5% (*n*/*n*) of the L-glutamate is polymerized with 95% (*n*/*n*) of D-glutamate (or vice versa) the polymerization rate is reduced to 1/3 of the value obtained for each enantiopure isomer. At equimolar proportions of D-glutamate and L-glutamate the polymerization rate is 1/17 that of the pure enantiomers (Idelson and Blout 1958)

sense of the polymer occurred after about four residues were added. Moreover, evidence accumulated that the helical configuration of a polymer was substantially less stable when the polypeptide contained both D- and L-residues. For the D,L-glutamate reaction mixture, a polymer in the preformed helical configuration composed of a single enantiomer reacted about five times faster with its own enantiomer than with the opposite optical isomer (Lundberg and Doty 1957).

A similar stereoselectivity in the polymerization of glutamate derivatives has been reproduced by others and was described for a variety of other amino acids including α-alanine, leucine, valine, isoleucine, and phenylalanine (Borchers et al. 2004) These data clearly indicate that the presence of a D-amino acid in an α-helix or β-sheet of mostly L-enantiomers has a distorting and destabilizing effect.

In order to investigate the stereoisomer distribution of tryptophan monomers in formed oligomers, the team of Pier Luigi Luisi at ETH in Zurich followed an innovative approach. One of the tryptophan enantiomers was isotopically labelled and after oligomer formation by a polycondensation reaction, ion monitoring mass spectroscopy was used to investigate the stereoisomer incorporation. The results show that homochiral oligo-tryptophan was 8.3 and 40 times more frequent in heptamers with $n = 7$ and decamers with $n = 10$, respectively, than would have been expected for a statistical distribution. Conversely, heptamers and decamers with approx. equal proportions of D- and L-enantiomers were underrepresented (Blocher et al. 2001).

It would be of great interest to see in future experiments whether mixtures of different amino acids also lead to a preferential formation of homochiral oligopeptides consisting of more than one type of amino acid, a result which we would expect to be demonstrated.

3.4.3.2 Enantiomeric Cross Inhibition of Nucleotides

As in proteins, the chiral monomer units of DNA and RNA are enantiopure, each of the nucleotides containing either D-deoxyribose or D-ribose. The importance of chiral purity during polymerization was also found by the experimental investigation of this genetic material, particularly in studies on the template-directed oligomerization of nucleotides (Bonner 1995a): the oligomerization of an activated guanosine mononucleotide proceeded readily if the monomers were of the same stereochemical configuration as the template, and was far less efficient if the monomers were of the opposite handedness (Joyce et al. 1984). More recent research activities shifted to a study of nucleotide analogues such as pyranosyl-RNA (Beier et al. 1999), threofuranosyl nucleic acids TNA (Schöning et al. 2000), and peptide nucleic acids PNA that had been proposed as initial pre-RNA genetic materials (Sutherland 2007). In Chap. 8 of this book we will have a closer look at these candidates in pre-RNA oligonucleotide chemistry. Anyhow, short tetramers of pyranosyl-RNA did not exhibit "enantiomeric cross inhibition" because the oligomerization of homochiral tetramers was not, or only weakly, inhibited by the presence of the non-oligomerizing diastereoisomers (Bolli et al. 1997). However, the rate of oligomerization was significantly reduced when only heterochiral tetramers were available for co-oligomerization (Borchers et al. 2004).

Due to these experimental results on tetranucleotides, the chiroselective replication of biopolymers became an attractive model for explaining homochirality in nature. Saghatelian et al. (2001) theoretically simulated a 32-residue peptide replicator (see also Siegel 2001) that was calculated to amplify homochiral products from a racemic mixture of peptide fragments. This approach used a chiroselective autocatalytic cycle. Similar amplifying models will be discussed in more detail in Chap. 10.

Summarizing, the enantiomeric cross-inhibitory effect was measured and present for peptides and for oligonucleotides, except short pyranosyl-RNA tetranucleotides. It provides evidence for the assumption that big chain length biopolymers would preferably arise in an enantioenriched environment during processes of chemical evolution and that life could arise in a greatly superior way from an environment possessing a certain enantiomeric excess. Life in its primitive form seems to require an ongoing source for the accumulation of chiral molecules on early Earth (Bonner 1995a). As a consequence of this, modern theories on the origin of life were supposed to follow *abiogenic* models explaining the origin of the molecular parity violation long before biological evolution. The next chapters will hence be dedicated to detailed and topical explanations of modern abiogenic theories on how life's asymmetry originated.

Was life's asymmetry transferred from chiral minerals to racemic carbon-based molecules? Or was it written into the atomic nuclei of these molecules in form of the weak force? Or did chiral fields, e.g. in interstellar space, violate molecular symmetry in prebiotic times long before the actual origin of life on Earth? Our journey through these theories will come to an end by presenting information on the chiral molecular ingredients of extraterrestrial samples.

Chapter 4
When Crystals Deliver Chirality to Life

We have already outlined that biomolecules like DNA, proteins (enzymes), and also lipids[1] are composed of homochiral monomers that do not tolerate any pollution by the "wrong" enantiomer. In order to understand the decisive asymmetry of evolutionary processes at life's beginning, we have to think of an asymmetric element put into the vicinity of a racemic or prochiral chemical substrate. In this chapter, a crystallization nucleus spontaneously generated in solution will serve as such an asymmetric element. According to this model, the chiral information will be spread from the chiral crystallization nucleus to the whole crystal and subsequently transferred to an organic substrate which itself might evolve asymmetrically into a self-organizing living system.

In the context of mainly inorganic crystal chemistry, it is refreshing to remark that snail shells are spiral systems that – as a matter of principle – are chiral and can be right-turning or left-turning. Interestingly, most snails construct their chiral shells deliberately; within most species, snail shells are nearly homochiral. Usually, right-helical shells are stereoselectively preferred and dominant, as Brunner (1999) investigated and documented beautifully. In the case of typical edible snails (*Helix pomatia*), right-helical shells have the majority: only one out of 20 000 species is constructed in the atypical left-helical way.[2] Other snail species show this particular stereoselectivity as well; in exceptional cases, as in the tower snail, left-helical shells are dominant. The Cuban tree snail is the exception, producing shells in a racemic right- to left-helical ratio.

At the University of Glasgow, biochemist Graham Cairns-Smith (1982, 1985) thought on the tricky question: "wherefrom does the shell, which is made up of the purely inorganic aragonite-modification of calcium carbonate, obtain the

---

[1] The homochirality of lipids is, for example, manifested in the molecular side chains of vitamin E and chlorophyll, both containing two tertiary carbon atoms in their enantiopure *R*-configuration (Spach and Brack 1983; Meierhenrich et al. 2001e).

[2] In the French Bourgogne, where edible snails are thought to be delicious and processed on a tons-per-year scale, exact knowledge of the right- to left-helical shell ratio was obtained (Brunner 1999).

U. Meierhenrich, *Amino Acids and the Asymmetry of Life*. Advances in Astrobiology and Biogeophysics,

information in which direction it has to turn?" The homochirality of the macroscopic inorganic calcite shell must have found a way into it. Cairns-Smith reasoned that organic proteins from the living and supple part of the snail might transfer the chiral information through the phase boundary to the inorganic calcite shell. This obvious transfer – and now the story becomes interesting – might be turned the other way around: would it be possible to transfer chiral information out of a mineralogical crystal to organic molecules that are in immediate contact with its surface? Consequently, wouldn't it be possible that the origin of biomolecular asymmetry and, furthermore, the origin of life have inorganic roots? Cairns-Smith called this migration process the "genetic takeover" from crystals to the first living beings and founded therewith a respected and distinguished branch of research on the origin of life.[3]

Inorganic crystals were abundant on the early Earth and they have always been in contact with organic molecules as well. That is reason enough for us to take a closer look at the symmetry and chirality of crystals. Indeed, crystals can be chiral and they were proposed to have served as asymmetric elements during chemical evolution.

In general, solids are composed of molecules that are packed together tightly to create a rigid, stable structure. In almost every case, such an arrangement of molecules is patterned. This orderly pattern is the crystalline structure, also called the lattice structure, of the solid. A *chiral* lattice structure can be formed in three ways.

1. Enantiopure compounds crystallize and form a chiral crystal lattice. (The crystallization of chirally pure molecules necessarily results in a chiral lattice structure.)
2. Achiral compounds crystallize and form a lattice structure that is chiral; this asymmetric crystallization is unpredictable and is the topic of cutting-edge research.
3. Chiral molecules in the racemic state undergo spontaneous resolution in which the two enantiomers segregate into a conglomerate of enantiopure crystals.

Chiral crystals, like any other asymmetric object, exist in two enantiomorphous forms of equal energy (Sakamoto 2004). In this book we will discuss formation way number 2 (Sect. 4.1) and way number 3 (Sect. 4.2) in more detail, since only here an asymmetry is developed and enhanced by the system. Chiral crystals obtained by formation way number 1 require enantiopure compounds and do not increase any enantiomeric excess, making them uninteresting in the scope of this book.

Quartz, the most common mineral, belongs to the second group (way 2) and possesses a chiral lattice structure. Bear in mind that the monomer of quartz, silicon dioxide (SiO_2, or silica) is achiral. Molten quartz is not optically active (Barron 2004). The quartz lattice, however, shows a helical structure twisting either to the right or to the left making two distinct enantiomorphous forms, which are represented in Fig. 4.1. Enantiomorphous quartz crystals are optically active; the optical

[3] Advanced parts of Cairns-Smith's theory however are highly controversial, particularly when he argues on protein molecules that served as catalysts in the growth of "crystal genes" on clay surfaces, later on breaking away from the clay to become three-dimensional living cells.

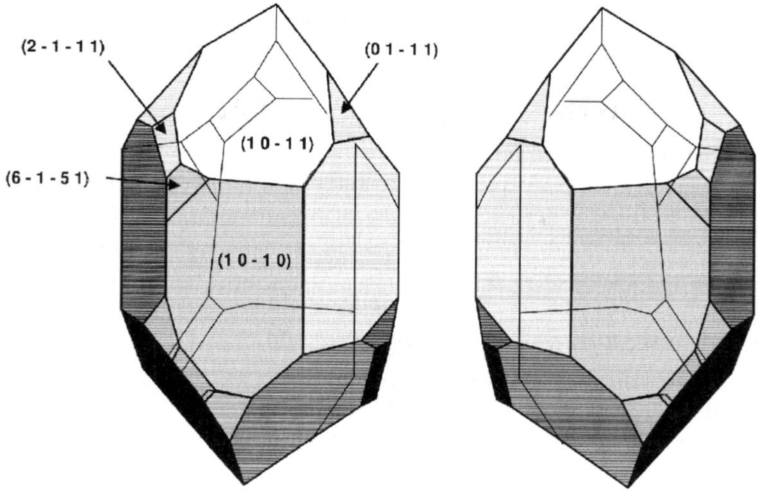

Fig. 4.1 Enantiomorph (−)-*l*-quartz (*left*) and (+)-*d*-quartz crystals (*right*) viewed along the a-axis. Miller indices that are used in crystallography to classify crystal surfaces are indicated in brackets for the left-handed quartz

rotation depends on the angle between optical axis and light, which means that the quartz crystal is anisotropic. We distinguish between (+)-*d*-quartz and (−)-*l*-quartz. Quartz is morphologically chiral (Herschel 1822; de Vries 1958).

Other chiral crystals exist in addition to quartz. Even if people believed that chiral crystallization is a very rare phenomenon, a survey of Koshima (2004) revealed that the statistical probability for the chiral crystallization of achiral compounds was around 8%. Sakamoto (2004) carefully surveyed 30 000 organic crystalline compounds and determined their chirality, showing that 19.4% of these compounds crystallized as chiral enantiomers. How can we identify them?

A chiral crystal must belong to a chiral space group. From a total of 230 space groups, 65 space groups are chiral, having translation and rotation symmetry elements only. Achiral crystals have symmetry elements as a mirror plane or a center of inversion. The most frequent chiral space groups are $P2_1$ and $P2_12_12_1$. When the space group of the single crystal of a compound is chiral, the crystal is designated as chiral.

4.1 Spontaneous Crystallization of Chiral Minerals

In order to prepare chiral crystals from achiral compounds, no special crystallization technique is required. Ordinary methods can be employed, such as cooling a hot saturated solution of a compound or evaporating the solvent to obtain crystals.

Determining if a crystal is chiral can be achieved by measuring the optical rotation in solution using a polarimeter (Koshima 2004).[4]

In 1990, Dilip Kondepudi, a former member of Ilya Prigogine's Brussels school of thermodynamics, surprised professional circles with a simple crystallization experiment on chiral symmetry breaking. At Wake Forest University in North Carolina, he and his team studied the crystallization of sodium chlorate $NaClO_3$. The crystals of sodium chlorate belong to the cubic space group $P2_13$ and are – as in the case of quartz – optically active although the molecular compounds of the crystal are not chiral. Kondepudi et al. prepared aqueous solutions of sodium chlorate by dissolving 100 g $NaClO_3$ in 120 mL of distilled water. After complete dissolution of the solute at 50°C, samples of 25 mL were transferred to several petri dishes and left for crystals to grow. After 4 to 5 days, the crystals bigger than 1 mm in size were separated and their optical rotation was measured by a pair of polarizers.[5] The crystals appeared blue when the polarizers were crossed and turn red on rotating the analyser clockwise or anticlockwise, depending on the crystal's handedness. For a 1 mm crystal, the change in colour from blue to red occurs at a rotation of about 7°. After precipitation of a total of 1000 crystals, 525 $NaClO_3$ crystals were identified to be of *l*-configuration whereas 475 crystals were of *d*-configuration. As expected, crystals were found in almost equal numbers in repeated experiments. Nothing new. The frequency distribution for crystal enantiomeric excesses was said to be monomodal, as demonstrated more than 100 years ago (Kipping and Pope 1898) and as illustrated in Fig. 4.2.

When the solution was simply stirred during crystallization with a magnetic stirrer, however, a dramatic effect was observed: as before, *l*- and *d*-crystals were separated and counted. Almost all of the $NaClO_3$ crystals (99.7%) had the same handedness, either *l*- or *d*-! In a total of 32 different crystallization experiments, 18 set-ups were predominantly of crystals in *l*-configuration whereas 14 were predominantly of *d*-configuration. The racemic state became unstable. Kondepudi et al. (1990) obtained a bimodal frequency distribution of the crystal's enantiomeric excesses as given in Fig. 4.3. A total of 11 829 crystals was counted. The direction of the obtained enantiomeric excess was unpredictable and random. In this context, the system showed stochastic behaviour. The observed phenomenon occurred without the need for careful control of temperature or other experimental parameters.

---

[4] If a crystal is composed of achiral compounds, its chirality disappears in solution. To circumvent this problem, light emitted from TFT-displays can be used in order to distinguish between the chirality of enantiomorphous crystals. This light is linearly polarized. Enantiomorphous crystals can thus be deposited directly on the screen of a notebook and regarded with a polarizing glass. The required angle of the glass to transmit the light gives information on the crystal's optical rotation and chirality.

[5] For determination of the $NaClO_3$ crystal chirality, linear polarizing laminated films from Edmund Optics can be used. One laminar was mounted on a light table (for instance, as used for photography) and the petri dish containing the $NaClO_3$ crystals was placed on top of it. Light shines from the bottom through the diffuser then through the polarizing film (called the polarizer) and finally through the sample. Now holding and rotating a second laminar (called the analyser) between the petri dish and one's eyes makes the optical activity of an individual crystal clearly visible (Personal Communications with Thomas Buhse and Dilip Kondepudi in December 2005).

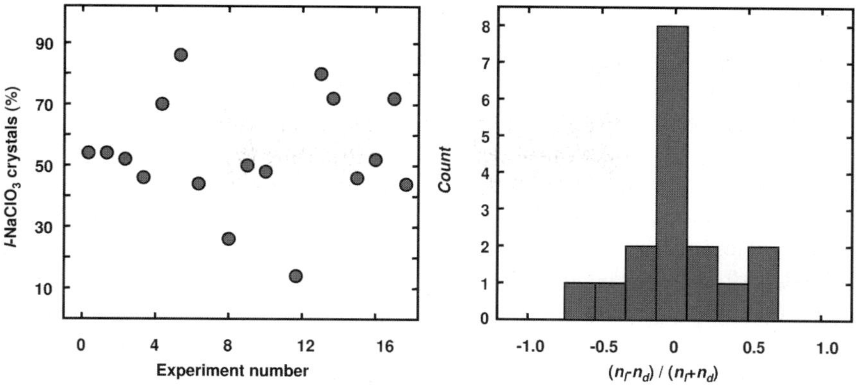

Fig. 4.2 Quantity of *l*-$NaClO_3$ crystals in percent (*left*) and monomodal frequency distribution for crystal enantiomeric excesses (*right*) after precipitation in non-stirred solutions (see Kondepudi et al. 1990; Kondepudi and Asakura 2001)

This kind of breaking of the chiral symmetry in the form of enantioselective crystallization was not induced by the presence of optically active compounds, external seeds, or other external asymmetric influences.[6] Also impurities in the reactants could be excluded for the observed results.

Presenting these intriguing results at his typically overcrowded lectures at international conferences, Dilip Kondepudi was often asked whether the direction of

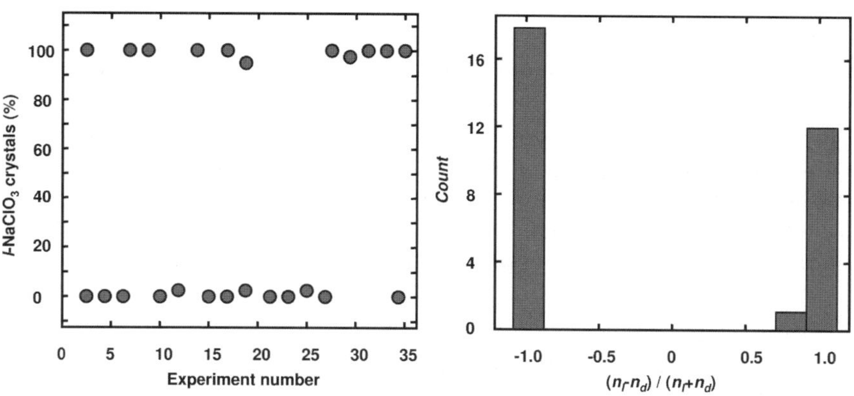

Fig. 4.3 Quantity of *l*-$NaClO_3$ crystals in percent (*left*) and bimodal frequency distribution for crystal enantiomeric excesses (*right*) after precipitation in stirred solutions (see Kondepudi et al. 1990; Kondepudi and Asakura 2001)

[6] In order to distinguish between *enantioselective* and *enantiospecific*, Hellwich and Siebert (2006) clarified that enantioselectivity means that one enantiomer out of several possible enantiomers is preferentially produced in a chemical reaction. A reaction is termed enantiospecific, if chiral starting materials are converted into chiral products. If the configuration of the starting material is known, then the product's configuration of an enantiospecific reaction can be predicted.

stirring in the form of left-stirring versus right-stirring of the stir bar might have induced the excess of either *d*- or *l*-enantiomorphous $NaClO_3$ crystals. He answered – smiling – that in stirred solution three distinct processes are of importance: primary nucleation, autocatalytic secondary nucleation, and crystal growth.[7] In the above experiment, secondary nuclei were rapidly produced from a primary nucleus. The chirality of the primary nucleus thereby determines the overall chirality of the system, a process that is chirally autocatalytic. The rapid production of secondary nuclei was confirmed by videotaping the collision of a $NaClO_3$ crystal and the stir bar. Kondepudi and colleagues assumed, however, that secondary nucleation alone is not sufficient to produce the observed homochirality. According to Kondepudi et al. (1990), this process must be accompanied by the suppression of nucleation of crystals of the opposite handedness in "a form of competition". But, what is that?

In Chap. 10, the Frank model will be introduced to explain the amplification of small enantiomeric enhancements. If we use the Frank model as a reference to better understand the formation of almost homochiral crystals by $NaClO_3$ crystallization in stirred solutions, we should note that mutual inhibition of two enantiomers is essential for the amplification of enantiomeric enhancement. In the $NaClO_3$ crystallization, the mutual inhibition is accomplished by the fast event of secondary nucleation in which the solute concentration rapidly drops below the saturation level so that no more crystals can be formed. In this context, secondary nucleation is, however, not understood in detail and Kondepudi's explanation remains debated (Buhse et al. 2000; Viedma 2005).

When the $NaClO_3$ solution was not stirred, there was no rapid autocatalytic production of nuclei observable: all of the nuclei were produced through primary nucleation and their handedness was random, producing a statistically equal number of *l*- and *d*-enantiomorph $NaClO_3$ crystals.

The observed chiral symmetry breaking also occurs in stirred solutions of sodium bromate $NaBrO_3$. Since most, if not all, crystals can generate secondary nuclei, Kondepudi and his collaborators expect this phenomenon to be observable in crystallization reactions of other achiral compounds that crystallize in enantiomeric forms, too (Kondepudi and Asakura 2001).

Later studies on the kinetics of the spontaneous crystallization of chiral crystals showed that small amounts of homochiral D- or L-lysine added as primary nuclei to a stirred solution of D/L-glutamic acid were capable of determining the chiral resolution of glutamic acid crystals (Buhse et al. 1999).

The symmetry breaking in spontaneous crystallization processes is not limited to the precipitation of chiral crystals from solution. The analogous effect was observed for the crystallization of a chiral polyaromatic hydrocarbon molecule from a melt. In this case, chiral 1,1′-binaphthyl served as the ideal system (Fig. 4.4): the racemization half-life measured at room temperature for a 1,1′-binaphthyl solution in benzene is relatively long and was determined to be 9 h. In the molten state above 158°C, the enantiomers interconvert rapidly with a racemization half-life of less than a second.

[7] In fact, the idea of inducing an inverse enantiomeric enrichment into a chemical system by stirring to the right or to the left is not trivial as it turned out very recently and as we will discuss in Sect. 4.3.

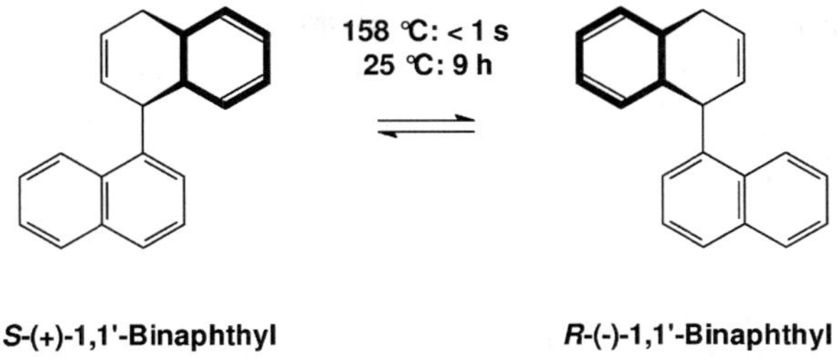

Fig. 4.4 Chemical structures of 1,1′-binaphthyl structures that interconvert slowly in benzene solution at room temperature and rapidly in the molten state above 158°C (Kondepudi and Asakura 2001)

The non-stirred crystallization of 1,1′-binaphthyl was observed from its melt by lowering the temperature below 158°C. This experiment resulted in the formation of chiral crystals following a monomodal frequency distribution for the enantiomeric excess with a maximum at zero. Crystallization from a melt stirred by a Teflon stir bar, however, showed – as in the case of enantiomorph $NaClO_3$ crystals – crystal symmetry breaking since the enantiomeric excess generated in almost every crystallization is greater than 80%. The frequency distribution of the enantiomeric excess in this case was bimodal in contrast to the monomodal distribution obtained in the non-stirred experiment (Kondepudi and Asakura 2001). This phenomenon seems to be a general one since other systems such as the chiral octahedral cobalt complex cis-$[Co(H_2O)(OH)(en)_2]^{2+}$ show equally a monomodal frequency distribution of enantiomeric excesses in non-stirred solution compared to a bimodal distribution by stirring (Asakura et al. 2000).

Such generation of highly enantioenriched solid systems from the racemic or even achiral liquid state shows us how, in nature, chiral asymmetry in crystals may spread during the processes of chemical evolution before the origin of biological life. Kondepudi et al. concluded that similar processes including autocatalysis and competition might help in understanding the possible origins of biomolecular homochirality. This model is, however, not yet capable of explaining the preference of a particular configuration of one enantiomer over the other.

4.2 Heated Debate: Which Amino Acid Enantiomer Precipitates First?

It is important to widen our view and to understand that not only minerals such as quartz and molecules such as 1,1′-binaphthyl can crystallize in chiral space groups. Amino acids themselves – white powders at room temperature and ambient pressure – can precipitate from aqueous solutions forming crystals. This precipitation occurs with nucleation and autocatalytic crystal growth processes having the potential to show noticeable enantioenriching properties (see Thiemann 1974).

As we have seen above, an equimolar mixture of two enantiomers in solution may crystallize either as a racemic compound with both enantiomers in equal quantities embedded in the crystal lattice or – less commonly – a spontaneous resolution separates each enantiomer in different enantiomeric crystals as in the case of Pasteur's tartaric acid precipitation followed by manual separation (see formation way number 3). Most amino acids, including aspartic and glutamic acid, belong to the most frequently observed first type and do not show spontaneous resolution by crystallization. Crystals of racemic amino acids are usually obtained as 50% *d*-enantiomers and 50% *l*-enantiomers.

In 2001, Cristobal Viedma (2001) from the University of Madrid, Spain, showed how the amino acids D,L-aspartic and D,L-glutamic acid as racemic compounds in aqueous solution crystallized at room temperature as distinguishable enantiomers with each enantiomeric molecule separate in its own enantiomorph crystal. In order to obtain such results, the author had added heterogeneous porous materials such as refractory porous bricks and porous paper sheets to influence the crystallization process allowing capillary action, concentration by evaporization and crystallization of the amino acids in a narrow upper zone of the porous materials. X-ray diffraction patterns were used for enantiomer identification verifying the spontaneous resolution of the enantiomers by this kind of crystallization. Viedma explained this result by assuming that the applied crystallization on a porous medium gives rise to high supersaturation levels. These levels might enable the system to go far from thermodynamic equilibrium where kinetic criteria dominate over thermodynamic criteria, and the formation of metastable and "anomalous" phases becomes possible. According to Viedma, this atypical behaviour might also be found in natural sedimentary environments and was thus proposed to be associated with the origin of life and its homochirality phenomenon. This is indeed an attractive model and we will wait for future experiments that will have to verify systematically whether other amino acids and chiral biomolecules such as sugars will show similar trends in atypical crystallization behaviour on porous media. Future research will also have to tackle the poorly understood crystallization behaviour in general: there are no firm rules to predict whether a particular pair of chiral partners will follow the behaviour of the vast majority of chiral molecules and crystallize together as racemic crystals, or as separate enantiomers (Fasel et al. 2006).

Since 2002, the research activities of Meir Shinitzky at the Weizmann Institute of Science in Rehovot, Israel, attracted tremendous attention in the field of the crystallization of an amino acid combined with the spontaneous appearance of an enantiomeric excess.[8] Shinitzky et al. (2002) studied solubilities of D- and L-tyrosine in water applying several analytical techniques such as UV-absorption, radioactive labelling & tracing, and optical rotation, which were specifically adapted to the system. Surprisingly, the authors measured unexpected differences in the solubilities of

---

[8] In this context, the spontaneous appearance of an enantiomeric excess is not necessarily the same thing as 'chiral symmetry breaking'. The situation can be compared with the non-stirred vs. stirred crystallization of $NaClO_3$ in which in both cases a non-zero *e.e.* can be obtained, but only the stirred one represents chiral symmetry breaking. More deeply reflections on these semantics can be found in Ávalos et al. (2004).

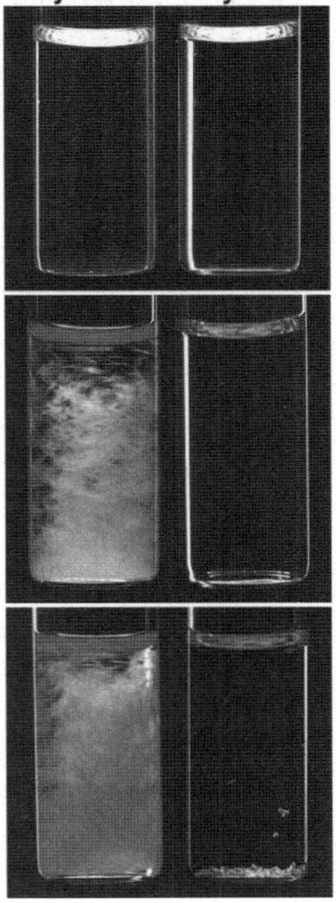

Fig. 4.5 Precipitation of D- and L-tyrosine enantiomers from 10 mM aqueous solution taken at 0, 6, and 24 h (Shinitzky et al. 2002). Illustration used with permission of the authors

the tyrosine enantiomers in water by all of these techniques. D-Tyrosine crystallized faster than L-tyrosine! The liquid phase above the precipitate became enriched in L-tyrosine! This effect is illustrated in Fig. 4.5.

Alain Schwartz from the University of Nijmegen, the editor of the journal 'Origins of Life and Evolution of Biospheres' in which these data were first printed, was equally surprised by these results and published a referee's comment simultaneously with the Shinitzky paper saying, "... in spite of the real prospect that future work will show this chiral discrimination to be an experimental artefact, I strongly encourage publication so that others may confirm, extend and amplify these results" (Schwartz 2002). And indeed, Wolfram Thiemann, Ute Jarzak, and I repeated the Shinitzky-precipitation experiment at the University of Bremen with various tyrosine sources confirming the higher crystallization rate for D-tyrosine.[9]

The different crystallization rates of the tyrosine enantiomers disappeared in D_2O. For the authors, it seemed possible that minute energy differences between D- and L-tyrosine, originating from parity violation or other non-conservative chiral discriminatory rules (see Chap. 5), could account for this observation. Shinitzky et al.

[9] Shinitzky's original finding was further supported by radioactive tritium-labeling of tyrosine and post-crystallization radioactive tracing of ^{3}[H]-L-tyrosine in the supernatant. The repetition of these experiments in the isotope laboratory at the University of Bremen did, however, not confirm the asymmetric counting rates obtained by Shinitzky et al. (2002).

concluded that the obtained results seem to violate the generally accepted axiom that chiral isomers are identical in their bulk energetic quantities.

Spectacular findings like this, however, require particularly robust and sophisticated data. Parity non-conserving energy differences between amino acid enantiomers had been calculated to be as small as 10^{-14} to 10^{-17} kT (see Chap. 5). Despite thorough research at the international level, an experimental proof of these energy differences has not yet been achieved. Shinitzky et al. (2002) investigated the amino acids D-tyrosine, L-tyrosine, and D,L-tyrosine from Fluka with >99% purity recrystallized in double distilled water. Subsequent to recrystallization, the purity of the amino acids was not controlled. Decomposition reactions of the amino acids were checked by thin layer chromatography only. Modern analytical techniques like enantioselective chromatography were not applied in order to detect and quantify possible contaminants.

In general, three fundamentally different routes (1-3) are used for the production of enantiomerically pure amino acids (see review of Breuer et al. 2004). 1) In extractive processes, amino acids are isolated as components of natural protein-containing material. 2) The alternative chemical method uses the Bucherer-Bergs variant of the Strecker-mechanism to produce racemic amino acids, followed by conversion into enantiomerically pure compounds by a number of biocatalytic methods. Finally, 3) biological methods such as biotransformations and fermentation methods allow the enantioselective synthesis of amino acids. We, therefore, have to assume that the supplier Fluka synthesized the different tyrosine enantiomers used in the Shinitzky precipitation experiment in different ways. L-Tyrosine can be isolated from biological protein-rich material, but also by one of the other ways, whereas D-tyrosine requires one of the above-mentioned chemical or biological synthesis pathways. As a consequence, the remaining pollutants – about 1% for each amino acid – are probably different in all products. These might be amino acids but also other organic and inorganic substances. As we know in physical chemistry, the growth of crystals (in this case, amino acids) and their surface properties are often dominated by impurities. Therefore, I assume, that the measured differences in the solubilities of the tyrosine enantiomers were caused by impurities (not necessarily amino acids!). Minute amounts of these impurities even in sub-trace abundances may have contaminated the L-tyrosine samples and worked as solubilizers.

In 2006, Scolnik et al., including Shinitzky, published that – indeed! – experimental tests of the hypothesis on parity non-conserving energy differences, where one enantiomer is compared to the other in solution, are hampered by the possible presence of "undetectable impurities" (Scolnik et al. 2006). Here, the authors correctly address the highly possible effect of impurities on their own previous experiment. Only the word "undetectable" might be incorrect. To my knowledge, however, the authors did not reject the above-mentioned paper published in the well-known Kluwer journal "Origins of Life and Evolution of Biospheres", which might have been a straightforward consequence.

Instead, Shinitzky et al. continue publishing (Scolnik et al. 2006) that they have now overcome the problem of possible impurities by measuring specific properties of synthetic polypeptides. Differences in structural transitions between poly-L- and poly-D-amino acids were reported. Surprisingly, the reader is informed that all of the four investigated polymers – each composed of 24 amino acids – show "purities >98%" only. This means, that out of 100 amino acids, two might be wrong. Consequently, in 4 polymers, two "wrong" amino acids might be implemented. In other words: two out of four polymers might be "wrong" (i.e. 50%!). Is it possible that these wrong polymers cause effects, which Scolnik et al. (2006) hurriedly interpret to possibly originate from parity non-conserving energy differences?

Recently, Meir Lahav who was employed at the same time as Meir Shinitzky by the world-famous Weizmann Institute of Science – ironically door to door working – challenged the original Shinitzky paper (Shinitzky et al. 2002) by reinvestigating the above mentioned tyrosine crystallization (Lahav et al. 2006) with D-Tyr, L-Tyr, and D,L-Tyr from different sources including samples directly from Shinitzky. Lahav et al. noted that samples provided by Shinitzky indeed displayed the effect he reported in his article, however, their results could not be reproduced with samples obtained from other sources. In contrast to the original work of Shinitzky et al., Lahav et al. focused on impurities in different samples of tyrosine enantiomers applying enantioselective chromatography in the well-known analytical lab of Volker Schurig at the University of Tübingen. Here, numerous other amino acids were identified in the tyrosine-samples and quantified as contaminants. The authors concluded that these contaminants – which were measured in quantities up to a few percent – caused the abnormal differential crystallization behaviour of the individual tyrosine enantiomers instead of any parity non-conserving energy difference.

Meir Shinitzky and his team replied to Lahav's criticism by insisting on the observation of different solubilities between D- and L-tyrosine. The new argument was that synthetic D,L-tyrosine was proven to show chiral enhancement in its supersaturated aqueous solutions. The precipitate and the supernatant were separated and their specific rotation indicated a highly significant chiral enhancement of approximately 60% L-tyrosine versus approximately 40% D-tyrosine in the supernatant (Deamer et al. 2007).

Most recently, Stanley Goldberg from the University of New Orleans entered into this scientific debate exposing another justification: he proved experimentally that a microbial spore which is often present in laboratory air is capable provoking different rates of the tyrosine enantiomers' crystallisation (Goldberg 2008).

It is not up to me to be the final judge of this scientific conflict. I would rather like to encourage experimentalists to study the chosen conditions of the opponents in more detail and to contribute to the solution of this divergence by supplying their own experimental data.

Very recently, Donna Blackmond and her colleagues at Imperial College London proposed a new approach in which crystallization could have played a part in selecting L-amino acids for proteins in living organisms. This approach is based on the equilibrium phase behaviour of amino acids (Klussmann et al. 2006). For some amino acids, such as serine, a tiny excess of one enantiomer can lead to that enantiomer's preferential precipitation in crystals, leaving the solution highly enriched in the other enantiomer. Blackmond et al. demonstrated that this enantiomer enrichment can be engineered for amino acids that don't display this behaviour by themselves. Small molecules such as achiral dicarboxylic acids added to the racemic mixture of amino acids cocrystallized and became thus incorporated into the amino acid's crystal. This cocrystallization promoted the extraction of one form of the amino acid from the solution (Klussmann et al. 2007).

A common characteristic of all these models is, however, that the resolution of enantiomers by spontaneous crystallization cannot be predicted at present (Koshima 2004). Furthermore, it is still a long way from optically active sodium chlorate or 1,1′-binaphthyl crystals to the distinct chirality of amino acids and other biomolecules. We will thus try to better understand the adsorption of organic molecules onto the surface of chiral crystals in the next paragraphs. Before doing so, we will have a closer look at the intriguing possibility that the direction of stirring might induce an enantiomeric enhancement in an appropriate chemical system.

4.3 Turning World: The Direction of Stirring

At the Biological Research Institute of Szeged in Hungary is the research team of Lajos Keszthelyi, which accumulated considerable international reputation through its contributions to the origin of biomolecular asymmetry. Here in Szeged, the idea became apparent that a chemical reaction starting with racemic or prochiral reactants performed in a right-stirred solution might give rise to contrary enantioenrichments in chiral products than the left-stirring of the same system. The following chemical reaction can be related to the prebiotic formation of (proto-) proteins and hence attracted the interest of the Szeged group: specific derivatives of amino acids called N-carboxyanhydrides (NCA) are known to form a polypeptide chain by a polymerization reaction. Kovacs and Keszthelyi (1981) studied the copolymerization of these derivatives in the case of D,L-glutamic acid and in the case of D,L-alanine. Reactions were performed twice in 50 mL vessels stirred by calibrated precision magnetic stirrers at identical speeds of 540 rpm but with opposite directions of rotation. The experimental set-up is depicted in Fig. 4.6. Kovacs and Keszthelyi carefully followed the polymerization kinetics, characterized the obtained polymers by determining their molecular weight, and obtained chiroptical information by measuring circular dichroism spectra. Astonishingly, counterclockwise stirring of the solution produced longer polymers. And indeed, an effect of the direction of stirring was found so that clockwise stirring seemed to cause a preference for the incorporation of D-enantiomers into the polymers. The authors admitted, however, that the differences suggested a trend but were not convincing enough to conclude that stereoselectivity was due to the direction of stirring in these polymerizations.

So we had to wait until 2001, when Josep Ribó and his colleagues (2001) at the University of Barcelona in Spain discovered and published an experiment in which the vortices in stirred solutions induced chirality in a chemical system. One has to admit that the selected system was special: homoassociates of achiral diprotonated porphyrins formed helical agglomerates during an aggregation process. The direction of vortex motion during aggregation determined the chirality sign of these homoassociates, which was controlled by means of circular dichroism measurements.

Fig. 4.6 Stirred not shaken: Experimental set up for clockwise (*right*) and counterclockwise (*left*) stirring of appropriate chemical systems. The obtained vortices are magnified

Such experiments, however, are interesting for better understanding the influence of macroscopic forces on bifurcation scenarios, but they have not yet been taken serious enough to explain the origin of biomolecular asymmetry, since the selected systems seem too distant from both molecules of prebiotic importance and prebiotic physico-chemical conditions.

Even if – and what a big if! – one could elucidate appropriate chemical reactions that seem reasonable to have occurred during chemical evolution and showed molecular symmetry-breaking due to the direction of stirring, one would have to answer the question of where the right- or left-vortex originated. Was it just random or might the direction of spiral water turning have been linked e.g. to the rotation of Earth?

In this context, interesting information was published on the well-known Coriolis effect. There is no doubt that the Coriolis effect, named after the French engineer Gaspard Gustave de Coriolis, is responsible for the strong tendency of cyclones and tornados to spin counterclockwise in the northern half of the globe and spin in the other direction in the southern half. Moreover, the Coriolis effect plays a role in the flow of rivers, but the question remains whether the Coriolis force is strong enough to be a detectable influence on a water vortex, such as when it spirals down a drain. It is common knowledge that when water flows out of a bathtub it forms a vortex around the drain. Is there any scientific evidence that the bathtub's or washbasin's vortex in one hemisphere of the Earth turns opposite to the vortex in the other?

According to Gardner (2005), natives along the equator in Kenya are perpetrating an amusing demonstration to tourists. Here is how Robert Goldenburg[10] described in Gardner (2005) what he saw when he visited Nanyuki in 1988:

"A standard feature of such stops is the demonstration of the Coriolis force by a local native equipped with a plastic dishpan, a hole punched out of its bottom. The man corks the hole, fills the pan with water and picks up a few twigs or pieces of straw, which he floats on the water surface. He then marches 20 paces south of the equator, followed by a dozen tourists, and pulls the plug. The twigs turn counterclockwise as the water runs out. He repeats the demonstration 20 paces north of the equator. The twigs swirl clockwise.

When we asked him to repeat the demonstration *on* the equator, he obliged, and the twigs did not move as the water drained out. Having bought trinkets and soda, I joined two other groups as he went through the entire demonstration twice more. If this 'Coriolis force' demonstration seems a matter of chance, I must report that the twigs behaved exactly the same way three times in a row. The man stated that the speed of the twigs' rotation correlates to 'latitude', the distance from the equator."

If this is a tricky swindle, how is it managed? Gardner (2005) guessed that in walking north or south the native secretly imparts a rotary motion to the water by a slight tipping of the dishpan. As for bathtub vortices, the question remains controversial.

4.4 Deposits of Enantiomorphous Quartz on Earth

For decades, academic studies have tried to answer the logical question: do chiral crystals on Earth occur in racemic ratios or in an enantioenriched, if not enantiopure, form. Is there any enantioenrichment to be measured in crystals, at least locally,

[10] The letter was originally published in Science News (June 17, 1989).

that might have caused a chiral bias in biomolecules today? The equal versus non-equal distribution of chiral crystals on Earth has been studied especially concerning *d*-(+)- and *l*-(−)-quartz enantiomorphs (Klabunovskii 2001). As we have outlined above, quartz is morphologically chiral and only one of the many chiral crystalline silicates to be found as minerals on the Earth's surface. Each of these minerals is assumed to exhibit, to differing degrees, analogous enantiomorphous composition (Tranter 1985a).

The asymmetric catalytic function of enantiomorph crystals on Earth was proposed as one of many scenarios to have introduced the first enantioenrichments into chiral organic molecules during chemical evolution (Klabunovskii 1963; Thiemann and Teutsch 1990; Wächtershäuser 1991). The mechanisms of this process are known and have been reproduced in the laboratory (Bonner et al. 1974; Bonner et al. 1975a; Bonner and Kavasmaneck 1976; Bonner et al. 1981; MacDermott 1995; Bonner 1995b; Soai et al. 1999).

The distribution of *d*-(+)- and *l*-(−)-quartz crystals on the surface of Earth integrated over all locations are be expected to be equal (Klabunovskii and Thiemann 2000). Lemmlein (1973) detected an equal abundance of *d*-(+)- and *l*-(−)-quartz in 10 000 samples from 19 locations. It should be mentioned that this statement is still a subject of hot debate (Klabunovskii and Thiemann 2000), because other authors have detected a worldwide excess of 50.5% *l*-(−)-quartz in 16 807 samples (Palache et al. 1962) and 49.83% *l*-(−)-quartz in 27 053 samples (Frondel 1978).

In *local* deposits, however, the formation of an excess of *d*-(+)- or *l*-(−)-quartz is considered to be possible if the crystallization of quartz proceeds in the presence of fossil materials with (+)- or (−)-rotation or if a seeding effect gives rise locally to a large preponderance of one enantiomorph. In quartz samples from Plakas, Greece, the occurrence of 60.6% *l*-(−)-quartz in 549 samples was reported, whereas 300 crystals collected in Samshvildo, USSR, showed the occurrence of 56.0% *d*-(+)-quartz (Klabunovskii and Thiemann 2000). Based on these data, the conclusion was drawn that the distribution of *d*-(+)-quartz and *l*-(−)-quartz over the surface of the Earth is random and any enantioenrichments in chiral organic compounds being generated under the influence of optically active quartz crystals would occur locally and be randomly distributed.

In the context of enantiomorph crystals and an eventual global excess of *l*-(−)-quartz, it is interesting to note that *d*-(+)-quartz and *l*-(−)-quartz do not necessarily show exactly the same electronic energy. As we will discuss in more detail in Chap. 5, the weak nuclear interaction gives rise to small energy differences called parity non-conserving energy differences that can be calculated for a chiral molecule and its mirror image enantiomer. On this background, the energy of an enantiomorph crystal varies by a tiny amount compared with its mirror image enantiomer. Quantum mechanical ab initio calculations for quartz crystals showed that, per SiO_2 unit, the *l*-(−)-quartz crystal is energetically stabilized by $2.24 \cdot 10^{-20}$ atomic units (a.u.) versus its *d*-(+)-enantiomorphous form. Interestingly, an excess of 1.4% of *l*-(−)-quartz crystals over *d*-(+)-quartz crystals requires 10^{15} SiO_2 units, which corresponds to a typical quartz crystal with a side length of 0.1 mm (MacDermott and Tranter 1989; MacDermott 1993). Nevertheless, the majority of involved scientists currently assumes that enantiomorphous forms of *d*-(+)-quartz and *l*-(−)-quartz are equally distributed on Earth.

Despite such discussions, chiral quartz crystals themselves are inert and not catalytically active, and therefore are unlikely to have been important prebiotic catalysts. Other chiral minerals might have a higher potential to serve as catalysts in prebiotic reactions as we will see in the next section.

4.5 Adsorption of Chiral Organic Molecules on Enantiomorphous Crystals

We have seen that crystals can be chiral and that such chiral crystals can originate from the crystallization of a racemic solution or melt in laboratory experiments and also in deposits of enantiomorphous crystals on Earth. Locally, high excesses of enantiomorphous crystals have been recorded. The next question will be whether organic molecules such as amino acids can enantioselectively interact on mineral surfaces of enantiomorphous crystals via adsorption or similar processes in order to transfer the chiral information from the crystal lattice to the organics on its surface.

The preferential adsorption of one amino acid enantiomer out of a racemate onto one of the morphologically *d*-(+)- or *l*-(−)-crystal surfaces of quartz minerals was first demonstrated to be feasible by Bonner et al. (1974) at Stanford University. The asymmetric adsorption phenomenon, however, required that non-aqueous solvents were employed. Experimental conditions rigorously excluded moisture, making this particular reaction therefore implausible in any realistic prebiotic environment.

Particular attention should be given to an experiment that was recently performed at the Carnegie Institution at Washington, DC. Robert Hazen and his team (2001) studied a geochemically plausible scenario for the origin of biomolecular asymmetry by selecting the common rock- and sediment-forming mineral calcite ($CaCO_3$) with its rhombohedral space group $R\bar{3}c$ to study its individual interaction with D- and L-amino acids. It was known before that particular enantiomers of amino acids can be selectively incorporated into opposite enantiotopic faces of centrosymmetric crystals (Weissbuch et al. 1984). A peculiarity of the calcite crystal is that it is not chiral at all but its scalenohedral faces possess surface structures that have the potential for chiral selectivity. The surfaces themselves are not chiral.

After rigorous surface cleaning, calcite crystals with surface areas of 6–36 cm^2 were dipped into a 0.05 molar racemic aspartic acid solution. The adsorbed aspartic acid was removed after 24 h from the calcite crystal by pipetting and subjected to enantioselective gas chromatography on a Chirasil-Val phase. As illustrated in Fig. 4.7, a chiral selectivity in the adsorption of D- and L-enantiomers was observed to be of 10% difference for the two enantiomers (partly up to 40%).

Accompanying computational work of Hazen and Asthagiri (2007) was performed recently on ab initio calculations of alanine and aspartate adsorption on calcite 214 surfaces. They found, in coherence with the above experimental data, that alanine has a 2-point interaction and, therefore, no chiral selection, whereas

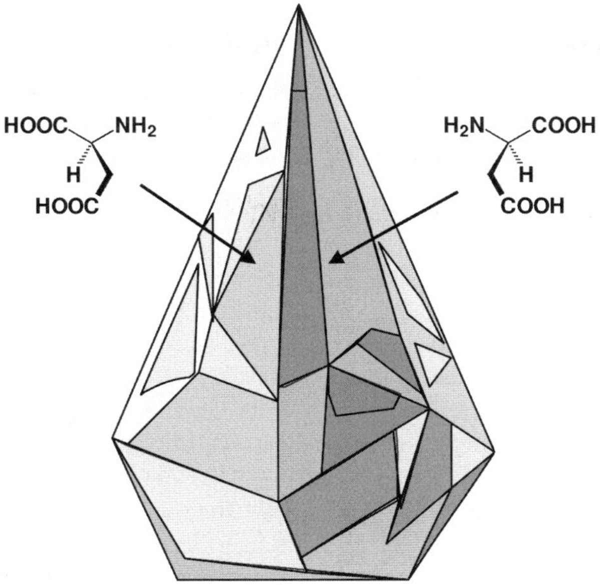

Fig. 4.7 Selective adsorption of L-aspartic acid (*left*) and D-aspartic acid (*right*) enantiomers on different scalenohedral calcite faces. Such a sorting process could explain why life selected L-amino acids for the construction of proteins (Hazen 2001)

D-aspartate has a strong 3-point binding to right-handed 214-type faces, but only 2-point binding to left-handed faces, hence the strong selective adsorption.

Hazen (2001) proposed that the results might be extrapolated to other amino acids and sugars, and concluded that polymerization reactions of amino acids on calcite surfaces represent a plausible geochemical mechanism for the production of homochiral polypeptides on the prebiotic Earth.

Besides calcite crystals, other minerals have been proposed to have violated molecular parity in primordial organics. In particular, pyrite crystals attracted our interest, since pyrite (FeS_2) has been linked to prebiological carbon fixation by providing both the energy source and the mineral surface for bonding the organic products of carbon fixation. Pyrite crystals are of crucial importance for surface metabolism in the frame of the popular "iron-sulphur world" proposed by the landmark theory of Günter Wächtershäuser (1992) for a chemoautotrophic origin of life on Earth. Günter Wächtershäuser is Honorary-Professor at the University of Regensburg and European Patent and Trademark Attorney in Munich. Interestingly, and hitherto ignored by the scientific community, pyrite crystals occur not only in the achiral cubic space group $Pa3$, but also in a triclinic space group $P1$. The triclinic space group was attributed to the low temperature formation of pyrite in sedimentary rocks or hydrothermal ore deposits. Triclinic crystals of space group $P1$ are enantiomorphous; they occur as l-pyrite and d-pyrite crystals (Wächtershäuser 1991). The achiral space group $Pca2_1$ classifies a third orthorhombic modification of pyrite crystals.

Wächtershäuser (1991) suggested that in a pyrite-pulled chemoautotrophic origin of life,[11] the organic constituents are seen as being produced in intimate contact with the pyrite surface and, in fact, concomitantly with the growth of the asymmetric pyrite crystal. This means that the homochirality of a given pyrite crystal was expected by Wächtershäuser to be transferred with a high enantioselectivity into the organic constituents formed on its surface.

But what have been the concrete experimental results of iron-sulphur world simulations in the laboratory up to today? The formation of a carbon-carbon bond was successfully proven to be possible in simulated iron-sulphur world conditions using activated acetic acid (Huber and Wächtershäuser 1997). However, hitherto no chiral molecule had been produced under such conditions even in the racemic state, not to mention any enantiomeric excesses induced by pyrite crystals in their chiral triclinic modification. We have thus no experimental proof yet that the chirality from the interior of a pyrite crystal might have been transferred to any prochiral or racemic molecule on its surface.[12] This beautiful suggestion remains a hypothesis only.

In addition to calcite and pyrite minerals, organophilic silica surfaces in the form of zeolites were proposed as host structures for the enantioselective interaction with prebiotic organic guest molecules. Particularly dealuminated zeolites with laevo/dextro-oriented channels were suggested as mineral surfaces involved in the asymmetric origin of life (Smith 1998). Experimental data on this proposal are, however, not yet available.

In summary, the supramolecular induction of an enantiomeric enhancement using natural and synthetic chiral crystals as hosts provides us with a method for transferring chirality to a prochiral substrate through host-guest interactions. Even though the factors and mechanisms that govern the product chirality and enantiomeric enhancement are not sufficiently clear, and the optical yields obtained are not satisfactorily high at present, it should be emphasized that the supramolecular induction of an enantiomeric enhancement exhibits several phenomena that are completely different from conventional chirogenesis performed in isotropic media and are thus worthy of future research (Wada and Inoue 2004).

[11] Chemoautotroph organisms chemically fix all their carbon constituents from carbon dioxide or other C_1-units in contrast to heterotroph organisms that are dependent on taking up organic carbon compounds as food.

[12] In November 1999, during a personal communication at the University of Bremen, G. Wächtershäuser himself did not attribute a causal function to chiral pyrite crystals to biomolecular homochirality.

Chapter 5
When Parity Falls: The Weak Nuclear Interaction

As we have seen, proteins (enzymes) are exclusively composed of amino acids in L-configuration. If we produce – in contrast to life's biochemical synthesis –, amino acids artificially in the laboratory, e.g., via the well-known Strecker mechanism, we will inevitably produce 50% of L-amino acids and 50% of D-amino acids. The enantiomers will be obtained in a racemic ratio and can be separated and quantified by enantioselective chromatography; the racemic mixture will give no signal in chiroptical measurements such as optical rotation dispersion ORD and circular dichroism CD. This is not surprising. It is the consequence of the symmetry of natural laws, the parity of fundamental laws of physics. Universal forces (or "interactions", as physicists prefer to say) such as the electromagnetic interaction (here photons are gauge bosons), the gravitational interaction (tentatively the boson is called graviton), but also the strong nuclear interaction (gluons as bosons) are symmetric in space. Here parity is not violated. This means that any event based on these forces (interactions) performed in a mirror would show identical results. Energy, colour, and weight, and also rate constants of the formation of the products are for example identical to these parameters in their mirror-image products. In the above reaction, during the Strecker synthesis no preference is given towards the production of L-amino acids or the production of D-amino acids. The two enantiomers will originate with the same probability and therefore be obtained highly symmetrically as a racemic mixture.

But attention: it is important that the above described symmetry of universal interactions is not a general principle even if it was assumed by physicists for quite a while in the past. The so-called weak force, the 4th fundamental interaction of physics that is carried by charged massive $W^{\pm}$ and neutral Z^0 bosons does violate parity! The weak nuclear force or weak interaction distinguishes undeniably between left and right. It can "feel the difference" between left and right (MacDermott 1993). The weak nuclear force is known to influence electrons inside atoms, causing them to travel helical paths that give all atoms a handedness. Due to its weakness – the weak interaction is hardly outreaching towards the exterior of atomic nuclei – it is usually marginal and does not manifest itself in everyday macroscopic events. Its most familiar effects are the β-decay of unstable atomic nuclei such as ^{40}K, ^{32}P, and ^{14}C and the associated radioactivity. The word "weak" derives from the fact that the field strength is some 10^{13} times smaller than that of the strong nuclear force, which holds together the protons and neutrons in the nucleus of an atom. The strong

U. Meierhenrich, *Amino Acids and the Asymmetry of Life*. Advances in Astrobiology and Biogeophysics,

nuclear force is often called the binding force of the nucleus and well known to all of us from nuclear power plants applying nuclear fission and also nuclear fusion as it is prospected for the international ITER project located in Cadarache, France. In radioactive β^--decay two particles are emitted: an electron e^- and an antineutrino $\bar{\nu}_e$ as illustrated in Fig. 5.1 for the general β-decay and the ^{14}C-case.

$$n \rightarrow p + e^- + \bar{\nu}_e$$

$$^{14}_{6}C \rightarrow {}^{14}_{7}N + e^- + \bar{\nu}_e$$

Fig. 5.1 The weak nuclear force is manifested in the radioactive β-decay: unstable nuclei having an excess of neutrons n emit electron and antineutrino via the β^--decay (*top*). The weak force is asymmetric causing that the emitted electron is longitudinally polarized. The unstable ^{14}C isotope serves as example of an isotope undergoing β-decay (*bottom*). ^{14}C contains 6 protons and 8 neutrons. Due to β^--decay, one of the neutrons becomes a proton changing the atomic number from carbon ($Z = 6$) to nitrogen ($Z = 7$) emitting – through mediation of a vectorboson W^- – an electron and an antineutrino

The big question that we will ask in the frame of this chapter will be: "Is there any link between the parity non-conservation of the weak nuclear force which manifests itself in the interior of atomic nuclei and the asymmetry of molecules in living organisms? Were universal physical processes such as weak nuclear forces involved in prebiotic pathways on the origin of life? And is there thus any connection between the asymmetry of a fundamental law of physics and asymmetric properties of biological life?"

5.1 The Fall of Parity: "I Couldn't Understand It"

In 1956, two young American scientists of Chinese origin Tsung Dao Lee at Columbia University in New York and Chen Ning Yang at the Princeton Institute of Advanced Studies carefully examined theoretically the radioactive β-decay of light nuclei and reasoned that the parity between left and right is violated here (Lee and Yang 1956).[1] In their original publication, they furthermore suggested

[1] According to the CPT-theorem, quantum field theory gives us today three 'symmetries' by which particles, antiparticles and their interactions can be characterized (Ellis 2003): The first is the charge C (plus or minus) and the second is parity P (left or right. But attention: Parity or space inversion is not the same as mirror reflection. The former sends a system with (x, y, z) coordinates in front of the mirror to an idealized image with (–x, –y, –z) coordinates, whereas for mirror reflection an object occupying the x, y, z plane will have an image with (–x, y, z) coordinates behind the mirror (Cintas 2007). However, often the result is equivalent). The third is time reversal T (forward or backward). By "time reversal" we mean nothing more than a reversal in the direction a particle or wave is moving. According to the CPT-theorem, a particle that is subjected to a combination of these three transformations shows unchanged behavior: Matter particles should behave the same as their antiparticles, if viewed in a mirror with their internal clocks reversed. A CPT-reversal of

experiments in order to test parity non-conservation in these interactions. This observation moved the fundaments of physics dramatically because until 1956 people thought that all the four fundamental natural forces are strictly symmetric. Lee and Yang were immediately honoured for their discovery with the Nobel Prize of Physics in 1957. To that time Lee was 30, Yang 34. The choice was inevitable.

The shock waves were still rumbling when, in 1957, the experimental verification of Lee and Yang's theoretical questioning of parity conservation in weak interaction was performed by Mme Chieng-Shiung Wu at Columbia University, New York. Together with Ernest Ambler and members of the National Bureau of Standards in Washington she selected the β-radiator ^{60}Co, which periodically emits electrons (β -particles) for her crucial and today highly cited experiment: ^{60}Co can be obtained by irradiation of ^{59}Co in a nuclear reactor with thermal neutrons. It is a suitable isotope to study the radioactive decay since its half-life is 5 years and therewith opportune for experiments. Add to this, it is feasible to orient Cobalt atoms in space in order to determine a direction to measure emitted decay products: Cobalt nuclei can be – similar to small magnets – compass-needle like oriented in a magnetic field. Usually these nuclei are not oriented and distributed in all directions. In a strong magnetic field ^{60}Co nuclei orient themselves in parallel to the field and in parallel to neighbour nuclei. The electron cloud in paramagnetic ^{60}Co ions – and this is the main advantage of ^{60}Co compared to other β-radiators – strengthens the magnetic field towards the nucleus. In order to eliminate the thermal motion of the particles, Wu's experimental set-up that is schematically depicted in Fig. 5.2 was cooled down to 0.01 K. Under such conditions, the majority of the cobalt nuclei is well oriented in the direction of the applied magnetic field.

If now, the radioactive decay were symmetric in space, the electrons (i.e., β^--particles) would be emitted in all directions including the direction of the ^{60}Co nuclei spins. If – on the other hand – one could observe an asymmetry in the angular distribution between 0° and 180° between the orientation of the parent nuclei and the momentum of the electrons it provides unequivocal proof that parity is not conserved in β-decay. Mme Wu and her team observed this asymmetry effect in the case of oriented ^{60}Co nuclei. Electrons were preferably emitted against the direction of polarization of the ^{60}Co nuclei (Wu et al. 1957). Here, mirror symmetry is not conserved. The weak nuclear interaction violates parity!

Physicists and physico-chemists were unwilling to accept the removal of symmetry from their theories.

---

a glass of milk of Lewis Carroll's Alice (in "Through the Looking-Glass and What Alice Found There" Alice asked her cat: How would you like to live in a looking-glass house, Kitty? ... Perhaps looking-glass milk isn't good to drink.") would mean that all charges would be reversed (making it antimilk), the structure would be mirror-reflected, and every molecular motion would reverse its direction (Gardner 2005). Until today, we assume indeed that CPT preserves symmetry; there is no violation of the CPT-theorem known, even if it was searched for intensively. However, particle interactions were shown to violate each of P and C individually, but also the CP combination of these transformations. This violation was proposed by Sakharov to be responsible for the dominance of matter over antimatter in the Universe today and is equally connected to the question "Can time go backward". Before the fall of parity, physicists believed that if you altered just one of the signs, the new sentence would still describe something nature could do.

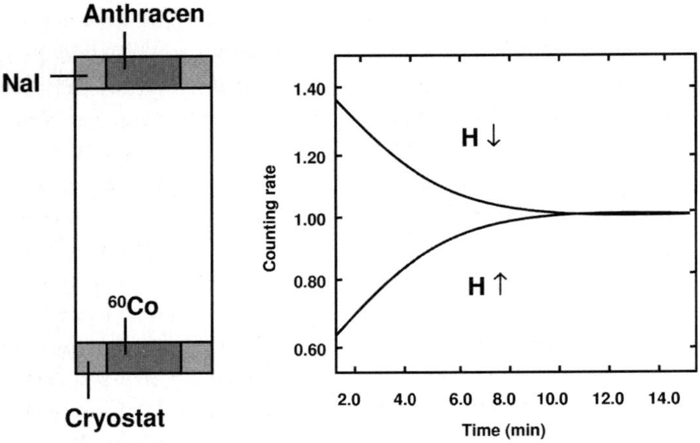

Fig. 5.2 Wu's experimental test of parity non-conservation provoked by β-decay of radioactive isotopes. (*Left*) Schematic drawing of the lower part of the cryostat: ^{60}Co nuclei were oriented in a magnetic field and their orientation was fixed in a cryostat at 0.01 K. The upwards-installed anthracen crystal was used to count the emitted electrons (β^--radiation); sodium iodide NaI was for counting equally emitted γ-radiation. (*Right*) The anisotropy of emitted electrons is depicted in function of the direction of the polarizing field pointing up and pointing down. If the polarizing field points down, the anthracen crystal captures more electrons. Electrons were preferably emitted against the direction of cobalt nuclei's polarization. Since the warm-up time for cobalt in the experimental set-up was about six minutes, the anisotropy of emitted electrons disappeared over time

Wolfgang Pauli (who died in 1958) is said to have bet large amounts of champagne against it when the suggestion of parity violation first appeared. "Now after the first shock is over," Pauli wrote to Weisskopf after the staggering news had reached him, "I begin to collect myself. Yes. It was very dramatic. On Monday, the twenty-first, at 8 p.m. I was supposed to give a lecture on the neutrino theory. At 5 p.m. I received three experimental papers on the first three test of parity. I am shocked not so much by the fact that the Lord prefers the left hand as by the fact that he still appears to be left-handed symmetric when he expresses himself strongly. In short, the actual problem now seems to be the question: Why are strong interactions right-and-left symmetric?" (Quack 2002; Gardner 2005).

The reaction of Richard P. Feynman is lively described by Martin Gardner: In April 1956, Gardner and Feynman participated at a conference on nuclear physics at the University of Rochester in New York State (Feynman 1997; Gardner 2005) where Feynman raised the question: "Is the law of parity sometimes violated?" Actually, Martin Block, an experimental physicist with whom Feynman was sharing a hotel room, had suggested the question to Feynman the night before. The answer to the theta-tau puzzle (to that time a fundamental question on two distinct types of neutral *K* mesons), said Block, might be very simple. Perhaps the lovely law of parity might not always hold. Feynman responded by pointing out that if this were true, there would be a way to distinguish left from right. It would be surprising, Feynman said, but he could think of no way such a notion conflicted with known experimental results. He promised Block he would raise the question at next day's meeting to see if anyone could find anything wrong with the idea. So the next day, at the nuclear physics meeting, when discussions started on the theta-tau puzzle, Robert Oppenheimer said, "We need to hear some new, wilder ideas about this problem." So Feynman got up, prefacing his remarks with, "I am asking this question for Martin Block." He regarded the notion as such an interesting one that, if it turned out to be true, he wanted Block to get credit for it.

Tsung Dao Lee and his friend Chen Ning Yang were present at the meeting. One of them gave a lengthy reply to Feynman's question.

"What did he say?" Block asked Feynman later.

"I don't know," replied Feynman. "I couldn't understand it".

"People teased me later," writes Feynman, "and said my prefacing remark about Martin Block was made because I was afraid to be associated with such a wild idea." I thought the idea unlikely but possible, and a very exciting possibility. Feynman made a 50-dollar bet with a friend that parity would not be violated. However Feynman lost that bet after the mentioned experimental proof of parity violation by Wu.

Freeman J. Dyson, a physicist now at the Institute for Advanced Studies in Princeton, called the reaction on the now-classic Lee and Yang paper the "blindness of most of his colleagues". Dyson is cited by Gardner (2005) with the words "A copy of the Lee and Yang paper was sent to me and I read it. I read it twice. I said, 'This is very interesting,' or words to that effect. But I had not the imagination to say, 'By golly, if this is true it opens up a whole new branch of physics.' And I think other physicists, with very few exceptions, at that time were as unimaginative as I."

Robert Frisch, the physicist who was co-discoverer of nuclear fission, reported that on 16 January 1957 he received the following air letter from a friend:

"Dear Robert: HOT NEWS. Parity is not conserved. Here in Princeton they talk about nothing else; they say it is the most important result since the Michelson experiment in 1887" (see Gardner 2005).

It is worth mentioning that the fall of parity in physics was the result of theoretical and mathematical work by Lee and Yang. They told the experimenters what to do. Lee and Yang themselves had no experience in the lab. Lee has never been working in a laboratory. Yang had worked briefly in a lab at the University of Chicago, where he was once kind of assistant to the Italian physicist Enrico Fermi. He had not been happy in experimental work. His associates had even made up a short rhyme about him (Gardner 2005):

Where there's a bang,
There's Yang.

Later on, it turned out that electrons are emitted during the β-decay with a velocity that is close to the speed of light. In our part of the universe, these electrons are always left-polarized, if we regard their spin towards their trajectory. The electrons' spin is antiparallel to their direction of propagation. The equally emitted antineutrino is right-polarized. The two types of electrons (left- and right polarized, where only left-polarized electrons are emitted via β^--decay) are related by mirror-image symmetry, which is, in principle, the same model as that which gives rise to optical activity in chemical compounds.

In contrast to antineutrinos and neutrinos, electrons can interact with matter and we will now have to continue asking the questions whether and how polarized 'homochiral' electrons generated by the β^--decay can transfer their inherent chirality towards any kind of organic environment. Did nuclear physics processes give ultimately birth to the asymmetry of life?

5.2 Asymmetric Radiolysis by Polarized Electrons

Our central question will now be whether this asymmetry inherent to the nuclear β-decay can have contributed to the selection of chiral molecules in life's biopolymers. Is it truly reasonable to assume that polarized electrons can interact with racemic or prochiral organic molecules such as amino acids inducing an enantioenrichment? In this case, a flux of spin-polarized β-electrons should react in a different and measurable way with *R*- and *S*-enantiomers.

Here, the interested reader will remark the strong analogy between polarized electrons as chiral driving force and 'chiral photons' in the form of circularly polarized light. The influence of 'chiral photons' on racemic organic molecules is indeed measurable which we will illustrate in the next chapter.

As mentioned before, polarized electrons (or positrons) emitted through the radioactive β-decay are highly energetic; their kinetic energy is not adapted to the range of excitation energies within molecules. Best results for the enantioselective interaction of polarized electrons with organic molecules were thus expected in the examination of relatively slow polarized electrons showing kinetic energies of some electron volt (eV) only. If we expect some enantioselective electron-molecule interaction, then it should be in this energy range. Production and handling of spin-polarized electrons in the low energy range is not particularly difficult and experiments of this kind have indeed been performed:

At the University of Edinburgh, Peter Farago studied the absorption of polarized electrons by individual enantiomers. He subjected enantiomers of D-camphor and L-camphor to 5 eV spin-polarized electrons. This experimental conception can be seen in close analogy to chiroptical circular dichroism measurements. Here and there a polarized beam (Farago: electrons; CD: photons) is differentially weakened by passing a left- or right-handed optically active medium. If any effect will be measured here, we may call it an "electronic circular dichroism" or an "electronic Cotton effect" (Rein 1992). And indeed, Farago was able to measure a differential influence: D- and L-camphor did not equally weaken the through-shining spin-polarized electrons! Differences were detected with high precision in the per mille range (Campbell and Farago 1987). Control experiments were performed carefully with racemic D, L-camphor and also with an unpolarized electron beam. The result was surprising: not because the effect was so small, but because the effect was much bigger than one could expect from quantum-theoretical calculations.

In similar experiments at the Biological Research Center in Szeged, Hungary, D- and L-tyrosine in alkali solution were irradiated separately with codissolved $^{90}SrCl_2$ as β-ray source. Unexpectedly, after 18 months, the absorption bands of D-tyrosine diminished compared to those of L-tyrosine, suggesting that the former had undergone more degradation than the latter (Garay 1968). At Stanford University, Garay's experiments were repeated in a way that the irradiated D- and L-tyrosine was additionally subjected to ORD and enantioselective gas chromatographic measurements. After 15 months, however, Bonner et al. did not at all notice the development of optical activity (Bonner et al. 1975b). Instead, Bonner et al. irradiated a solid film of the amino acid *rac*-leucine with 'left-handed' longitudinal polarized electrons but also with unnatural 'right-handed' longitudinal polarized electrons produced in a specifically developed linear accelerator source (Bonner et al. 1976). In agreement with Garay's finding, it was quantitatively observed that natural left-polarized electrons bring about the degradation of D-leucine more extensively than that of L-leucine, whereas the unnatural right-polarized electrons act in strictly the opposite sense. The significant enantiomeric excess generated in the product ranged from $-$ 0.60% to + 1.42%. It is worth mentioning that the natural left-handed electrons produced an asymmetric degradation favouring the formation of an excess of the L-leucine enantiomer occurring in biomolecules, while antinatural right-polarized electrons do precisely the opposite.

Even higher – not to say dramatic – enantiomeric excesses were reported by Wolfram Thiemann and coworkers at the Jülich Nuclear Research Center, Germany (Darge et al. 1976). In the experiment's design, it was assumed that asymmetric processes in aqueous solution are dominated by solution radiochemistry involving symmetrical free radicals that might overwhelm and obscure any asymmetric effect. Because of this, ^{32}P-phosphate in the form of $NaH_2{}^{32}PO_4$ was used as buffer and "internal" β-emitter together with the racemic amino acid D,L-tryptophan in water at $-25°C$. The low temperature was chosen in order to avoid asymmetric reactions caused by biological (micro-) organisms. The mean energy of the β-rays was 0.57 MeV with a 93.9% degree of polarization. After 12 weeks of asymmetric interaction, optical rotation was measured at 220 nm. Comparing the data with control experiments it was concluded that 33% of the racemic tryptophan had been decomposed and from the optical rotation of $+0.7 \pm 0.4$ mdeg as the mean value of sixteen independent runs, the authors concluded that a remarkable 19% enrichment of the D-enantiomer in the residue had been produced. We should not overlook that Bonner and coworkers (1979) tried to repeat this experiment using enantioselective gas chromatography to quantify the tryptophan enantiomers after subjecting them to β's of $KH_2{}^{32}PO_4$ at $-196°C$. They were not able to reproduce the effect of asymmetric radiolysis (Blair and Bonner 1980). Darge et al. (1979) immediately replied to this criticism that ORD measurements are capable of detecting the effect of the sum of all potential chiral products while the gas chromatographic method applied by Bonner and coworkers quantified exclusively the enantiomeric composition of the partially destructed residual tryptophan. This gas chromatographic method focused on the destruction of racemates whereas ORD measurements at 220 nm were said to be furthermore sensitive towards the equally probable chance of stereoselective *synthesis* triggered by left-polarized electrons as a source of optical activity.

The mechanism of the deceleration of polarized electrons in matter is dominated by collision and ionization processes. Details remain hitherto unknown in lack of experimental but also theoretical work. We know that the original polarization of the electrons is at least partly maintained during deceleration. Meiring (1987) was able to calculate asymmetries in radiolysis cross-sections for both β^-- and β^+-radiation, and to derive the connection of these cross-sections with decomposition rate constants. At Kyoto University, Akaboshi et al. (1979) contributed to elucidate the reaction mechanism by irradiating individual enantiomers of the amino acid α-alanine with β-electrons produced by radioactive ^{90}Yttrium-metal. The products were analyzed by electron spin resonance (ESR), a technique capable of characterizing radicals. As assumed, radicals were identified and quantification showed that D- and L-alanine enantiomers remained in distinguishable concentrations. This asymmetry was confirmed by ESR studies of Conte (1985), however, not observed with non-polarized electrons obtained from the Van de Graaff generator (Akaboshi et al. 1982) pointing on a real asymmetric effect. The interaction of Cerenkov radiation, the bluish light emitted as high-energy charged particles including β-particles, with liquid chiral enantiomers of (−)-*R*- and (+)-*S*-2-phenylbutyric acid was carefully studied by Garay and Ahlgren-Beckendorf (1990). Also here, a small asymmetry in the interaction was identified.

At this point we can conclude from all of the above experiments that – even if we are confronted with conflicting results – left spin-polarized ('chiral') electrons are emitted by the radioactive β^--decay and that such electrons are in principle capable of interacting differently with *R*- and *S*-enantiomers. The hitherto measured effects are small and/or difficult to reproduce and we are still waiting for the scientific breakthrough. We will leave it an open question whether or not the direct asymmetric radiolysis of racemic organic molecules by polarized electrons is a viable mechanism for the origin of biomolecular asymmetry. We can expect for the future that (a) more precise beams of polarized electrons showing higher polarization rates and fluxes and (b) better resolution of enantiomers with modern analytical techniques will provide even more convincing information on the possible direct asymmetric radiolysis of organic matter by polarized electrons.

Since the instant kinetic energy of spin-polarized electrons emitted by the β^--decay and other natural decay processes is usually some 100 keV and therefore orders of magnitude too high for any enantiomer discriminating absorption by organic molecules, the Vester-Ulbricht Process was proposed to circumvent this problem.

5.3 The Vester-Ulbricht Process

Tilo L. V. Ulbricht and Frederic Vester from the Universities of Cambridge and Saarbrücken were excited by reading reports of Lee and Yang, predicting the violation of parity in physics manifested by the weak nuclear force and its experimental proof by Wu.[2] Vester obtained this information at a cocktail-party as he lively describes (Vester 1974). He and Ulbricht, whom he met the next day at Yale University, immediately thought that Lee and Yang's finding might be the ultimate key to understand the origins of biomolecular asymmetry, clarifying equally our understanding of the origin of life on Earth.

As commonly known, instable elements such as ^{40}K with a half-life of 1.5×10^9 years and ^{14}C with a half-life of 6000 years are present in Earth's crust. These elements may have served as radiation source for electrons that are – as we know – exclusively left spin-polarized. Ulbricht and Vester simply proposed already in 1962 to decelerate ultra-fast spin-polarized electrons obtained by the β-decay before studying an eventual enantioselective interaction of the decomposition products with organic matter (Ulbricht and Vester 1962); a suggestion that attracted tremendous interest and triggered an armada of research activities until today.

5.3.1 Circularly Polarized Bremsstrahlung in the VU Process

It is known that if electrons pass matter they are decelerated, due to which they emit electromagnetic radiation that is called Bremsstrahlung (indeed, this German term describes the particular kind of electromagnetic radiation also in English language). In the context of the VU process it is important to note that Bremsstrahlung produced by spin-polarized electrons is equally polarized (Goldhaber et al. 1957)![3] The helicity of the spin-polarized electron is transferred to the helicity of the Bremsstrahlung, which is nothing else than circularly polarized electromagnetic radiation. The interaction of circularly polarized electromagnetic radiation with racemic or prochiral organic molecules was – even to that time – known to induce enantiomeric excesses. 'Chiral photons' can be used for asymmetric photochemical reactions such as asymmetric photolysis and asymmetric synthesis (we will develop a deeper understanding of asymmetric photochemistry in the next chapter). Ulbricht and Vester proposed the chain of the above-mentioned consecutive processes to explain the chiral asymmetry in biomolecules (Fig. 5.3). Today, this sequence is known as the Vester-Ulbricht (VU) process.

Worldwide, an ultimate experimental verification of the Vester-Ulbricht process was searched intensively. Each of the individual steps of the consecutive Vester-Ulbricht process found independent experimental confirmation. Electrons emitted during β-decay are due to the weak nuclear interaction longitudinally polarized (Wu et al. 1957) and Bremsstrahlung emitted from β-rays is circularly polarized (Goldhaber 1957). Long before the VU hypothesis it had become apparent that asymmetric photochemistry of circularly polarized light in ultraviolet wavelengths with organic matter yields optically active products (Kuhn and Braun 1929).

[2] To that time, Lee was not aware of the fact that not only parity in physics is violated but that biology's molecular symmetry is broken as well.

[3] At the high-energy end of the Bremsstrahlung spectrum emitted photons are almost completely circularly polarized.

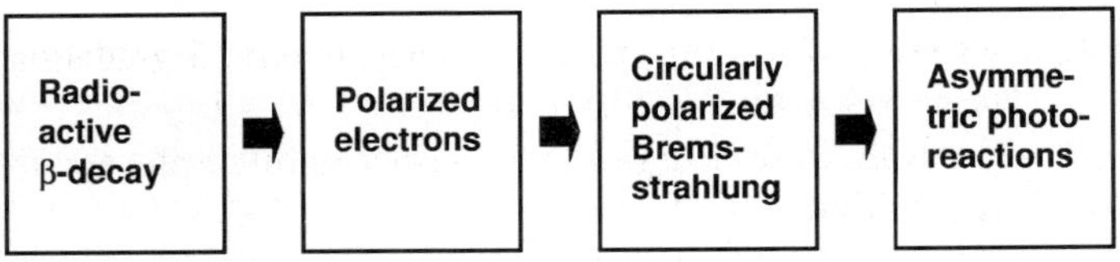

Fig. 5.3 The Vester-Ulbricht process. The natural β-decay of radioactive isotopes emits spontaneously and continuously longitudinally polarized electrons (or positrons) that decelerate by the interaction with matter producing circularly polarized Bremsstrahlung. This circularly polarized electromagnetic radiation can be interpreted as 'chiral photons' and is known to interact with racemic or prochiral organic molecules via asymmetric photoreactions in order to induce an enantiomeric excess

5.3.2 In Search of the Ultimate VU Experiment: California Dreaming?

The complete consecutive Vester-Ulbricht process, however, has not yet been realized successfully in the laboratory. It had been searched for intensively with long-time irradiation experiments at various β-ray sources like ^{104}Rh illuminating different enantiomers such as the amino acids alanine, proline, leucine, valine, and others (Goldanskii and Khrapov 1963). In none of these experiments optical activity was induced, nor could differences in the action of the electrons on individual enantiomers be observed.

At Stanford University, William A. Bonner (1974) subjected aqueous solutions and solid samples of amino acids such as D, L-leucine for 1.34 years to β-radiation. Racemic mixtures but also individual enantiomers were irradiated separately. While slight differences were noted in the percent radiolysis of solid D-leucine (12.7%) and solid L-leucine (16.2%), no significant enantioenrichment was found by enantioselective gas chromatographic analysis of irradiated D,L-leucine (Bonner and Flores 1975).

It is interesting to know that ^{14}C-labeled racemic amino acids had been produced with high specific activities long before the discovery of the Fall of Parity in the laboratory at the Lawrence Radiation Laboratory at the University of California. In subsequent years these crystalline samples had undergone self-radiolysis due to the pure β-rays from their ^{14}C-decay and the accompanying γ-ray Bremsstrahlung. After having been informed on the parity violation in physics and the resulting asymmetry of β-electrons, Melvin Calvin and coworkers examined the residual amino acid samples (then 12 to 24 years old) for possible asymmetric degradation employing ORD measurements. Despite high total radiation dosages and radiolysis of the extent of 3–80% no optical rotation was observable within the sensitivity of the spectropolarimeters of 0.002° (Bonner et al. 1976). Two years later, these crystalline amino acid samples were analyzed by enantioselective gas chromatography. Also here, the D/L ratios of the radioactive D,L-amino acids were within experimental error 50:50 indicating that these amino acids suffered no asymmetric degradation, despite self-radiolysis of up to 67% (Bonner et al. 1978).[4]

Today, we have various reasons to assume that these unsuccessful results are not surprising. The reasons are manifold.

1. The energy of ultra-fast electrons dissipates during the electron's movement in matter via collision and ionization processes towards particles and molecules. Only a small part of approximately 10^{-3} of the initial β's kinetic energy is emitted as Bremsstrahlung.
2. The energy of the Bremsstrahlung is only partly located in the photochemically active range of the spectrum; roughly 10^{-3} parts of the Bremsstrahlung are in the near ultraviolet.
3. The degree of circular polarization decreases linearly with photon energy and is small at photoenantioselective energies (MacDermott and Tranter 1989). Approximately 10^{-4} parts of Bremsstrahlung produced by polarized electrons in the low energetic range are circularly polarized. Only this circularly polarized part of the light is active for asymmetric photochemical reactions.
4. The photolytic absorption of the circular polarized component of the Bremsstrahlung by organic molecules is about 10^{-2} as we will see in the next chapter.

The multiplication of these factors lets us estimate that any asymmetry obtained by the Vester-Ulbricht process should be about 10^{-12} and thus very small and extremely difficult, if not impossible, to observe (Rein 1992).[5] Of course, this experimental difficulty of the involved laboratories requiring high precision does not necessarily exclude the Vester-Ulbricht scenario from having contributed to prebiotic evolution. Prebiotic amino acids might well have been subjected to natural permanent asymmetric radiation of polarized electrons but also circularly polarized Bremsstrahlung. One requirement would have been that the obtained difference in prebiotic enantiomers is big enough to rise out of the statistical fluctuations of the chiral components. Autocatalytic reactions (we will discuss them in Chap. 10 in more detail) might have been capable of favouring the amplification of a very small excess of one particular enantiomer; even the excess that had been determined by the asymmetry of the weak nuclear interaction might have been sufficient for asymmetric amplification.

---

[4] I would be more than curious to analyze these ^{14}C-labeled amino acid samples from the Laurence Radiation Laboratory or analogous well conserved ^{14}C-labeled amino acids of comparable age by means of enantioselective techniques today.

[5] For the specialist reader, I would like to mention that we distinguished for didactic reasons between (a) the direct asymmetric radiolysis of racemic organic molecules by polarized electrons and (b) the Vester-Ulbricht process. This distinction has pedagogic reasons in a way that scientific phenomena are exposed starting from the simple (Sect. 5.2) and moving then to the more complex model (Sect. 5.3). However, this distinction is 'artificial' because an effect of longitudinally polarized electrons on racemic organic molecules might be (a) due to a direct electron-particle interaction or (b) by circularly polarized Bremsstrahlung via photon-particle interaction. Often, the experimentalist is unable to distinguish (see Bonner et al. 1976; Akaboshi et al. 1982).

Discussing the Vester-Ulbricht hypothesis and the hitherto lack of its experimental verification, we should not overlook that ionizing β-radiation could also cause the racemization of amino acids. If the rate of asymmetric radiolysis were lower than the rate of radioracemization, such radioracemization would inevitably diminish the effectiveness of the Vester-Ulbricht mechanism of engendering optical activity (Bonner et al. 1982; Bonner 1984). Effects of radioracemization or secondary symmetrical degradation effects might dominate over asymmetric reactions and thus be responsible for the lack of observing an asymmetric degradation during radiolysis of racemization-susceptible amino acids such as D,L-leucine with longitudinally polarized electrons.

5.3.3 What Are the Next VU Experiments?

Since the question of a possible relationship between asymmetry at the elementary particle level and chirality at the biomolecular level (a) remains unanswered until today and (b) is clearly of outstanding interest with regard to the question of the origin of life, I would like to urge others to extend earlier experiments and to develop new ones having a bearing on the matter. A strategy for a successful experiment might be to adapt experimental parameters to interstellar-like conditions in which amino acid structures had recently been identified (Bernstein et al. 2002; Muñoz Caro et al. 2002). In particular, the experimental study of asymmetric synthesis and radiolysis reactions of α-methyl-α-amino acids with the β^+-radiator ^{26}Al might have the potential to offer important results. Various α-methyl-α-amino acids were identified in meteorites showing considerable enantiomeric excesses (see Chap. 8) and these amino acids are stable against radioracemization (Bonner 1984). The ^{26}Al emitter of spin-polarized ('homochiral') positrons (van House et al. 1984) has largely contributed to the early thermal evolution of comets and small icy bodies (Meiring 1987; see also Chap. 9), in which amino acid structures were identified as well. Here, the experimentalist should take into consideration that positrons obtained by the β^+-decay show opposite handedness (right-polarization) compared to left-polarized β-electrons (Conte 1985). The β^+-radiator ^{26}Al should therefore induce the opposite handedness into organic molecules such as amino acids compared to a β^--radiator as e.g. ^{32}P. Furthermore, longitudinally polarized protons (Bonner et al. 1982) or supplementary and hitherto overlooked particles acting as external chiral driving forces might be examined carefully in future times.

5.4 Parity Non-Conserving Energy Differences

The Yamagata-Rein hypothesis, that dates back to a publication of Yamagata (1966), with its challenging theoretical background, has provided an alternative fresh look on the origin of biomolecular asymmetry and thus attracted the interest of chirality's scientific community since decades (e.g. Keszthelyi 1984; Borchers et al. 2004). D- and L-enantiomers are in fact not the exact mirror images of each other but differ in scalar physicochemical properties. The parity non-conserving neutral current weak interaction causes parity-violating forces that give rise to different bond strengths in enantiomer molecules! If the electromagnetic interaction, which is mirror-image invariant, were the only force that binds electrons to atomic nuclei, the electronic binding energy of D- and L-enantiomers were identical. But the asymmetric weak interaction slightly interferes with the electromagnetic interaction stretching its tiny asymmetry from atomic nuclei towards atoms and molecules (Rein 1992).[6]

5.4.1 The Electroweak Interaction

In the 1970s, Steven Weinberg (1980) and Abdus Salam (1980) independently unified electromagnetic and weak interaction resulting in the electroweak interaction. Electromagnetic and weak interaction should therefore be regarded as one ensemble. For this important achievement Weinberg and Salam obtained the Nobel Prize of Physics in 1979. This unification is similar to the way in which James Clerk Maxwell 150 years ago unified electric and magnetic forces towards the electromagnetic interaction. To that time, Maxwell did not see and interpret the weak interaction as additional partner-force, since the weak interaction is active in the innermost interior of atoms only, a region that was studied more intensively at the beginning of the 20th century.

Today, the electroweak interaction has become part of standard theory. It is important to know that the weak interaction fits – despite of its asymmetry and despite of its weakness – into general electrodynamics. Electromagnetic Coulomb forces that are known among students for determining atomic and molecular relations are no more only influenced by the virtual exchange of photons, which are as γ-quanta the gauge bosons of the electromagnetic interaction, but also by the vector bosons Z^0 of the weak interaction. The uncharged (neutral) Z^0 vector boson had indeed been measured experimentally. It is now entirely implemented in electroweak theory and gives rise to the weak neutral current.

It is evident that the electromagnetic component and the weak component, which we just have combined to the electroweak interaction, are an unequal pair. The weak part is largely hidden behind the electromagnetic interaction as a small correction being restricted to minute distances and possessing extremely small absolute values. But with respect to optically active and chiral molecules, the 'new' quality of the weak component has to be considered, especially its capability to differentiate between right- and left-forms.

[6] As a consequence, atoms can be optically active and rotate the plane of polarization of linearly polarized light by a tiny angle that already had been measured experimentally.

5.4.2 Characteristics of Electromagnetic and Weak Interaction

In fact, theoretical calculations were performed to precisely determine the energy of chiral molecules compared to their corresponding mirror-images. Calculations for the individual enantiomers resulted in the astonishing consequence that the energy of a given enantiomer is indeed non-identical to the energy of its own mirror-image. The obtained differences are called parity non-conserving (PNC) energy differences ΔE^{PNC} or sometimes parity-violating energy differences (PVEDs), as eye-catching illustrated by the number plate of one of its protagonists (Fig. 5.4). The value of this asymmetry is, of course, extremely small in comparison with the involved Coulomb interaction. The weak part of the binding energy V^{PNC} is – in contrast to the Coulomb interaction – characterized by the following parameters.

1. The Fermi coupling constant determines the force of the weak interaction, which is consequently much smaller than the Coulomb interaction V^{Coul} determined by charges of interacting particles. The weakness of the weak force is based on the W and Z bosons that are relatively heavy or – which means the same – that the effective reach of the weak nuclear current is extremely small.
2. The strength of the weak interaction is no more in inverse proportion to the distance of particles such as we know it from the Coulomb interaction, but active on extremely small distances, smaller than the radius of a proton.
3. The characteristic of the weak part of the binding energy V^{PNC} is that it bears the spatial asymmetry in mind. This spatial asymmetry is described and can be calculated by an interaction between spin and momentum of the interacting particles

Fig. 5.4 The North Carolinian PVED number plated Lincoln Town Car steered by Alexandra MacDermott from Wake Forest University (WFU) before she moved together with her husband Roger Hegstrom to Clear Lake University in Texas. Photo: Alexandra MacDermott

called helicity. The helicity itself is considered a pseudo-scalar value, since its sign changes by mirror-reflection of the coordinate system.

5.4.3 *Difference Matters: Energies of Enantiomers*

The small, weak, and pseudo-scalar potential V^{PNC} contributes to the attracting forces and the binding energy of molecules in a way, that a tiny amount of energy is added to one enantiomer whereas the same amount is subtracted from the energy of the mirror-image enantiomer giving rise to definite ΔE^{PNC} values. Usually – and as we have outlined in Chap. 2 – it has been assumed that all scalar values of enantiomers, with the exception of its capacity to rotate the plane of linearly-polarized light, are strictly identical. The enthalpy, the entropy, and hence the standard Gibbs energy of an isomerization reaction between the *R* and *S*-enantiomers of a chiral molecule were assumed furthermore to be exactly zero at all temperatures, including the temperature 0 K. The equilibrium constant *K* for this isomerization was assumed to be exactly 1. But now we have to bear in mind, that chiral molecules are little more (or little less) stable than their mirror-image enantiomers!

Based on this knowledge on the physical basics of chemistry, symmetry, and chirality, we will now go a step further and ask the question whether the atoms of a left-handed molecule are more strongly (or less strongly) bound compared to the atoms of a right-handed molecule. We may try to measure the equilibrium constant *K* to very high precision. If *K* were found to be systematically different from 1 within high experimental confidence, the symmetry would be proven by experiment to be absent and thus the absolute handedness would become observable (Quack 2002). The literature is full of exiting examples of chiral molecules such as chiral hydrogen peroxide H_2O_2 (Mason and Tranter 1983b; Bakasov et al. 1996; Berger and Quack 2000; Gottselig et al. 2001), chiral disulfane H_2S_2 including its deuterium and tritium isotopomers (Gottselig et al. 2001), chiral dichlorodisulfane ClSSCl (Berger et al. 2001), the chiral archetypes bromochlorofluoromethane CHBrClF, and the deuterated CDBrClF (Quack and Stohner 2001), or chiral Buckminsterfullerenes (MacDermott 1993) for which parity non-conserving energy differences have been calculated. Even for transition-states involved in the first induction of chirality in amino acid-forming reactions, non equivalent electronic binding energies were calculated for the enantiomers of α-amino-propionitrile (Tranter 1985b). We will not discuss amino acid or ribose molecules as first examples since the applied ab initio methods calculating the spatial electron distribution are complex. Our first introducing case will concern energy differences of the trouble-free ethylene molecule $H_2C{=}CH_2$.

Ethylene is composed of two carbon atoms and four hydrogen atoms. It has 16 electrons and a plane of symmetry. In order to make the ethylene molecule chiral, we will twist it by 10° around the carbon-carbon axis (Rein et al. 1979; Rein 1992). A torsion angle $\varphi = 10°$ twist in one direction will give the *P*-helical ethylene (right-helix), a $-$ 10° twist in the other direction will result in *M*-ethylene. The twisted ethylene molecule is now chiral, composed of two enantiomers. At this instant, a suitable software programme for ab initio calculations can be fed with bond distances and angles of the chiral ethylene in order to calculate (and minimize) the total energy. The programme will provide a precise picture of the electron distribution. Such calculations provide the advantage that we can calculate total energies for various molecular conformations. We are not limited to conformations that are preferably realized in nature, but we can study conformations not used naturally. For example we can change the ethylene conformation by twisting the ethylene molecule by the torsion angle φ. This was done and indeed for the two twisted ethylene enantiomers a parity non-conserving energy difference of $\Delta E^{PNC} = 4 \cdot 10^{-20}$ a.u. $= 10^{-18}$ eV was found (Rein 1992). The two twisted ethylene enantiomers are of different energy. The *P*-helix of twisted ethylene was favoured by 10^{-18} eV to *M*-ethylene and was thus calculated to be more stable. This is exciting and encouraging, even if such energy differences are too small to be directly measured for example with available spectroscopic methods.

With a certain approximation, we can generally use an energy difference ΔE^{PNC} for enantiomers that is specified by Eq. 5.1. The energy difference can be given in atomic units (a.u.), defined as the doubled ionization energy of hydrogen (27.2 eV). Alternatively, ΔE^{PNC} can of course be given in eV.

$$\Delta E^{\mathrm{PNC}} \approx 10^{-20} \cdot Z^5 \mathrm{a.u.} = 3 \cdot 10^{-19} \cdot Z^5 \mathrm{eV} \tag{5.1}$$

Equation 5.1 includes a large exponent of five on the atomic number of protons Z in the nucleus, which results in the case of carbon ($Z = 6$) in approximately four orders of magnitude. Particularly for higher elements this exponent was assumed to help in making the energy difference experimentally accessible. Later, we will see that chiral cobalt- ($Z = 27$) and iridium- ($Z = 77$) complexes have been studied to measure ΔE^{PNC} values experimentally.

In spite of the extremely small ΔE^{PNC} values, we have to face the consequence that *R*- and *S*-enantiomers of the same molecule are now by definition really diastereoisomers, not enantiomers. Remember that diastereoisomers show distinguishable scalar properties! The strict enantiomer of an L-amino acid is the D-amino acid made of antimatter, but the 'classical' D-amino acid is not energetically equivalent (Mason 1984). Based on the weak interaction in atoms' nuclei also atoms themselves become optically active! The 'strict' enantiomer of an atom with its electrons is an antimatter atom including positrons. This gives rise to a small electroweak optical rotation, which was measured experimentally for the heavy metal atoms Tl, Pb, Bi, and Cs in the gas phase (Emmons et al. 1983) in accordance with quantum mechanical calculations.

Hence the two natural enantiomers of a chiral molecule differ slightly in many ways, including energetically. According to the Yamagata-Rein hypothesis such differences in the electronic energy might have been important for the deterministic generation of biomolecular asymmetry in prebiotic evolution on the Early Earth. The condition is that L-amino acids must have been favoured compared to D-amino acids and that D-sugar molecules are favoured to L-sugars.

Values of ΔE^{PNC} do influence rate constants whenever chiral molecules (a) are formed, (b) racemize, (c) decompose, or otherwise react to give achiral products. The activation energies will in general be slightly different for the D- vs. L-enantiomers (Hegstrom 1984). A tiny difference in the energy of a chiral molecules compared to its mirror image enantiomer also influences the stability of chiral molecules in chemical reactions. L-Amino acids might be little more stable than D-amino acids or vice versa. They are not equally stable if we consider the contribution of the weak interaction to the electromagnetic interaction carefully. And this is what we have to do, independently on the size of the effect.

5.4.4 Calculated Energy Differences of Amino Acid Enantiomers

Ab initio quantum mechanical calculations that include the asymmetric electroweak interaction have been performed in order to determine the parity non-conserving energy difference ΔE^{PNC} of the binding energy in the electronic ground state for major biomolecules such as amino acids. Molecular structures of amino acids are more complicated than twisted ethylene but they are still manageable. Stephen Mason and his former assistant George Tranter (who moved since then to Oxford University and afterwards to industry) performed studies on the molecular wave function of α-alanine with its 48 electrons at London King's-College (Mason and Tranter 1983a, 1983b).

As we have seen in Chap. 2, α-alanine is the simplest chiral amino acid with its asymmetrically substituted α-carbon atom as stereogenic center. The enantiomer α-*R*-alanine corresponds to α-D-alanine and α-*S*-alanine is identical to α-L-alanine. In the α-*S*-alanine zwitterionic structure depicted in Fig. 5.5 the planar carboxylate-group is not fixed and can rotate around the carbon-carbon bond linking this group to the amino acid's central α-carbon atom.

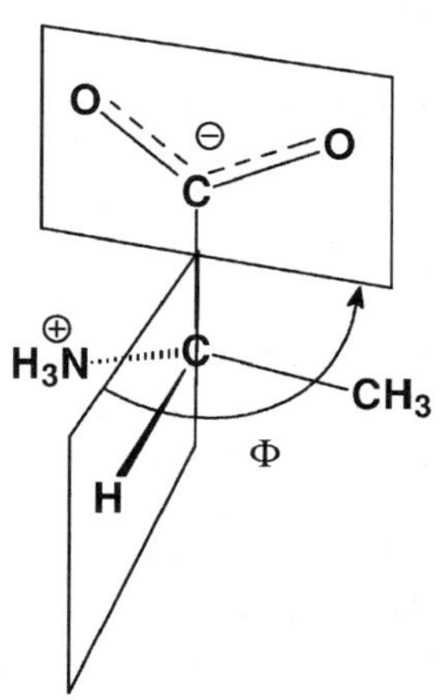

Fig. 5.5 Chemical structure of the amino acid α-*S*-alanine indicating the torsion angle Φ of the plane of the carboxylate-group COO^- and the fixed C-C_α-H-plane. In the solid state of aggregation the conformation shows Φ = 62°. In liquid solution, the confirmation is determined by $\Phi = 0$

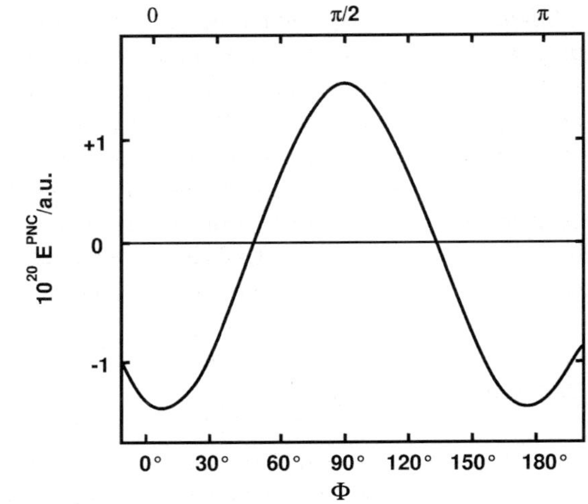

Fig. 5.6 The parity non-conserving energy shift as function of torsion angle Φ calculated for α-*S*-alanine. Corresponding data for α-*R*-alanine can be obtained by horizontal flipping of the curve

The torsion angle Φ can range from 0 to 180°, resulting in different conformer structures of α-alanine. Mason and Tranter calculated value and sign of the parity non-conserving energy shift for α-alanine as a function of this torsion angle Φ. The result is depicted in Fig. 5.6.

The parity violating energy shift is a function of the torsion angle Φ. If E^{PNC} is negative, the bonding energy is stabilized which is the case for α-*S*-alanine with values of Φ smaller than 50° and bigger than 130°. The parity non-conserving energy difference ΔE^{PNC} is defined by Eq. (5.2).

$$\Delta E^{PNC} = E^{PNC}(S) - E^{PNC}(R) \qquad (5.2)$$

In this range of Φ, E^{PNC} is indeed smaller than zero, which means that α-*S*-alanine (L-alanine) is energetically preferred compared to its mirror image enantiomer the α-*R*-alanine (D-alanine). For the intermediate range of $50\,° < \Phi < 130°$ the opposite is true and D-alanine is more stable than L-alanine. But this intermediate range is less realized in all possible alanine confirmations since the medial conformation of alanine is in the L-enantiomer preferring range. In aqueous solution – the most important matrix for prebiotic reactions of amino acids – alanine has a torsion angle $\Phi = 0°$.

So one concluded that L-alanine – as it is used in biomolecules – is energetically preferred in comparison with D-alanine, the 'unnatural' enantiomer at least in the important aqueous solution state. This preference is very tiny and was quantified with ΔE^{PNC} (α-alanine) $= 2.5 \cdot 10^{-20}$ a.u. = $0.7 \cdot 10^{-18}$ eV (see Fig. 5.6). For chemists, this tiny value corresponds to $6.8 \cdot 10^{-14}$ Joule/mol, an extremely small amount of energy. Quack et al. insist on the fact that more recent and improved ΔE^{PNC} calculations performed by his team at ETH Zurich gave rise to ΔE^{PNC} values that are at least one order of magnitude larger (Bakasov et al. 1996).

What about other amino acids? The extrapolation of the preference of the L-enantiomer in the case of α-alanine by the parity non-conserving energy difference toward other amino acids seems to be possible. In the case of twisted glycine, nearly identical results were obtained (Mason and Tranter 1983a; Rein 1992). Glycine, the simplest amino acid is relatively easy to calculate with quantum mechanical tools. Similar to twisted ethylene, glycine is chiral over most of the conformational range spanned by the rotation of the carboxylate plane about the bond to the α-carbon atom. Magnitude and sign of the parity non-conserving energy differences were calculated to give a similar sinusoidal dependence on the torsion angle for both glycine and L-alanine. Subsequent calculations for other zwitterionic amino acids in the aqueous solution state equally preferred the L-enantiomers relative to their D-enantiomers, as there are ΔE^{PNC} (valine) $= 4.6 \cdot 10^{-20}$ a.u., ΔE^{PNC} (serine) $= 1.7 \cdot 10^{-20}$ a.u., and ΔE^{PNC} (aspartate anion) $= 2.9 \cdot 10^{-20}$ a.u. (MacDermott and Tranter 1989).

So how about the typical amino acid confirmation in proteins (enzymes) that are, as we know, important biocatalysts composed of 20 different amino acids in L-configuration? Quantum mechanical calculations show that fragments of such proteins (polypeptides) composed of L-amino acids are little more stable than polypeptides composed of D-amino acids. This stabilization was calculated for both α-helices[7] and for β-sheet structures. A polypeptide chain composed of L-amino acids is stabilized by $-1.8 \cdot 10^{-19}$ eV per monomer unit compared to a mirror-image polypeptide chain of D-amino acid monomers. A L-polypeptide of n residues should therefore be stabilized relative to the D-polypeptide by $-1.8\, n \cdot 10^{-19}$ eV, but there may be additional contributions from cross-terms between residues, since these do not necessarily fall off with increasing residue separation (MacDermott and Tranter 1989). This calculation is valid for both α-helices and for β-sheet structures.

We will discuss and illustrate in Chap. 10 whether or not mechanisms involved in prebiotic evolution were capable of amplifying such a tiny difference in electronic energies of enantiomers.

Now, we will turn towards quantum mechanical ab initio calculations of sugar molecules, since we know that biopolymers of living organisms such as the nucleic acids DNA and RNA exclusively implement monomers of D-deoxyribose and D-ribose enantiomers into their molecular skeletons. Consequently we will now ask, "Are D- or L-sugar enantiomers preferred by ΔE^{PNC}-values, too?"

5.4.5 Calculated Energy Differences of Sugar Enantiomers

In biochemical reactions, sugar molecules such as glucose are synthesized and used as D-sugars (D-glucose). They are chemically related to the structure of D-(+)-glyceraldehyde (see Chap. 2). Sugar molecules are optically active and rotate the plane of linearly polarized light to the right. Some exceptions are known,

[7] The name α-helix is not based on the use of α-amino acids as molecular building blocks of proteins. The name α-helix is derived from the classification of X-ray diffraction diagrams of organic fibers. Linus Pauling and Robert B. Corey, biochemists at the California Institute of Technology, were the first to discover and name this helical structure.

as for example the inversion of (+)-saccharose, which decomposes by catalysis to form (+)-glucose and (−)-fructose. (+)-Saccharose exhibits a specific rotation of $[\alpha]_D = +66,5°$, (+)-glucose shows $[\alpha]_D = +52,7°$, and (−)-fructose gives $[\alpha]_D = -92,4°$. The chemical kinetics of this reaction can easily be followed by a polarimeter, since the optical rotation is dominated by the synthesized (−)-fructose and decreases during the ongoing reaction. This reaction is ideal for practical students training; it requires simply cane sugar and hydrochloric acid solution as well as a polarimeter (Försterling and Kuhn 1985).

Sugars in D-configuration are also used in the nucleic acids DNA and RNA in the form of D-ribofuranose. Ribose has 80 electrons and quantum mechanical calculations have been performed in order to determine sign and value of parity non-conserving energy differences. For the two ribose confirmations that mainly occur in nature, a reduction of the D-enantiomer's energy of 10^{-20} a.u. was calculated, whereas the L-enantiomer was affected with a higher energy. Consequently, the D-ribose enantiomer used in ribonucleic acid was calculated to be energetically favoured compared to non-biological L-ribose (Rein 1992). In the case of this particular sugar, however, values found in the literature are less coherent. MacDermott and Tranter (1989) calculated that D-deoxyribose is indeed stabilized due to parity non-conserving energy differences by ΔE^{PNC} (deoxyribose) $= -1.2 \cdot 10^{-20}$ a.u., whereas D-ribose in its preferentially adopted C3-endo solution confirmation shows a ΔE^{PNC} (ribose) = + $5.8 \cdot 10^{-20}$ a.u. value favouring non-biological L-ribose (see also MacDermott 1993). Nevertheless, the authors argued that the ultimate reason for the evolutionary selection of D-ribofuranose in life's genetic material is not yet clear. Prebiotic synthesis of ribose via the unselective formose reactions leads e.g. to many different stereoisomers and biological D-ribose might have been selected later during evolution by its parent compound, i.e., D-(+)-glyceraldehyde. Subsequent ab initio calculations were therefore performed with glyceraldehyde, the most simple sugar molecule. Glyceraldehyde was discussed to be the precursor molecule of more complex sugar structures in prebiological evolution and is thus a particularly interesting case to study. D-(+)- and L-(–)-glyceraldehyde are used as standards for the D/L-nomenclature. The energetic preference of D-(+)-glyceraldehyde by 10^{-20} a.u. compared to its mirror-image enantiomer was indeed confirmed by numerical calculations (Rein 1992).

Based on the weak nuclear current, D-(+)-glyceraldehyde shows an intrinsic energetic advantage such as D-deoxyribose, L-alanine, L-valine, L-serine, and the L-aspartate anion are preferred energetically as well. What makes the Yamagata-Rein hypothesis interesting for us is that these mirror image isomers are exactly the specific enantiomers used in biomolecular structures such as proteins and nucleic acids. We have just to keep in mind that their energetic preference is extremely small[8] and we will have to discuss whether this determinate intrinsic bias can be amplified by suitable mechanisms.

[8] In order to obtain larger parity non-conserving energy differences one may implement more heavy atoms such as phosphor (cf. the sugar-phosphate backbone of DNA), silicon, and sulphur (MacDermott 1993) in highly chiral environments. To cite some examples, larger values for parity non-conserving energy differences were calculated for the $(O_8)^{2-}$ helix (ΔE^{PNC} = + $55.4 \cdot 10^{-20}$ a.u.) and a similar $(S_6)^{2-}$ helix (ΔE^{PNC} = + $748 \cdot 10^{-20}$ a.u.).

Did "electroweak bioenantioselection" (MacDermott and Tranter 1989) contribute to the asymmetric origins of life? Our main problem to accept this captivating theory is that experimental evidence remains difficult to obtain. The value of the parity non-conserving energy difference of a typical chiral molecule is about 13 orders of magnitude lower, than measurable energetic differences as for example in the fine structure of atomic spectra. This is equivalent to raise the budget of national finances by 1 cent. A Minister of Finance would never consider this raise as an evolutionary advantage in economy and envisage a tax reduction (Rein 1992).

An experimental verification of ΔE^{PNC} values, however, is of great importance, since competing research teams have difficulties to reproduce the calculation of the parity non-conserving energy differences for the amino acids glycine, alanine, serine, and cysteine (Cintas 2001 and references therein; Wesendrup et al. 2003). Promising experimental trials for an ultimate ΔE^{PNC}-approval were undertaken and will thus be presented in the next paragraph.

5.4.6 Never Say Never: Measuring Parity Non-Conserving Energy Differences

More than 30 years ago, immediately after its theoretical prediction, it was assumed that the tiny parity non-conserving energy difference ΔE^{PNC} might be experimentally measured for chiral molecules verifying the above outlined wonderful theoretical approaches of Yukio Yamagata, Dieter Rein, Stephen Mason, and others. Besides measurements of specific heat and magnetic susceptibility, spectroscopic methods such as ultraviolet, infrared, generally optical, microwave, and nuclear magnetic resonance (NMR), but also Mößbauer, and Raman spectroscopy (see Bolik et al. 2007) were applied. Often, isolated molecules in the gas phase under well defined conditions but also physico-chemical nucleation processes were studied intensively for individual chiral compounds and their corresponding mirror image enantiomers with the aim to detect a small differential shift (see e.g. Quack 1989, 1993; Wang et al. 2000, 2003). Particular attention was drawn to reactions where very small differences of energy can become manifest by amplification or cascading. Polymerization and precipitation were considered as such reactions, but also the possibility of cascading differential properties of enantiomers by irradiation-induced chain reactions.

In a remarkable experiment, Otto Merwitz studied differences in the radiochemical behaviour of enantiomers at the Jülich Nuclear Research Center in Germany. He irradiated D-, D,L-, and L-β-phenylalanine with unpolarized γ-radiation from a ^{60}Co source. The absorption of γ-quanta by the substrate provokes the dissociation of phenylalanine under release of carbon dioxide CO_2. The decarboxylation reaction is illustrated in Fig. 5.7.

As depicted, the amino acid β-phenylalanine was 100% labelled with ^{14}C-isotopes at its carboxylate-carbon atom. By utilizing this kind of ^{14}C-labelled enantiomers it was possible to very precisely quantify the radiochemically-released

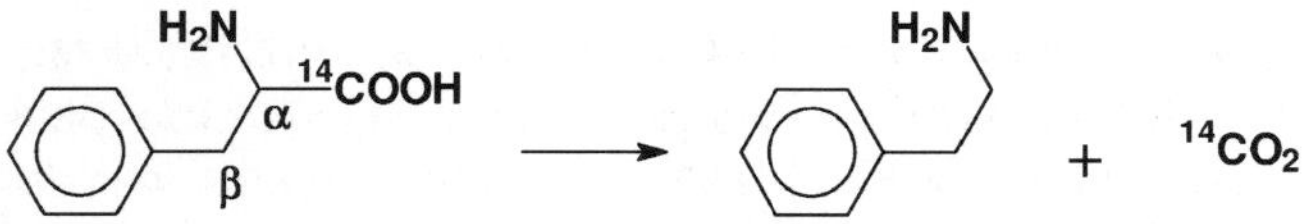

Fig. 5.7 γ-Quanta induced radiolysis of ^{14}C-labeled β-phenylalanine enantiomers producing $^{14}CO_2$

$^{14}CO_2$ by measuring its counting rate of ^{14}C in the gas phase. In Table 5.1, the counting rate is given for the individual enantiomers D- and L-β-phenylalanine and the racemic mixture D,L-β-phenylalanine.

The results indicated that the counting rate of $^{14}CO_2$ and therewith the cleavage rate of the D-β-phenylalanine enantiomer was higher by a maximum factor of 2.7 compared to the L-enantiomer (Merwitz 1976). The D-enantiomer decarboxylated faster. Values for the racemic mixture were recorded in almost exactly halfway between the values for the individual enantiomers. The differential behaviour of the enantiomers was valid for doses of the electromagnetic radiation between $5 \cdot 10^2$ rad and $5 \cdot 10^4$ rad. Below 50 rad the counting rates could not be measured with sufficient accuracy. Above 10^4 rad the differences became smaller.

For further clarification of this crucial experiment, subsequent differential radiolysis experiments with ^{14}C-labelled (Nordén et al. 1985) and ^{13}C-labelled (Merwitz et al. 1998) leucine enantiomers in the solid-state point in a very similar way to the preferential decarboxylation of the D-enantiomers. The decarboxylation of D-leucine was proven to be more effective than that of L-enantiomer. One should note that the above asymmetric results were not assumed to be caused by circular polarization of the photons. Nordén et al. (1985), however, could not completely exclude the possibility that (a) asymmetric crystal defects or (b) impurities from the original preparations of the amino acids survived the recrystallization and sublimation caused the asymmetric behaviour.

Are these results the ultimate experimental proof for parity non-conserving energy differences as claimed by the author? Discussions continued (Schleser et al. 1991; Merwitz et al. 1991) and are ongoing until today; the final answer to this question cannot be given yet.

Table 5.1 Dose dependent counting rates of $^{14}CO_2$–cleavage after irradiation of ^{14}C-labeled β-phenylalanine with unpolarized electromagnetic radiation

Enantiomer	Counting rate at 50 rad [cpm/g][a]	Counting rate at $5\cdot10^2$rad [cpm/g][a]	Counting rate at $5\cdot10^3$rad [cpm/g][a]	Counting rate at $5\cdot10^4$rad [cpm/g][a]	Counting rate at $5\cdot10^5$rad [cpm/g][a]
D-β-phenyl-alanine	$1.3 \cdot 10^4$	$2.4 \cdot 10^4$	$3.0 \cdot 10^4$	$4.6 \cdot 10^4$	$1.5 \cdot 10^5$
D,L-β-phenyl- Alanine	$9.0 \cdot 10^3$	$1.6 \cdot 10^4$	$2.2 \cdot 10^4$	$3.5 \cdot 10^4$	$1.4 \cdot 10^5$
L-β-phenyl-alanine	$5.7 \cdot 10^3$	$9.0 \cdot 10^3$	$1.2 \cdot 10^4$	$2.5 \cdot 10^4$	$1.3 \cdot 10^5$

[a] Units of cpm for the counting rate are indicated per gram of labeled β-phenylalanine

Alternatively, crystallization reactions that form macroscopic crystals via autocatalytic growth might offer the potential to amplify tiny energetic differences between enantiomers. They were thought to hold the key to get experimental access to ΔE^{PNC}. As we have discussed earlier, the atomic number Z of the element in the asymmetry center contributes to ΔE^{PNC} to the power of 5 Eq. (5.1). Combining these two understandings, Szabó-Nagy and Keszthelyi (1999) studied crystallization experiments with sodium ammonium tartrate, Δ- and Λ-tris (1,2-ethanediamine)cobalt(III) and Δ- and Λ-tris(1,2-ethanediamine)iridium(III) complexes. The cobalt- and iridium-compounds are chiral octahedral complexes (see Chap. 2 and Fig. 5.8). They were chosen as candidates for the study of crystallization processes since cobalt ($Z = 27$) and iridium ($Z = 77$) provide relatively high atomic numbers compared to carbon atoms ($Z = 6$).

Racemic aqueous solutions of the octahedral complexes were carefully produced and after four weeks of water evaporation at room temperature half of the dissolved molecules were precipitated as polycrystalline material. The crystalline material was dried and again dissolved in water. Circular dichroism spectra of dissolved sodium ammonium tartrate crystals that had been running in parallel were measured to be precisely zero. However, in the case of the cobalt-complex the distribution of the circular dichroism signals for the crystalline material was shifted from zero by a value of $-2.3 \cdot 10^{-4}$! Even higher records were obtained for the iridium-complexes. The distribution of the crystalline material is shifted in the same direction as for the cobalt-complexes by a value of $-39 \cdot 10^{-4}$. CD shifts were larger when less material was crystallized, as theoretically expected. Chiral impurities and bacterial contaminations were excluded for the interpretation of these results. After multiple repetitions and a careful statistical evaluation the authors assumed that the recorded shifts in the observed autocatalytically growing crystals, which increased with the atomic number Z, are real and caused by ΔE^{PNC}. No CD shift was found in the case of sodium ammonium tartrate with carbon in the stereogenic center. Assuming that these differences point to experimental evidence for the existence of ΔE^{PNC}, Szabó-Nagy and Keszthelyi proposed the ΔE^{PNC} as the determining agent for the origin of homochirality of biomolecules. To my knowledge this is until today the most hopeful and also most cited experiment for the experimental ΔE^{PNC} determination.

Nevertheless, we would have to add that even if these reproducible effects in the crystallization behaviour of specific chiral metal complexes were found, how can one guarantee that their origin is intrinsically from parity violation and not from some parasitic effects arising from intermolecular interactions in the macro-

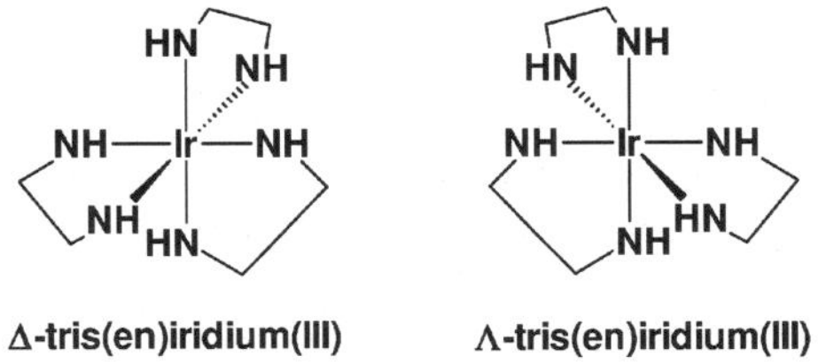

Fig. 5.8 Chemical structure of the chiral octahedral Δ- and Λ-tris(1,2-ethanediamine)iridium(III) complexes, crystallized by Szabó-Nagy and Keszthelyi (1999)

scopic experiments? Minor impurities could play major roles. The contemporary environment has been virtually saturated with L-amino acids, and so potential chiral contamination is a continuing concern (Deamer et al. 2007). How can one surely guarantee the absence of effects of this type and of other parasitic effects in a macroscopic experiment? The other difficulty is to relate quantitatively the macroscopic observations expressed in a CD value of $-2.3 \cdot 10^{-4}$ for the octahedral cobalt crystals and a CD value of $-39 \cdot 10^{-4}$ for the octahedral iridium crystals to the theory of parity violation. This seems to remain extremely difficult, at present (Quack 2002).

In view of the above calculations and experiments, we will have to close this chapter with the conclusion that whether or not the effects of the weak interaction on molecular dynamics are sufficiently large to be able to dominate random conditions or fluctuations has not yet been met with certainty. We do wait for the ultimate experimental evidence of the ΔE^{PNC} determination in the future.

5.4.7 Implications of Parity Non-Conserving Energy Differences

The weak nuclear interaction is a universal force. If parity non-conserving energy differences based on the weak nuclear interaction were involved in life's choice for the homochiral structure of its biopolymers, then life based on the same chemical elements as ours (amino acids, ribose, etc.) anywhere in the Universe would contain molecules of the same chirality as observed on Earth. If so, then "left-life" might extend throughout the universe. "Right-life" would be possible only in an antimatter world where the weak force could go the other way (Gardner 2005).

With all other sources of homochirality such as magnetochiral anisotropies or circularly polarized electromagnetic radiation, the sign of the handedness will be accidental. The same or the opposite chirality as on Earth would be equally possible on life-supporting planets in other Solar Systems (Borchers et al. 2004).

If, for example, astronauts find right-handed amino acids on Mars, it could be excluded that asymmetry on the particle level could be a factor in determining the handedness of organic molecules. Accordingly, in the upcoming chapters we will focus on the determination of enantiomeric enhancements in extraterrestrial samples. Particularly, meteorites were subjected to enantioselective analyses of their chiral organic ingredients including amino acids (see Chap. 8), but also simulated cometary matter was shown to contain a wide variety of amino acids (see Chap. 7). Authentic cometary matter will be analyzed enantioselectively in the near future after landing on a comet; the same is true for the enantiomer-separating analysis of Martian organic molecules (see Chap. 9).

Chapter 6
Chiral Fields: Light, Magnetism, and Chirality

In this chapter we will ask the inspiring question: "Can the asymmetry of a chiral field – such as a specifically oriented magnetic field, radiation field, or the combination of both – be transferred to racemic or prochiral organic molecules in a manner that an enantiomeric enhancement is induced, that is to say, transferred from the mass-less chiral field into the organic molecule?" And indeed, the answer to this question is "yes". In order to understand systematically such phenomena, we will discriminate between (a) the combination of a magnetic field with unpolarized light, known as magneto-optical effects and (b) circularly polarized light as a chiral field to induce enantiomeric enhancement by a process often called photochirogenesis.

To date, four absolute asymmetric photochemical pathways for the generation of the enantiomeric enrichment have been postulated (Fig. 6.1). In this chapter, each

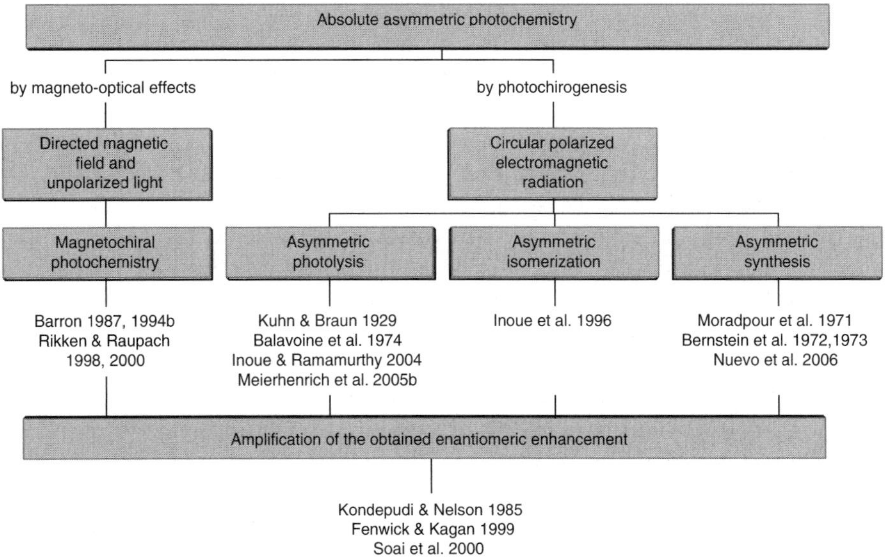

Fig. 6.1 Chiral fields as vehicles for asymmetry: four pathways to induce an enantiomeric enhancement via absolute asymmetric photochemistry. We distinguish between magneto-optical effects and pure photochirogenesis. Circularly polarized light is a true chiral field and offers possibilities to induce the *e.e.* via well-defined processes called photolysis, isomerization, and synthesis

of them will be explained and discussed individually. I would like to mention that new concepts and recent results on absolute asymmetric reactions induced by photochemical processes appear to be particularly encouraging, and modern theories suggest that complex organic molecules could evolve in the ice mantles of dust grains in interstellar and circumstellar space, where magnetic fields and electromagnetic radiation are abundant. Consequently, we will take a close look at this aspect.

6.1 Magneto-Optical Effects

A highly captivating idea is to investigate the potential of a magnetic field to be involved in a photochemical reaction pathway as the means of producing enantiomer enhancement. The magnetic field of Earth or any interstellar magnetic field might then be considered important to the origin of life's molecular asymmetry.

Let us begin with a very general but simple question: "Are there any known interactions between a magnetic field and physico-chemical processes?" Thiemann (1984) answered this question with a clear "yes". He outlined several examples from the domain of magnetochemistry, starting with Zeeman-splitting of spectral lines [used in the analytical tool nuclear magnetic resonance (NMR) spectroscopy] and covering paramagnetic spin resonance, the Faraday effect [leading to magnetic optical rotation dispersion (MORD) and magnetic circular dichroism (MCD)], and others. But all this, of course, does not answer the question of magnetic fields inducing stereoselective effects in chemical reactions. Since the magnetic field alone is defined as a pseudo-chiral force, we have to think at least about a combination of two pseudo-chiral fields to expect a chiral effect on chemical structures.[1]

According to Thiemann (1984), an immediate and simple approach to the problem lies in the following idea: if we fix a symmetric and achiral substance such as benzene between the poles of a strong magnet, we observe a Cotton effect and a rather strong magnetic circular dichroism, due to the Faraday effect (see Chap. 2). The benzene becomes pseudo-optically active. The effect is perfectly symmetric, i.e., it changes sign, upon reversal of the field. If we then switch the magnet off, the effect, manifested by the MCD signal, disappears immediately. Would it be possible to "freeze this optical activity", i.e., to transform this pseudo-chirality into a "true", i.e., permanent, chiral structure? Can a magnetochiral anisotropy – an effect linking chirality and magnetism – give rise to an enantiomeric excess in a photochemical reaction driven by unpolarized light in a parallel magnetic field? From whatever point one is looking at it, the research in this area covers a vast field and deserves attention. Hypotheses and rather wild speculations on such a magnetochiral interaction have always been around in the literature, but only since the advent of a few fundamental papers by Lawrence Barron of the University of Glasgow has a solid foundation, and hence a firm justification, for such endeavours been laid down. Barron (1994a, 1994b) and others (Rhodes and Dougherty 1978; Rikken and Raupach 1998) pre-

[1] The definition of true versus false chirality lies not in the scope of this book and can be found in Barron (2004).

dicted theoretically that an enantiomeric enhancement could indeed be induced in chiral or prochiral racemic systems by the interaction with two "pseudo-chiral" fields such as *un*polarized electromagnetic radiation and a parallel magnetic field.

A few experimental trials have been published based on hypotheses of this kind, making use of an interaction of magnetic fields in asymmetric chemical reactions. Among these, some did not yield the wanted results (see Teutsch and Thiemann 1986, 1990), and some were greeted with enthusiasm due to their rather dramatic positive results, but were soon withdrawn from the panel because Zadel et al. (1994) admitted to having intentionally manipulated the results (see also Chap. 1).

Until recently, however, convincing experimental verifications for this theorem, though searched for intensively, was hard to find. The first evidence was provided in June 2000 by reports on the first photoresolution of a chiral system, a Cr^{III}-tris-oxalato complex in aqueous solution. Rikken and Raupach (2000) published the sound experimental observation of a magneto-optical effect that provided the link between magnetism and chirality. According to their report, the sign of the enantiomeric excess in the octahedral complex was determined by the relative orientation of the light beam to a fixed magnetic field.

The Cr^{III}-tris-oxalato complex was chosen for this study because this chiral octahedral complex is unstable in aqueous solution, continuously dissociating and reforming. Interestingly, the dissociation-process is accelerated by the absorption of light. In 2000, the Cr^{III}-tris-oxalato complex was irradiated by Rikken and Raupach with an unpolarized laser beam (sic!) oriented in parallel to a static magnetic field (Fig. 6.2). Here, a small excess of one enantiomer in the order of *e.e.* = 1.0×10^{-5} was observed. Reversing the magnetic field direction resulted in the detection of an equal concentration of the mirror-image enantiomer.

The observed process consists of a polarization independent refractive index difference between left- and right-handed systems, which is proportional to $B \times k$, where k is the wavevector of the light and B is the magnetic field. In the above photochemical reaction with unpolarized light in a magnetic field, this magnetochiral anisotropy was shown to give rise to a small enantiomeric excess linear in $B \times k$. The absolute magnetochiral dichroism *MD* value is regarded as the difference between the absorption coefficient of unpolarized light parallel to a static magnetic field $\varepsilon_{\uparrow\uparrow}$ and the absorption coefficient of unpolarized light antiparallel to a static magnetic field $\varepsilon_{\uparrow\downarrow}$ Eq. 6.1. The claimed effect based on magnetochiral dichroism yielded relatively low enantiomeric excesses and required a high magnetic field of about 15 T, compared with the terrestrial field of $5 \cdot 10^{-5}$ T.

$$|MD| = \varepsilon_{\uparrow\uparrow} - \varepsilon_{\uparrow\downarrow} \tag{6.1}$$

Soon after publication of these original data, questions concerning the magnetochiral anisotropy arose: does the magnetic field convert the unpolarized electromagnetic radiation into circularly polarized light due to the well-known phenomenon of the Faraday rotation? And would this circularly polarized light be able to introduce an enantiomeric excess into the chiral octahedral complex? Is it possible to explain the described and now observed magneto-optical effect by the overlap of two well-known physico-chemical effects, namely the mentioned Faraday rotation

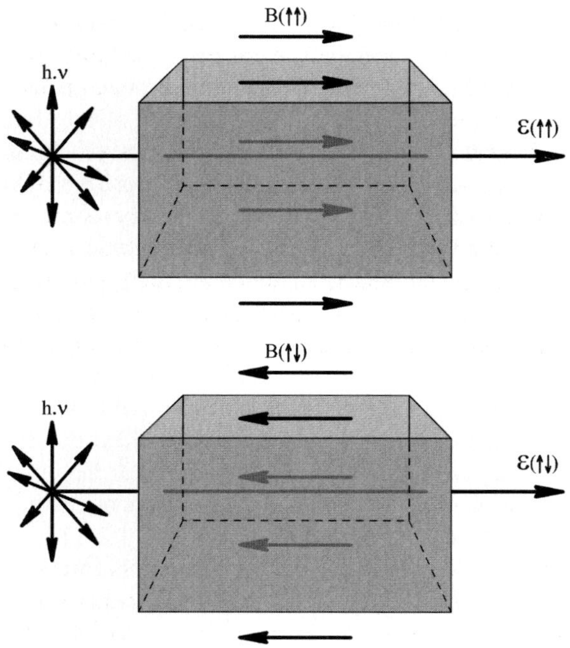

Fig. 6.2 Magneto-optical effect: scheme to generate an enantiomeric enrichment in a chiral organic molecule in solution by performing a photoreaction with unpolarized light in a static magnetic field. In the upper image, the unpolarized light beam $h\nu$ passes the aqueous solution in parallel to a static magnetic field $B_{\uparrow\uparrow}$. In the lower image, the unpolarized light passes antiparallel to the magnetic field $B_{\uparrow\downarrow}$. The change of the light's orientation provokes the opposite enantiomeric excess in the dissolved chiral molecules (see Barron 2000a). Illustration: Stéphane Le-Saint, University of Nice-Sophia Antipolis

and the asymmetric interaction of circularly polarized light with matter? In fact, Rikken and Raupach (1998) and Barron (2000a, 2000b) carefully distinguish between two mechanisms that might introduce an enantiomeric excess into racemic organic compounds or into prochiral educts: A 'cascaded' and a 'pure' mechanism.

1. *'Cascaded' mechanism.* The magnetic field 'converts' the unpolarized light first into circularly polarized light as a result of the Faraday effect and the obtained circularly polarized light interacts with the molecule producing circular dichroism, which is well known to result in an excess of one of the enantiomers. The whole procedure is called a 'cascaded' mechanism. This sequence of effects was, however, calculated to be too small to end in the observed enantiomeric excesses, because a magnetic field of 1 T may introduce an enantiomeric excess of only $e.e. = 5 \times 10^{-7}$.
2. *'Pure' mechanism.* Due to this, a so-called 'pure' mechanism for the observed magnetochiral anisotropy was taken into consideration, capable of producing much larger enantiomeric excesses, because it involved less electric dipole interactions. This means that the observed magneto-chiral dichroism is a phenomenon distinct from Faraday rotation.

For the above-described experiment, Rikken (2004) concluded that the cascading mechanism did not contribute significantly and that the true magnetochiral effect dominated the cascaded one. In other systems, however, he admitted that the cascading mechanism might be dominant.

In future experiments, the applications of enantioselective magnetochiral photochemistry to other chiral systems with higher relevance to prebiotic systems might provide additional convincing arguments that this effect played an active role in the origin of biomolecular homochirality. According to Barron (2000a), magnetochiral dichroism is – in addition to circularly polarized light and the electroweak interaction – a serious candidate for the source of handedness in biomolecules. This model, however, has not yet won wide acceptance in explaining biomolecular homochirality.

6.2 Photochirogenesis

Is it reasonable to wonder if a collection of 'chiral photons' –sometimes loosely referred to as 'chiral light' – is capable of inducing enantioenrichments into prochiral or racemic molecules? Or, in other words, can one imagine a migration of chiral information from pure and mass-less electromagnetic radiation to matter?

In theory, the effect of circularly polarized light represents indeed a "true-chiral" influence on racemates of organic molecules (MacDermott 1993; Barron 1994a, 1994b) and hence photochemical reactions applying circularly polarized light have been examined extensively in many systems in the laboratory.

Reactions that induce an enantiomeric enhancement in prochiral or racemic systems with circularly polarized light are often called *absolute asymmetric*, as there is no net chirality in the molecular reactants. This discriminates asymmetric photochemistry from ordinary photochemistry of chiral molecules, which can simply be induced by non-polarized light (Rau 2004). Absolute asymmetric photochemical reactions induce enantiomeric enhancement by an absorption process. This requires that two different circularly polarized light absorbing ground state species, i.e., enantiomers or diastereoisomers, are present. Here, we will concentrate on enantiomers. Treatment of photoreactions of diastereoisomers will not be covered, because (a) non-polarized light is sufficient to modify the optical activity of a sample and (b) such reactions are not of relevance for the origin of biomolecular asymmetry on Earth.

6.2.1 Asymmetric Photolysis of Racemic Organic Molecules

Since 1929, irradiation of racemic or prochiral precursor molecules with circularly polarized light has been a proposed scenario for the origin of biomolecular asymmetry. Werner Kuhn, a pioneer in the field of asymmetric photolysis, achieved at the University of Heidelberg the first enantioselective photodecomposition (i.e., photolysis) by irradiating solutions of ethyl-α-bromopropionate and N,N-dimethyl-α-azidopropionamide (Kuhn and Braun 1929; Kuhn and Knopf 1930). These molecules were – even for chemists – rather exotic and non-representative of prebiotic conditions during chemical evolution.

In 1974, the research team of Henri Kagan at the University of Paris-Orsay achieved world-wide reputation by the irradiation of racemic camphor (Balavoine et al. 1974), resulting in the highest photochemical induced enantiomeric excess of 20% by carrying the photodecomposition to 99% completion. With the aim of investigating and verifying photochemical theories for the induction of biomolecular homochirality via electronically excited states, topical research focuses on the effect of circularly polarized light on racemic organic molecules such as amino acids.

6.2.2 Asymmetric Photolysis of Amino Acids

In 1977, molecules of assumed relevance for prebiological pathways, such as the amino acids alanine (Nordén 1977) and leucine (Flores et al. 1977), were decomposed enantioselectively. Irradiation was performed at 212 nm in acidic solution provoking a particular electronic transition that is called the (π^*, n)-transition. The corresponding photolysis mechanisms were recently developed (Nishino et al. 2001, 2002). It was demonstrated for each amino acid that the enantiomeric purity achievable during photolysis of a racemate is dependent on (a) the *anisotropy factor* (g) that is the ratio between the circular dichroism ($\Delta\varepsilon$) value and the extinction coefficient (ε), and (b) the extent of reaction (Balavoine et al. 1974). The anisotropy factor g was formerly called the *dissymmetry factor* and is given by Eq. (6.2).

$$g = \frac{\varepsilon_l - \varepsilon_r}{\varepsilon} = \frac{\Delta\varepsilon}{\varepsilon} = \frac{\Delta\varepsilon}{(\varepsilon_l + \varepsilon_r)/2} \tag{6.2}$$

Due to its dependence on ε, the anisotropy factor is a function of the wavelength. In order to determine g at a given wavelength, one can choose between two possibilities.

1. One measures the differential absorption $\Delta\varepsilon$ of one individual enantiomer (*R*- or *S*-enantiomer) against right- and left-circularly polarized light, designated in Eq. (6.2) as r and l.
2. One determines g according to Eq. (6.3) by recording the differential absorption $\Delta\varepsilon$ of a pair of enantiomers (*R*- and *S*-enantiomer) against one kind of circularly polarized light, e.g. individual *R*- and *S*-enantiomers against left-circularly polarized light or individual *R*- and *S*-enantiomers against right-circularly polarized light (Rau 2004).

$$g = \frac{\varepsilon_S - \varepsilon_R}{\varepsilon} = \frac{\Delta\varepsilon}{\varepsilon} = \frac{\Delta\varepsilon}{(\varepsilon_R + \varepsilon_S)/2} \tag{6.3}$$

Table 6.1 Anisotropy factors (g) of selected amino acids

Amino acid	λ [nm]	g
Alanine	> 200	0.007
Alanine (pH 1)	213	0.029
Glutamic acid	> 200	0.008
Leucine (pH 2)	213	0.0244
Leucine (pH 1)	215	0.028
Tyrosine	266	0.024

Anisotropy (g) values are given at room temperature and indicated wavelength λ

Anisotropy factors for selected amino acids are given in Table 6.1 (see Rau 2004 and references therein). Among the proteinaceous amino acids, it is leucine that shows a relatively high anisotropy value of $g = 0.0244$ (Flores et al. 1977) and has therefore often been selected for enantioselective photolysis experiments.

For the last 25 years, the highest enantiomeric excess achieved for enantioselective photolysis of an amino acid has remained 2.5% in the case of leucine. Very recently, the challenging purely photochemical induction of higher enantiomeric enrichment of amino acids was intensively revisited with the help of innovative ideas, concepts, and instrumentations using four main approaches.

1. *pH-dependence*. The Yoshihisa Inoue ERATO research team at Osaka University studied the pH-dependence of the chiroptical properties, particularly the anisotropy g (and therefore the ability to induce a high *e.e.*), of the amino acid leucine. Solutions with different acidity were subjected to right- and left-circularly polarized photons and the resulting enantiomeric excess was determined subsequently. It was shown that enantioselective photodecomposition depends strongly on the degree of protonation of the amino acid's carboxylic group, such that protonated carboxylic groups at pH values close to 1 generate a higher anisotropy. Based on this effect, which was promised also to be studied for other amino acids, a Norrish Type II mechanism was proposed, including γ-H abstraction and Norrish type II cleavage of the leucine molecule. The obtained enantiomeric excess was, however, only close to 0.2% (Nishino et al. 2001).
2. *Two-photon excitation*. In order to increase the anisotropy g, two-photon excitation processes were studied. Femtosecond circularly polarized laser pulses were applied in a specific norbornadiene-quadricyclane system to make use of an additional anisotropy g^* derived from the excitation of an electronically excited state. For the chosen system, however, the induction of enantiomeric enrichments during irradiation involving two-photon excitations turned out to be very similar to that obtained by one-photon excitation. Nevertheless, the principal approach is innovative and might provide advanced capabilities.
3. *Elliptically polarized light*. The potential of right- and left-*elliptically* polarized light in asymmetric photolytic reactions was systematically studied by William A. Bonner at Stanford University subjecting acidic solutions of leucine to irradiation. It was shown that a decomposition up to 93% justified asymmetries close to 3% enantiomeric excess. As theory predicts, irradiation with "clean"

right-*circularly* polarized light or left-*circularly* polarized light in parallel experiments ended with 4% enantiomeric excess and thus a higher degree (Bonner and Bean 2000).

4. *Irradiation in the solid state*. Studying the interaction of racemic organic molecules with left- and right-circularly polarized light in aqueous solution represents the conventional attempt at enantioselective photolysis. Solitary (π^*, n)-electronic transitions of the amino acids' carboxylic groups with values close to 212 nm (5.85 eV) were obtained here, because water absorbs wavelengths below 200 nm making higher energetic electronic transitions (at lower wavelengths) inaccessible. Taking this into account, leucine molecules in their *solid state* were exposed by our research team to circularly polarized synchrotron radiation (Nahon et al. 1997) of variable polarization (Alcaraz et al. 1999; Nahon and Alcaraz 2004) and energy obtained from a newly developed electromagnetic planar/helical crossed undulator. Using this concept, additional high-energy electronic transitions such as (π^*, π_1)-, (π^*, π_2)-, and even (σ^*, σ)-transitions of amino acids (Fig. 6.3) could be excited below 200 nm providing more effective optical anisotropies. Following this approach, we took the energy-dependence of the anisotropy *g* into consideration, which will now be outlined in more detail.

To establish the most suitable conditions for enantioselective photodecomposition, the absorption spectrum of leucine with its electronic transitions was studied between 120 and 250 nm in the vacuum ultraviolet photon range.

In order to determine the direction of a possible photon-induced enantiomeric excess, chiroptical investigations of amino acids were taken into account, also in the

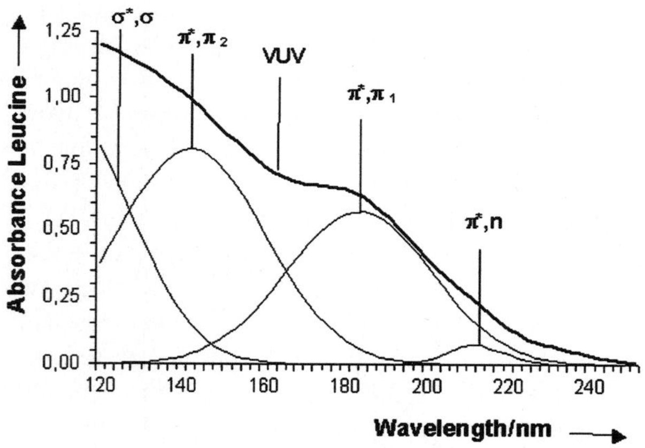

Fig. 6.3 Vacuum ultraviolet absorption spectrum and electronic transitions of the amino acid leucine recorded in a solid sample film with synchrotron radiation at beamline SA-61 in the Laboratoire pour l'Utilisation du Rayonnement Électromagnétique LURE, Paris. The (π^*,n)-transition of leucine was observed at 211 nm, its (π^*,π_1)-transition at 183 nm, the tentative (π^*,π_2)-transition at 142 nm, and its potential (σ^*,σ)-transition at 121 nm. Enantioselective photodecomposition of leucine can be considered to be sufficiently active precisely at each of the determined electronic transition energies

vacuum ultraviolet range. Extreme care was required to obtain artefact-free, solid-state circular dichroism spectra. This is because circular dichroism spectra in the solid state can inevitably be accompanied by parasitic signals that originate from macroscopic anisotropies of a sample, such as linear dichroism (LD) and linear bifringence (LB) (Kuroda 2004).

Circular dichroism spectra of D-leucine were recorded in the vacuum ultraviolet region of the spectrum by the team of Søren Hoffmann at the University of Århus, using the specifically dedicated synchrotron beamline UV1-ASTRID. The obtained circular dichroism spectrum is given in Fig. 6.4. It shows a negative band at 207 nm for its (π^*,n)-electronic transition and, surprisingly, an intense positive band at 191 nm corresponding to the (π^*,π_1)-transition. The circular dichroism spectrum in the solid, here microcrystalline, state is noticeably different from previously recorded circular dichroism spectra in solution. The intense band at 191 nm, allowing for relatively high enantiomeric excesses by irradiation at this wavelength with circularly polarized light, attracted our particular interest.

Based on the signs of the obtained CD data one can deduce that the (π^*,n)-electronic transition of D-leucine is induced preferentially by right-circularly po-

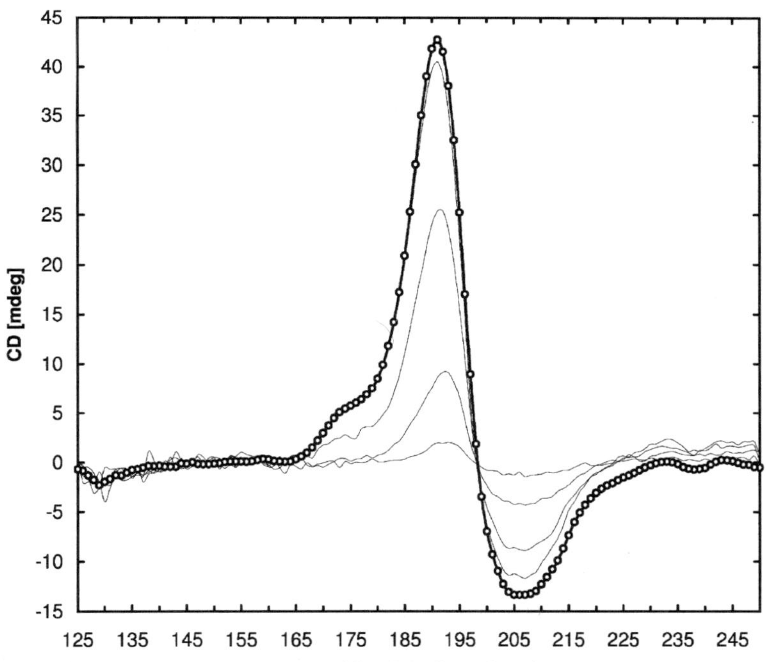

Fig. 6.4 Vacuum ultraviolet circular dichroism spectrum of the amino acid D-leucine, recorded in a solid sample film with synchrotron radiation at the Synchrotron Facility ISA, University of Århus (Denmark). Film thickness 1 μm (*bold line*), film thickness 0.2, 0.4, 0.6, and 0.8 μm (*thin lines*). D-leucine micro-crystals were immobilized on UV-transparent MgF_2 windows (Meierhenrich et al. 2005b)

larized light and the positive (π^*,π_1)-electronic transition of D-leucine is preferentially excited by left-circularly polarized light. Therefore, irradiation of D,L-leucine at the 207 nm (π^*,n)-electronic transition with left-circular polarized light is assumed to result in an enantiomeric excess of D-leucine. At the 191 nm (π^*,π_1)-transition, left-circular polarized light should lead to an enantiomeric excess of L-leucine, and so on. Using right-circular polarized light, inverse enantioenrichments are foreseen to be obtained (Meierhenrich et al. 2001a).

With these experiments performed in the solid state, we made the asymmetric 191 nm electronic transition accessible for enantioselective photolysis reactions. Moreover, solid-state conditions were chosen, because we assumed that amino acids that are present in interstellar environments (see upcoming Chaps. 7–9) will typically occur there under low-temperature, low-pressure, and low-gravity conditions in the solid state and not primarily in the liquid state.

Following these chiroptical measurements, the assumption was born that the origin of life's molecular asymmetry is due to irradiation of racemic solid state amino acids in space. In certain interstellar environments, an originally symmetric mixture of amino acids might have become asymmetric by irradiation with 'chiral photons'. In order to find further arguments for this hypothesis, solid state racemic leucine was exposed in the laboratory to individually left- or right-circularly polarized synchrotron radiation, representing astrophysical conditions. The breakthrough of these experiments came indeed with the idea of irradiating an amino acid in its solid state. Previous experiments were restricted to the liquid state.

How can one obtain the required circularly polarized light in the vacuum UV at precisely defined energies? We used the polarizing undulator[2] called Ophelie (Nahon et al. 1997) installed in the Super-ACO storage ring at LURE as a source of circularly polarized light showing admirable circular polarization rates above 90%. These rates are given in Fig. 6.5.

Irradiation of solid state D,L-leucine was performed within the SU-5 beamline (Fig. 6.6, Nahon et al. 2001) with the energy of the circular polarized synchrotron radiation first set to 182 nm, closely matching the desired 191 nm (π^*,π_1)-transition, and second to 170 nm, so that information about the energy dependence of the process could be gained.

After 3 h of irradiation resulting in photodecomposition of approx. 70% of the starting material, chemical analysis of the remaining leucine residue was performed by enantioselective gas chromatography. The application of a new analytical technique enabled our group to identify the selective degradation of left-handed leucine exposed to right-circularly polarized radiation. Due to different molecular absorption coefficients of the two leucine enantiomers, the highest obtained enantiomeric excesses in the residues was +2.6% in D-leucine (Meierhenrich et al. 2005b). The sign of the induced *e.e.* was dependent on the direction of circular polarization as illustrated in Fig. 6.7.

[2] An undulator consists of a periodic arrangement of dipole magnets creating an alternating static magnetic field. Electrons that traverse this alternating field are forced to undergo oscillations and emit electromagnetic radiation called synchrotron radiation.

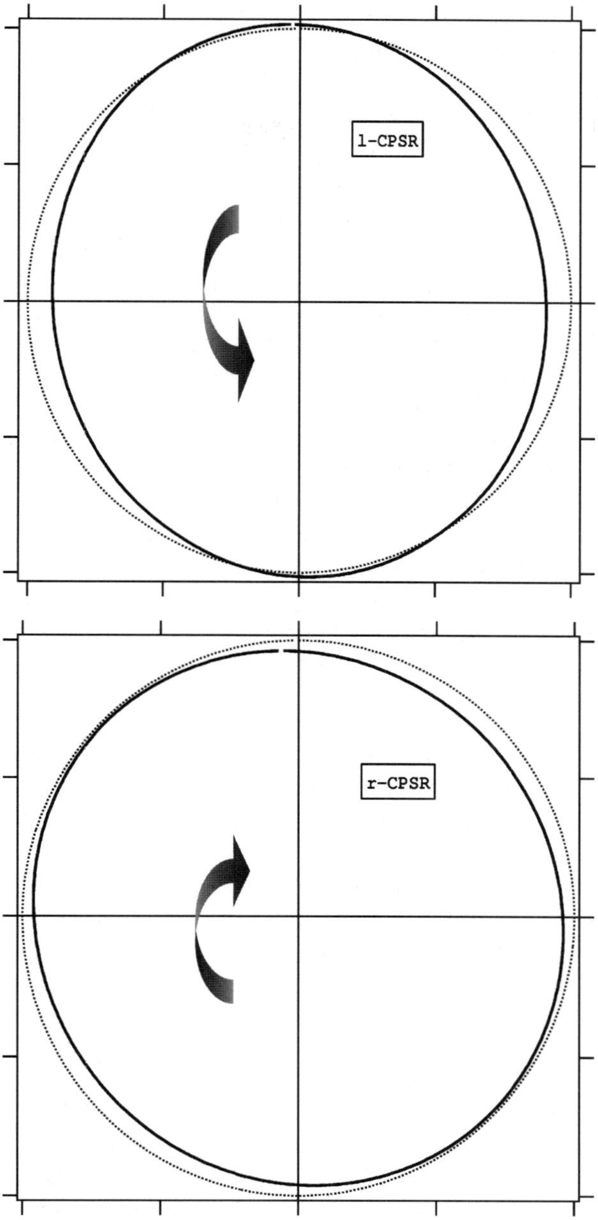

Fig. 6.5 Circular polarization rates of the applied synchrotron radiation at 182 nm. The absolute circular polarization for left-circularly polarized synchrotron radiation (l-CPSR) is 94% (*top*) and 91% for right-circularly synchrotron radiation (r-CPSR), respectively (*bottom*). Thin lines illustrate the perfect circular polarization case (see Nahon and Alcaraz 2004)

Fig. 6.6 Synchrotron beamline SU-5, part of the Laboratoire pour l'Utilisation du Rayonnement Electromagnétique LURE, Centre Universitaire Paris-Sud, France, was constructed to produce and analyse circularly polarized radiation (Nahon and Alcaraz 2004). Here, samples of amino acids were irradiated allowing the induction of a significant enantiomeric excess in the solid state. *Top*: High vacuum chamber for irradiation of amino acids in the solid state. *Top left*: Towards photodiode detector. *Top center*: Sample holder for 7 MgF_2 windows on the surface of which amino acids are deposited which can be moved individually into the synchrotron light. *Top right*: Towards synchrotron source. *Bottom left*: Location of amino acid samples on top of turbo molecular pump. *Bottom center*: Polarimeter, capable of determining 4 Stokes parameters. *Bottom right*: Photodiodes. Today, this polarimeter is installed on the new VUV beamline DESIRS at SOLEIL (Gif sur Yvette, France) for which a new variable polarization undulator has been designed

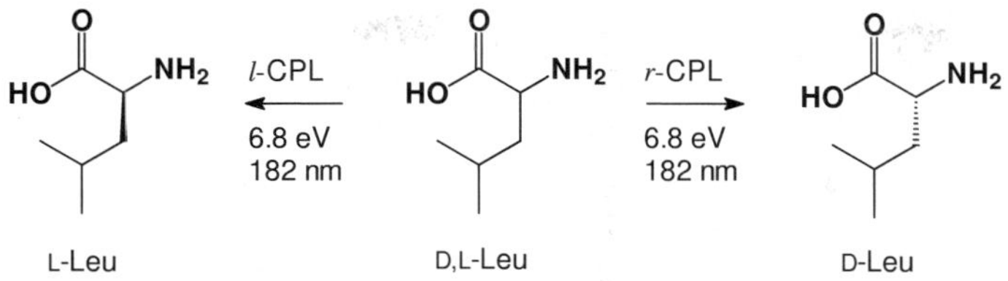

Fig. 6.7 Tipping the biomolecular balance: D,L-Leucine subjected to irradiation with left-circularly polarized light at 182 nm resulted in an enantiomeric excess of L-leucine, while irradiation with right-circularly polarized light gave an excess of D-leucine

In a sample with equal amounts of D-and L-leucine, right-circularly polarized light destroyed slightly more of the L-leucine, while left- circularly polarized light destroyed slightly more of the D-form. Irradiation of solid-state D,L-leucine with circularly polarized light at a higher energy, $\lambda = 170$ nm, did induce a minor value for the enantiomeric excess, which does not change the sign by switching photon helicity. This observation is coherent with the fact that the circular dichroism signal of leucine (see Fig. 6.4) at 170 nm is vanishing.

The detection of enantiomeric excesses in samples of leucine, induced by vacuum ultraviolet asymmetric photochemical reactions such as enantioselective photolysis under realistic interstellar/circumstellar conditions, was demonstrated to be feasible. These experimental data support the assumption that ice mantles of tiny interstellar/circumstellar dust grains can experience asymmetric reactions when irradiated by circularly polarized ultraviolet light. In space, this type of irradiation may have led to small but significant local enrichment of L-amino acids, as measured in meteorites (see Chap. 8). This excess could have been enough to lead to autocatalytic processes for the formation of life based on L-amino acids. The results have far-reaching consequences on the understanding of the origin of life on Earth and its evolution, suggesting that the biomolecular asymmetry of amino acids was induced in interstellar space, long before the origin and biological evolution of life on Earth took place. Afterwards, these asymmetric amino acids might have been delivered via (micro-) meteorites, interplanetary dust particles, and/or comets to Earth, where they triggered the appearance of life.

Nevertheless, a more systematic study of asymmetric vacuum ultraviolet photolysis reactions in the solid state for *different* amino acids would help us to better understand the processes leading to their asymmetric formation and their role in prebiotic chemistry. Using this concept, the variation of the photon's energy enables the determination of the anisotropy factor (g) as a function of the wavelength, and from this value the conditions required to obtain enantioenrichments can in turn be determined. Worldwide, there are only two polarimeter-connected light-sources for circularly polarized synchrotron radiation available. One is based in Tsukuba, Japan, the other in the synchrotron in the South of Paris. We will continue to use this unique technique in the new generation synchrotron SOLEIL, which just finished construction in Gif sur Yvette, France.

Enantioselective photolysis reactions, however, suffer in principle from the logical but inconvenient condition that in order to produce a significant enantiomeric excess which can then be amplified by appropriate mechanisms, high amounts of the amino acids have to be photodecomposed, i.e., destroyed, as determined by the anisotropy factor $g (g = \Delta\varepsilon/\varepsilon)$. The higher the anisotropy factor g of the chiral molecule (g is function of the wavelength and the state of aggregation) and the higher the extent of reaction ξ, the higher the enantiomeric excess that can be induced in the system. This relationship is illustrated in Fig. 6.8 (see Balavoine et al. 1974).

As g-values for amino acids are generally small (cf. Table 6.1), high enantiomeric enrichments are only reached at a large ξ, where most of the starting material has disappeared. So Fig. 6.8, in some ways, is "deceptive" (Rau 2004).

Enantioselective sensitized photochemical reactions, in which an optically active sensitizer transfers energy to a prochiral or racemic substrate through non-covalent interactions in the excited state (Inoue 2004), might justify higher anisotropies and circumvent this problem. Similar to catalytic and enzymatic asymmetric syntheses, the photosensitization necessitates only a tiny amount in catalytic quantities of the optically active sensitizer. This is what distinguishes this particular strategy from other asymmetric photochemical reactions, which employ circularly polarized light, chiral complexing reagents, chiral auxiliaries, chiral supramolecular hosts, and chiral crystal lattices, since the chiral interaction occurs exclusively in the excited state (Inoue 2004). Suitable optically active sensitizers for enantiodifferentiating photoreactions involving amino acids of prebiotic relevance, however, have not yet been identified. Topical research activities focus on this kind of challenging photoreaction.

Another common criticism of the above scenario should be mentioned: the relevance of models for the asymmetric photolysis of amino acids in the origin of biomolecular asymmetry were recently questioned because the non-aliphatic amino acids tryptophan and proline show opposite Cotton effects in circular dichroism spectroscopy compared to aliphatic amino acids and thus would be generated as

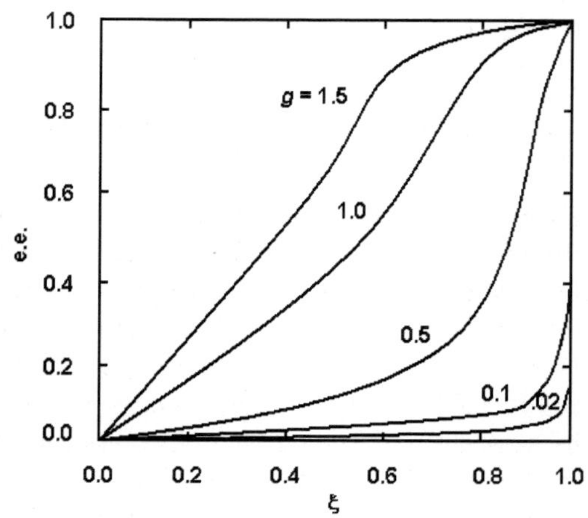

Fig. 6.8 In some ways "deceptive": Enantiomeric excess *e.e.* as a function of extent of reaction ξ and anisotropy factor g

D-enantiomers (Cerf and Jorissen 2000). This would, of course, be problematic! However, the referred circular dichroism spectra were exclusively recorded in aqueous solution, a state of aggregation that is – as outlined above – not relevant to astrophysical and interstellar conditions.

In order to further investigate this argument we subjected solid amorphous films of the amino acids L-valine, L-alanine, L-leucine, and L-proline to circular dichroism measurements. Synchrotron light between 130 and 330 nm was used at the Institute for Storage Ring Facilities at Århus University. Newly recorded circular dichroism spectra showed that not only L-valine, L-alanine, and L-leucine exhibit characteristic bands in the vacuum ultraviolet at 182 nm, but also L-proline showed a signal of the same sign exactly matching this wavelength (energy).

Circular dichroism spectra given in Fig. 6.9 were recorded for amorphous films of the individual amino acids after vacuum sublimation. The corresponding bands are red-shifted compared to the circular dichroism spectrum of leucine depicted in Fig. 6.4, because in Fig. 6.4 microcrystalline D-leucine was used instead of an amorphous film. For more detailed information on the comparison of circular dichroism spectra of microcrystalline and amorphous films, see Kuroda (2004).

The obtained chiroptical data clearly demonstrate that at least the four tested amino acids show very similar circular dichroism spectra in the vacuum ultraviolet region in the solid state. Furthermore, diamino acids were experimentally and theoretically, i.e., based on ab initio calculations, shown to possess nearly identical circular dichroism spectra (Bredehöft et al. 2007). In contrast to the skeptical comments of Cerf and Jorissen (2000), photolysis reactions of amino acids in the solid state thus remain suitable possibilities for the induction of an enantiomeric enhancement into prebiotic and interstellar amino acid structures.

What about the mentioned tryptophan? In the next chapter (Sect. 7.3.3), we will discuss experimental studies on the recruitment order of amino acids at the beginning of biological evolution. These studies come to the unisonous conclusion that complex amino acids such as tryptophan seem to be recruited late in proteins. In contrast, the above tested amino acids L-valine, L-alanine, L-leucine, and L-proline are believed to be recruited very early during the origin of life, which may underline the above hypothesis.

6.2.3 Asymmetric Photoisomerization

In contrast to asymmetric photolysis, asymmetric photoisomerizations include processes in which chiral organic molecules undergo enantioselective isomerization induced by circularly polarized light. The pioneering work on this topic was performed by Inoue et al. (1996) studying the direct photoderacemization of *E*-cyclooctene. *E*-Cyclooctene is chiral and composed of the (−)-*R*-*E*-cyclooctene and (+)-*S*-*E*-cyclooctene enantiomers. It was selected for this kind of experiment because it shows extraordinary high optical rotation with $[\alpha]_D$ values of −426 for the *R*-enantiomer. A photoreaction of racemic *E*-cyclooctene induced by circularly polarized light at $\lambda = 190$ nm in pentane solution provokes a $E \rightarrow Z$ isomerization

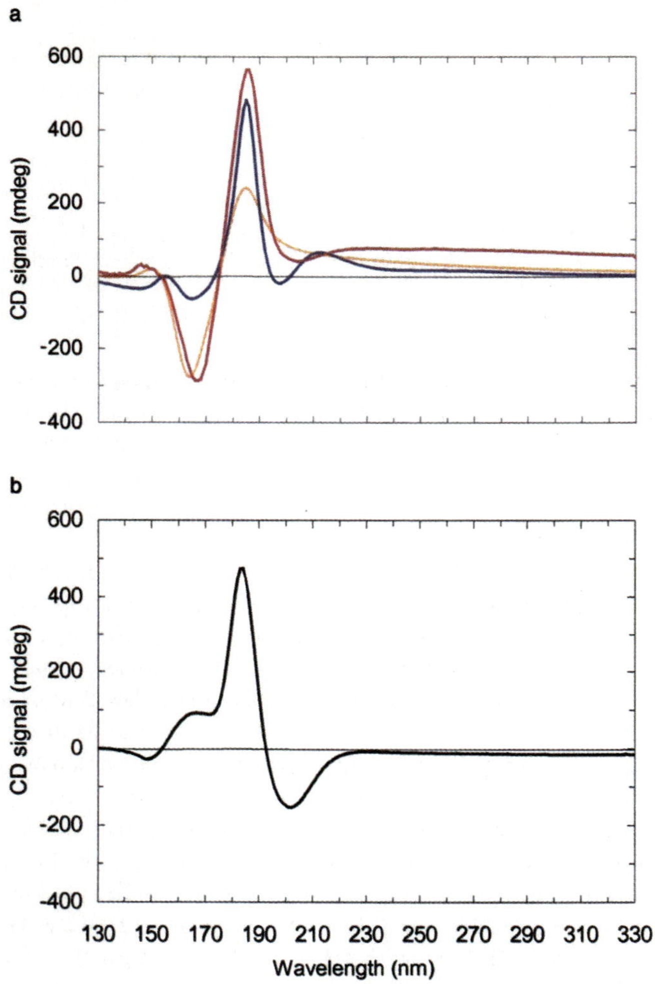

Fig. 6.9 Vacuum ultraviolet circular dichroism spectra of solid-state amino acids. **a**, CD spectra between 130 and 330 nm of solid amorphous L-valine sublimate (*red line*), L-alanine sublimate (*blue line*), and L-leucine sublimate (*yellow line*) immobilized on MgF_2. **b**, L-proline sublimate. A CSA standard was used for instrument calibration purposes; D-enantiomers gave CD spectra of opposite sign. The band of L-proline matched the band of L-alanine, L-valine, and L-leucine

producing the achiral *Z*-cyclooctene as given in Fig. 6.10. Interestingly, during this photoreaction, the optical activity of the reaction mixture rises, passes a maximum, and disappears again. However, over long periods, a photostationary state will be reached due to the inverse $Z \rightarrow E$ isomerization, which is also active under irradiation (Rau 2004). Very recently, functional groups were incorporated along the cyclooctene skeleton leading to changes in its photochemistry.

It was proposed that these results give additional hints, that indeed photoreactions involving circularly polarized light are capable of inducing enantiomeric excesses.

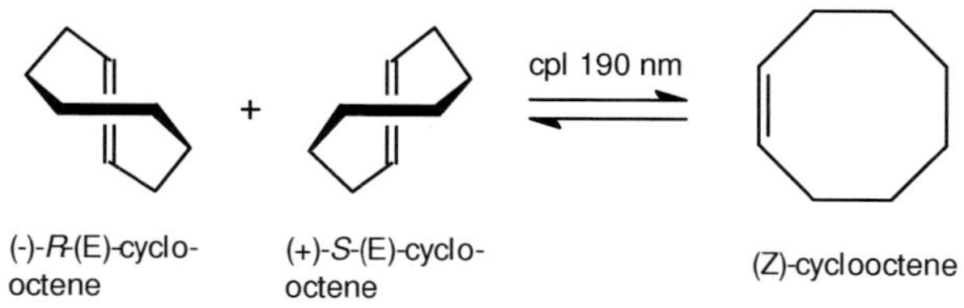

Fig. 6.10 Photoisomerization of racemic E-cyclooctene towards achiral Z-cyclooctene. Triggered by a photoreaction with circularly polarized light, the initially racemic reaction mixture develops optical activity

Similar reactions with amino acids and other biomolecules were assumed to provide further insights into the crucial role of 'chiral light' in chemical evolution.

6.2.4 Asymmetric Synthesis

Another strategy to circumvent the problem of excessive photodecomposition during induction of an enantiomeric enhancement by asymmetric photolysis reactions is the spontaneous asymmetric photoformation, i.e., asymmetric synthesis, of amino acid structures, which is particularly interesting under interstellar conditions with circularly polarized light.

In contrast to asymmetric photolysis and asymmetric photoisomerization, the pure synthesis of optically active molecules in non-racemic yields induced by circularly polarized light has remained a difficult task to achieve. The first successful attempts were reported 30 years ago. In these experiments, photocyclization of alkenes in solution, performed in the presence of iodine, led to the formation of polyaromatic hydrocarbon molecules. By irradiation with circularly polarized light, the chiral hexahelicene given in Fig. 6.11 was synthesized with optical yields below 2% (Moradpour et al. 1971; Bernstein et al. 1972).

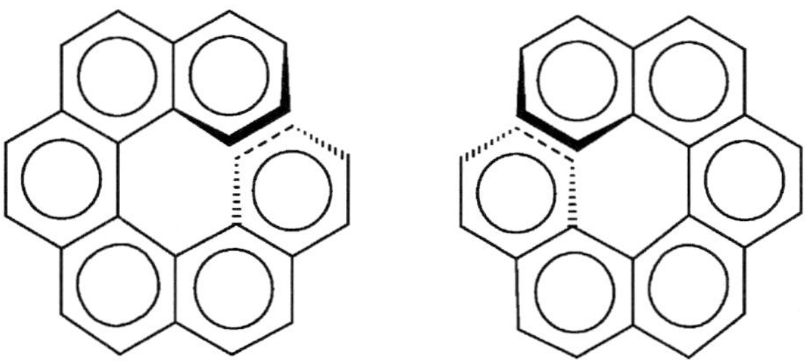

Fig. 6.11 Chemical structure of hexahelicene. M-hexahelicene is depicted on the *left*, the P-enantiomer is given on the *right*. This particular chiral molecule possesses no stereogenic center; it had been synthesized with an enantiomeric enhancement by a photochemical reaction using circularly polarized light at $\lambda = 313$ nm

The reaction's dependence on wavelength and the substituents' structure has also been examined, and the mechanism was claimed to involve selective excitation of the enantiomeric conformers (Bernstein et al. 1973). Photoproduction of amino acids has been reported to be possible with the help of initial photon acceptors (Sagan and Khare 1971; Khare and Sagan 1971).

Very recently, two groups demonstrated contemporaneously the spontaneous photoformation of a variety of amino acid structures under interstellar conditions (Bernstein et al. 2002; Muñoz Caro et al. 2002). Since the two groups used unpolarized light for the photoreaction the obtained amino acids were chiral but racemic (see comments in Shock 2002). More recently, analogous experiments were performed with circularly polarized light as an asymmetric driving force in order to directly generate enantioenriched amino acids. These experiments show preliminary successes in a way that chiral amino acids and diamino acids were synthesized with circularly polarized light under simulated interstellar conditions. The obtained results will be presented in Chap. 7.

Now, we will continue asking the question if there are natural sources of circularly polarized electromagnetic radiation and, if so, whether they are sufficient in intensity, energy, and polarization to be promising candidates for inducing biomolecular asymmetry.

6.2.5 'Natural' Sources of Circularly Polarized Light

Circularly polarized electromagnetic radiation, which is required for the photochemical introduction of enantiomeric excesses into racemic mixtures of organic molecules or into prochiral educts, was indeed found to be formed by various 'natural' sources. On Earth's surface, circularly polarized light (both in the ultraviolet and visible region of the spectrum) is extremely weak in intensity and handedness.[3] Therefore, possible astronomical sources of circularly polarized light were searched intensively and described recently. Circularly polarized light was identified in galaxies, stars, the interstellar medium, and planets by photo-polarimetric measurements with Earth-based telescopes.

6.2.5.1 Double Reflections on Planets' Surfaces

The circular polarization of scattered sunlight from planets and satellites is a general phenomenon. Observations from Earth-based telescopes have been performed in order to determine circular polarizations of scattered sunlight from Venus, Mercury,

[3] Electromagnetic radiation emitted by the non-oriented dipoles of the sun is unpolarized. Sunlight scattered by Earth's atmosphere is linearly polarized and non-asymmetric. Only multiple scattering (Mie scattering) in the atmosphere gives circular polarization. However, the circular component is (a) low (0.1%), (b) in the infrared spectral range, ineffective for chemical reactions, and (c) of different sign in the morning and in the evening. Sunlight is therefore considered insufficient for enantioselective photochemistry.

the Moon, Mars, Jupiter, Uranus, and Neptune. In the 30 years since the fundamental work of James Kemp and colleagues at the University of Hawaii, fractional circular polarizations (S) have been measured in electromagnetic radiation from several planets. The obtained global values (S_0) were compared with the fractional circular polarization of each planet's northern hemisphere (S_N) and southern hemisphere (S_S). Data from Venus, the Moon, and Jupiter (Kemp et al. 1971a, 1971b) confirmed approximately that

$$S_0 = 1/2(S_N + S_S). \tag{6.4}$$

The absolute amount of fractional circular polarization $|S|$ was determined to be a function of the phase angle (φ) between the Sun-planet and Earth-planet directions. In the case of Jupiter, S_N and S_S pass zero at $\varphi = 0°$. Venus which, like Jupiter, has a dense atmosphere shows a polar scattering effect with the same signs of S_N and S_S as Jupiter, for the same sign of φ. Kemp and Wolstencroft (1971) proposed a scattering, non-magnetic mechanism for the generation of circular polarization at gaseous surfaces (for magnetic mechanisms cf. Lang and Willson, 1983). This model is consistent with the observed signs of the polar effect on Jupiter and Venus. What Kemp et al. called the elliptical scattering power S/φ (or $\delta S/\delta\varphi$) was determined to be notably smaller for Venus than for Jupiter, at the same wavelength in the deep red.

The model for the generation of circular (or elliptical) polarization of scattered electromagnetic radiation for planets and satellites with little or no atmosphere and condensed, solid, and rough surfaces (e.g. Moon or Mercury) differs from the gas-layer case described above. The relevant mechanism here is based on double reflections of light from crystal grains exposed on a surface and was developed by

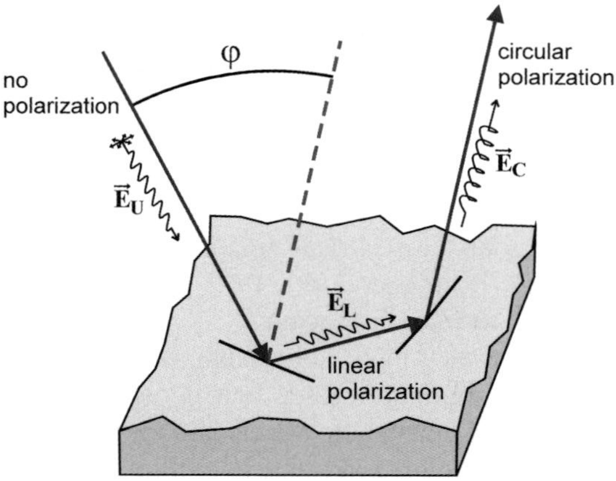

Fig. 6.12 Double refection of non-polarized light on a surface can create circular polarization. After the first reflection, the unpolarized light expressed by the electric field vector E_U becomes partially linearly polarized (electric field vector E_L). After the second reflection, the light becomes circularly polarized (electric field vector E_C). Amount and direction of the circular component are functions of the phase angle φ between the trajectories of incoming and double reflected light

Bandermann et al. (1972). As depicted in Fig. 6.12, the first reflection linearly polarizes the ray and the second, from an adjacent grain, elliptically polarizes it by virtue of the material being slightly absorbent. The Bandermann model uses geometrical optics and Fresnel's law. It predicts that the hemispheric values S_N and S_S have opposite signs for the gas-layer case model in comparison with Venus and Jupiter for all φ between 0° and a critical phase φ_c. As the magnitude of φ increases toward 90°, S changes sign. The data for the Moon, at $\varphi = +90°$, are consistent with this model since $+90° > \varphi_C$.

Thus, data obtained for fractional circular polarizations are understood in detail in sign and amount in the cases of Venus, Jupiter, and the Moon. With Mercury, the situation is less certain. The fractional circular polarization data obtained for Mercury showed an asymmetry in S_0, which is difficult to interpret. The authors took several potential error effects into account but concluded that this asymmetry is real (see also Meierhenrich et al. 2002b).

Nonetheless, energy and zero-sum of the reflected circularly polarized light from planets was almost certainly insufficient for consideration as a chiral field capable of tipping the biomolecular balance towards the selection of L-amino acids.

6.2.5.2 The Bonner-Rubenstein Neutron Star Hypothesis

The subsequently developed Bonner-Rubenstein hypothesis considers another extraterrestrial mechanism for the molecular parity violation observed in the terrestrial biosphere. According to assumptions of Rubenstein et al. (1983), circularly polarized electromagnetic radiation is emitted from the poles of an extremely rapidly rotating neutron star in a way that, at one pole, right-circularly polarized synchrotron radiation is emitted, whereas the other pole emits left-circularly polarized synchrotron radiation. The neutron star itself formed as the result of a supernova explosion. This circularly polarized light, which was proposed to emit mainly in the ultraviolet and visible wavelengths, might have interacted with interstellar organic matter, the precursor of the solid bodies in the universe such as our planet. By this mechanism, the racemic constituents of the organic mantles of the stellar cloud grains could be asymmetrically photolysed, producing an excess of the corresponding more stable enantiomer in the grain mantles. This model has been further developed by Bonner and Rubenstein (1987), Bonner (1991, 1992 and 1995a), and Bonner et al. (1999a), and has resulted in a topical discussion.

Stephen Mason (1997, 2000) from London King's College argued against this model, considering the 'Kuhn-Condon zero-sum rule' (Kuhn 1930, Condon, 1937), that the integral of the circular dichroism over the whole wavelength range is zero. In case, the neutron star would emanate the required radiation over the entire spectrum and no enantiomeric enhancement could be produced by photochemical interactions with organic molecules. Bonner et al. (1999b) replied that only the circular dichroism band within the characteristic absorption spectra of prebiotic molecules is relevant for the incriminated enantioselective photochemistry by synchrotron radiation.

Despite this discussion, circularly polarized synchrotron radiation seems to be difficult to observe and detect in supernova remnants of pulsars (Bailey et al. 1998; Bailey 2000). Such detection has not yet been performed successfully. Consequently, considerable doubt arose against the hypothesis that synchrotron radiation of neutron stars induced enantiomeric enrichments in racemic or prochiral interstellar molecules (see e.g. Keszthelyi 2001).

6.2.5.3 Light Scattering on Aligned Dust Grains

As an alternative to the ideas of double reflections on planets' surfaces or rapidly rotating neutron stars, circular polarization of ultraviolet photons caused by specific interstellar scattering processes was proposed as a plausible original source of biomolecular handedness. This admissible hypothesis came into mind all of a sudden in 1998, when strong infrared circular polarization was observed by Bailey et al. (1998) in Orion OMC-1 using the 3.9-m Anglo-Australian Telescope including a polarimetry system. The team of Jeremy Bailey measured levels of circular polarization as high as 17%. Two years later, Bailey (2000) reported the theoretical background, namely that the OCM-1 region is known to have grains aligned by a magnetic field. Aligned spheroidal grains can cause circular polarization of incoming unpolarized light by Mie scattering. In general, Mie scattering describes the scattering of electromagnetic radiation by spherical particles, named after its developer, the physicist Gustav Mie (1908). In contrast to Rayleigh scattering[4], Mie's solutions to scattering embraces all possible ratios of diameter to wavelength. In 2003, a 23% circular polarization was discovered by Bailey (2003) in NGC 6334V.

Bailey's results were confirmed by observations such as the extended search for circularly polarized infrared radiation in additional areas of the Orion OMC-1 molecular cloud reported by Buschermöhle et al. (2005). This team used the United Kingdom Infrared Telescope (UKIRT) at Manua Kea Observatory in Hawaii. At 2.2 μm, circularly polarized light was detected as depicted in Fig. 6.13. Distinct regions of the OMC-1 molecular cloud show positive circular polarization depicted in red colour, whereas in other regions negative circular polarization dominates, shown in blue. Most of the areas show very low circular polarization close to the zero-value given in green colour. In order to provide a suitable explanation for the observed interstellar circular polarization of light, the authors discuss that in the first step, partial plane polarization may result from scattering or from dichroism. Circular polarization is then produced in cases of multiple scattering such as scattering of linearly polarized radiation, scattering of unpolarized radiation by aligned aspherical grains or dichroism involving multiple clouds or so-called twisted magnetic field lines.

We have to take into consideration that one cannot directly observe the circular polarization in ultraviolet wavelengths required for asymmetric photolysis of chiral molecules. Observations of circularly polarized light were only possible in the less energetic infrared region of the spectrum, because of the high dust obscuration in

[4] As indicated in Chap. 2, Rayleigh scattering describes the scattering of electromagnetic radiation by particles that are much smaller than the wavelength of the electromagnetic radiation.

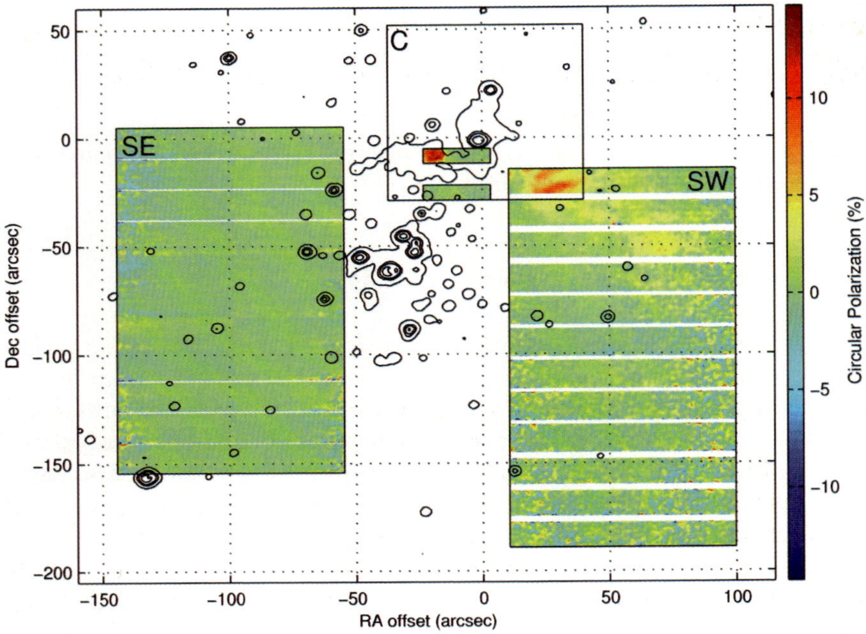

Fig. 6.13 Seeing *red*: Map of circular polarization in the Orion OMC-1 molecular cloud recorded in the near infrared at 2.2 μm. North is to the top and east is to the left. The square above center (labelled C) designates the area observed by Chrysostomou et al. (2000). The two observed fields are centered to the southwest (SW) and southeast (SE) of BN. Two smaller rectangular areas that lie entirely in area C are also shown. The observation of circularly polarized light is not feasible in the ultraviolet range of the spectrum, due to high dust extinctions. The illustration was first published by Buschermöhle et al. (2005). Reproduced with permission of the authors and the American Astronomical Society (AAS)

Orion's massive star formation regions. However, calculations of the scattering of light from aligned grains show that substantial circular polarization could still be produced at ultraviolet wavelengths. When an ultraviolet flux was present, it was calculated to exhibit substantial levels of circular polarization as well. So it is plausible to assume that there are locations in these massive star-forming complexes where light from an ultraviolet emitting star could scatter off the aligned grains.

The circular polarization in OMC-1 shows regions of both signs but the polarization structure is on a larger scale than the expected size of a protostellar disk – or our solar system. Consequently, if this source were responsible for the selection of life's handed molecules, the same direction of circular polarization would have affected all planets and also comets of our solar system.

With respect to the analysis of eventual enantioenrichments in samples of extraterrestrial origin such as meteorites, comets, or Mars (for details see Chaps. 8 and 9), this means, that Mie scattering models on aligned interstellar dust grains can indeed be assumed to induce an asymmetry. Since the circular polarization was observed on an extremely large scale, such processes would generate the same handedness of chiral molecules within our Solar System, including comets.

Chapter 7
Key to the Prebiotic Origin of Amino Acids

Amino acids are the crucial molecular components of life. They are the chiral building blocks for the molecular architecture of proteins (enzymes) and certainly played a key role in both the appearance of life and the emergence of biomolecular asymmetry (homochirality) on Earth. Proteins of living organisms are composed of 20 basic amino acids. It is thought that certain, if not all, amino acids had already been present on Earth nearly four billion years ago, long before the appearance of life. How did these chiral amino acids, the key molecular ingredients triggering biological evolution and life as we know it, originate?

Modern research distinguishes between three alternative models: (1) The formation of primordial amino acids might have occurred in the early Earth's atmosphere as successfully simulated in the laboratory by the well-known Urey-Miller experiment. (2) Hydrothermal systems are proposed to have the potential of generating a variety of amino acids before life started on Earth. And (3) The origins of amino acid formation may lay in the depths of space, since a pool of hitherto 89 amino acids has been identified in meteorites. Moreover, the foundation for the molecular asymmetry of amino acids may have taken place in well-defined regions of interstellar space: this scenario proposes that primordial amino acids were – frozen within tiny particles of ice – exposed to highly energetic, "asymmetric" ultraviolet radiation.

The aim of this chapter is to expose the current research activities on life's molecular building blocks and to present new data supporting the interstellar formation of amino acid structures including their asymmetry.

7.1 Amino Acid Formation Under Atmospheric Conditions

Based on the ideas of Alexander Oparin in Russia and Harold Urey in the United States in the first half of the 20th century, it was assumed that the atmosphere of the early Earth was strongly reducing and dominated by hydrogen H_2, methane CH_4, and ammonia NH_3. These gases could have been captured from the primordial solar nebula. The giant planets like Jupiter and Saturn had retained these gases since that time, as they were too massive to release H_2 into space. One concluded that

U. Meierhenrich, *Amino Acids and the Asymmetry of Life*. Advances in Astrobiology and Biogeophysics, © Springer-Verlag Berlin Heidelberg 2008

terrestrial planets, including Earth, had atmospheres of similar composition early in their histories before they had time to evolve to further stages.

Based on this supposition, Stanley L. Miller, member of Nobel laureate Harold Urey's research team at the University of Chicago, simulated such an atmosphere in the laboratory. The atmosphere composed of the above gases was processed by spark discharges imitating lightning. A 250 mL liquid water fraction taken from the simulation system showed the generation of a variety of amino acids and other organic molecules after several days of treatment (Miller 1953, 1998; Miller and Lazcano 2002). Here, a reducing atmosphere with methane (CH_4) as the carbon-providing molecule was assumed. Several racemic amino acids were formed under these conditions as products of spark discharge, photoprocessing, or heat.

However, the dominant view in recent years has been that the primitive atmosphere consisted of a weakly reducing mixture of CO_2, N_2, and H_2O, combined with lesser amounts of CO and H_2. This is because the atmosphere of the early Earth is considered to be largely produced from volcanic outgassing. There is a considerable literature stating that the hydrogen emitted from volcanoes is released as H_2O rather than H_2, and most of the carbon is released as CO_2 rather than CO or CH_4 (Kasting and Brown 1998). The weakly reducing mixture of the early Earth's atmosphere and particularly the high partial pressure of CO_2 of about 0.2 bar is consistent with the most likely mechanism for balancing the smaller solar luminosity at the time of formation of the earth, i.e., 30% lower than at present. This required an increased greenhouse effect in Earth's atmosphere, probably due to the accumulation of CO_2 to a minimum amount of 0.2 bar (Walker et al 1981; Kasting 1992).

The problem with the Urey-Miller mechanism is that amino acids could not be generated in a weakly reducing atmosphere. The HCN intermediate, required for amino acid formation, forms only from spark discharges in gas mixtures containing N_2 and CH_4. However, it does not form in atmospheres containing predominantly N_2 and CO_2, in which the nitrogen atoms formed by splitting N_2 almost invariably combine with oxygen atoms to yield NO (Kasting and Brown 1998). Stanley Miller himself reported that glycine was essentially the only amino acid synthesized in the spark discharge experiments with CO and CO_2 (Miller 1998). Other research teams of Akiva Bar-Nun and Sherwood Chang investigated the problem and found that, at most, only traces of amino acids form under such realistic weakly reducing conditions (Bar-Nun and Chang 1983). As expected for Urey-Miller types of atmosphere simulating systems, the amino acids produced remained racemic, since no asymmetry was introduced via spark discharges or other carriers into the system.

Furthermore, the generation of amino acids under the simulated atmospheric conditions of the early Earth was performed with ultraviolet light as the energy source (Sagan and Khare 1971; Khare and Sagan 1971). In this case, the photochemical reaction of a gaseous mixture containing CH_4, C_2H_6, NH_3, H_2O, and H_2S yielded a variety of amino acids. However, the presence of ethane with its preformed carbon-carbon bond was required as the main ingredient in the gas mixture. Hydrogen sulfide was also necessary as an acceptor for the long wavelength photons. Without ethane no amino acids were generated.

7.2 Amino Acid Formation Under Hydrospheric Conditions

Alternatively, hydrothermal systems were considered as an environment that could allow for the prebiotic formation of organic molecules (Hennet et al. 1992; Holm and Andersson 1998; Proskurowski et al. 2008). For the generation of hydrocarbons, lipids, amides, esters, and nitriles under such conditions, a Fischer-Tropsch type mechanism was proposed (Holm and Charlou 2001). However, the synthesis of amino acid structures in such an environment remains problematic. Under simulated "hydrothermal" conditions, selected starting compounds can react and lead to amino acids such as glycine, alanine, aspartic acid, serine, leucine, and others. However, the selected starting chemicals, i.e., formaldehyde, ammonia, and cyanide cannot be considered as being "prebiotic". Starting with these compounds of the Strecker synthesis was considered to be "equivalent to starting with glycine" (Miller 1998).

The team of Kensei Kobayashi at Yokohama National University tried to identify relicts of D-amino acids in different deep-sea hydrothermal systems pointing to their eventual abiotic origin. However, amino acids were only found in their L-configuration, suggesting their biogenic origin rather than any hints of abiotic chemical synthesis (Takano et al. 2003a, 2003c, 2004a).

Thus, by both atmospheric and hydrospheric simulation experiments, amino acids can obviously be synthesized under ideal, however not realistic, prebiotic conditions of the initial mixture of precursor molecules. This points to the need of formulating an alternative mechanism of prebiotic relevance for the spontaneous generation of amino acids.

7.3 Amino Acid Formation Under Interstellar Conditions

As nearly 90 chiral amino acids have been discovered in carbonaceous meteorites (see appendix; Kvenvolden et al. 1970; Cronin and Pizzarello 1997; Engel and Macko 2001; Ehrenfreund et al. 2001b; Meierhenrich et al. 2004), a significant contribution of extraterrestrial amino acids synthesized in interstellar environments under asymmetric conditions and delivered to Earth in the form of cometary and interplanetary dust offers an attractive alternative to a purely terrestrial (atmospheric and hydrospheric) origin of chiral organic molecules such as amino acids. Therefore, we remain particularly curious to topical research results and discussions on the formation of amino acids under interstellar conditions; these conditions need to be outlined.

7.3.1 Photochemistry in the Interstellar Medium

The interstellar medium and the circumstellar medium are harsh environments where predominantly simple gas molecules and interstellar dust particles are processed by the bombardment of hard ultraviolet photons and cosmic rays. Dense clouds in the interstellar medium possess kinetic temperatures as low as 10 K and densities from $10^3 - 10^6$ particles cm^{-3}. They are the birthplace of stars and planetary systems. As introduced in Chap. 1, solid dust particles in dense clouds, with a typical size of 0.1 μm, accrete an ice layer made of H_2O, CO, CO_2, CH_3OH, and NH_3 as the main components (Gibb et al. 2000, 2001, 2004). These ices undergo energetic processing by ultraviolet photons and cosmic ray particles (Muñoz Caro et al. 2001; Mennella et al. 2001). Ultraviolet irradiation is responsible for the dissociation of the ice molecules, creating free radicals, which recombine to form organic molecules (Greenberg 1986; Kerridge 1999; Schwartz and Chang 2002; Muñoz Caro and Schutte 2003; Muñoz Caro et al. 2004).

In order to gain information on the chemical and chiral structure of interstellar organic molecules, interstellar processes can be (a) simulated in the laboratory on Earth. Such simulation experiments will be presented in this chapter. Furthermore, one may envisage supplementary chemical analysis of (b) meteorites, (c) comets, and (d) organic molecules on the surface and subsurface of the planet Mars. These ambitious projects, which are at the center of topical research activities of the European and the US Space Agencies, will be presented in the upcoming chapters always concentrating on the phenomenon of chirality.

7.3.2 Simulation of Interstellar Chemistry in the Lab

At Leiden Observatory for Astrophysics, The Netherlands, the research team of J. Mayo Greenberg specialized in the synthesis of simulated interstellar and circumstellar ices. A specifically developed low pressure-low temperature chamber for the representative simulation of interstellar and circumstellar processes was installed there. A schematic view of this chamber is depicted in Fig. 7.1, a photo of the original instrument is given in Fig. 7.2 (a more detailed description can be found in Muñoz Caro (2003)).

The chamber basically consists of a high vacuum chamber with a pressure of approx. $1 \cdot 10^{-7}$ mbar. Inside this chamber, a gas mixture is deposited on a cold finger at 12 K, forming an ice layer, and irradiated. The system is pumped by a turbo pump backed up by a diaphragm pump. The low temperatures typical of dense clouds, down to 10 K, are achieved by means of a closed-cycle helium cryostat. The cold finger consists of a sample holder, in which an IR-transparent caesium iodide (CsI) window is mounted. The deposition and irradiation of the ice is monitored by means of infrared spectroscopy. Deposition of the molecules and subsequent photoprocessing occurs on the surface of this cold finger. In the simulation experiment, the cold finger was simulating the interstellar dust particle.

After pumping and cooling the system down to simulated interstellar conditions, a gas mixture consisting of water, ammonia, methanol, carbon monoxide, and carbon dioxide (H_2O, NH_3, CH_3OH, CO, and CO_2) of molar composition 2:1:1:1:1 was deposited onto the cold finger's surface. The composition of the gas mixture

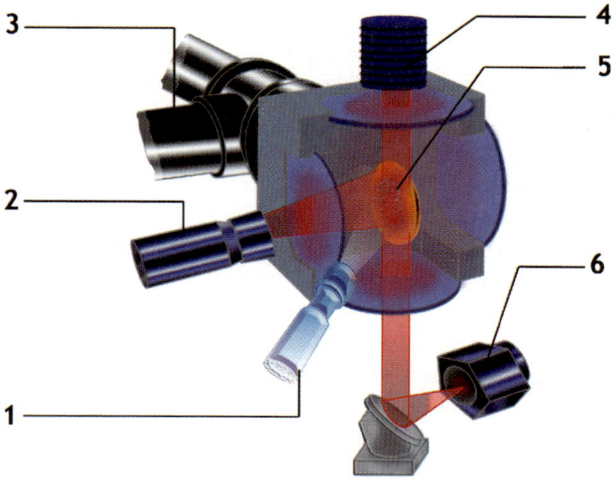

Fig. 7.1 Shedding light on interstellar processes: scheme of the low pressure-low temperature chamber for simulating the deposition and photoprocessing of interstellar and circumstellar ices by the adaptation to the physico-chemical interstellar and circumstellar conditions. **1** Deposition tube for gas molecules, **2** Vacuum UV lamp that emits mainly at Lyman-α (121 nm), **3** vacuum pump, **4** light source of IR spectrometer, **5** CsI window for sample condensation, **6** detector of IR spectrometer. Illustration: Stéphane Le-Saint, Université de Nice-Sophia Antipolis

Fig. 7.2 Space simulation chamber at Leiden Observatory, The Netherlands. The ice sample is located inside the vacuum chamber on an Al-block, or CsI window, located on the tip of the cold finger at $T = -261°C$ and irradiated by vacuum ultraviolet photons. These conditions recreate interstellar ices present in the pre-solar nebula during the formation of the solar system

is similar to that found close to protostellar sources (Gibb et al. 2001) and dense molecular cloud ices (Gibb et al. 2004). Interstellar ice abundances adopted for the starting ice mixture before irradiation were based on those derived from observations with the Interstellar Space Observatory (ISO) by Gerakines et al. (1999) and Ehrenfreund et al. (1999), excluding water ice which was increased by a factor of 2, as it is known that water is overabundant in comets. Two deposition lines were used for the gas mixture with a gas flow of 10^{16} molecules sec^{-1}.

The system was separated by a MgF_2 window from a UV-light source, i.e., a microwave-stimulated hydrogen discharge lamp, that emits 1.5×10^{15} photons sec^{-1} (Weber and Greenberg 1985) from 7.5 eV (165 nm) to 10.5 eV (117 nm) with the highest flux peaking at 9.8 eV (121 nm at the Lyman-α line) (Gerakines et al. 1995; Muñoz Caro and Schutte 2003). The typical irradiation and deposition time was 24 hours. Please note that the number of deposited molecules was higher than the number of photons induced into the system.

In order to follow the photochemical reactions during ice deposition, irradiation, and heating, infrared (IR) spectra were measured in situ with the help of a FTIR spectrometer.

After irradiation, the samples were heated at 1 K min^{-1} to 40 K and then at 4 K min^{-1} up to room temperature. During the heating process, the photochemically formed radicals start to move in the simulated interstellar ice and can be assumed to recombine in order to form oligomers and polymers. After heating, the above-described samples to room temperature, a small amount of residue remains, showing a yellowish colour as it is depicted in Fig. 7.3.

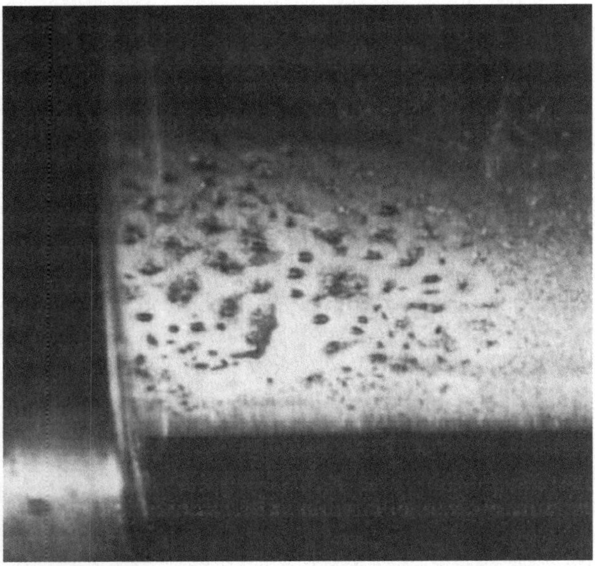

Fig. 7.3 Producing a comet in the lab: yellowish residue obtained after simulating interstellar and circumstellar ices on the surface of the cold finger (here: an aluminium block of 30·40 mm surface). After extraction, hydrolysis, and chemical derivatization, a wide variety of different chiral amino acid structures was identified therein

In the residue, non-chiral saturated organic compounds containing carboxylic groups but also hexamethylenetetramine (HMT) (Agarwal et al. 1985; Briggs et al. 1992; Bernstein et al. 1995) and structurally related compounds such as methyl-HMT, hydroxy-HMT, and others (Meierhenrich et al. 2001f; Muñoz Caro et al. 2004) were identified. In addition, non-chiral N-heterocycles and amines were determined in the residues serving as potential precursors of biological cofactors (Meierhenrich et al. 2005a). Previous work claimed the identification of trace amounts of a small number of amino acids in photo-processed interstellar/circumstellar ice analogues (Briggs et al. 1992; Kobayashi et al. 1999a). Except for glycine, however, such analytical results were not confirmed by isotopic labelling of the initial gas mixture, a necessary verification to exclude contamination with trace amounts of organic compounds.

7.3.3 Fresh in the Ice: Amino Acid Structures

7.3.3.1 Enantioselective Analysis

For the identification of chiral molecules, the irradiated samples of interstellar and circumstellar ice analogues were subjected to enantioselective analysis at the CNRS Centre de Biophysique Moléculaire in Orléans, France. An enantioselective gas chromatograph coupled with a mass spectrometer (GC-MS) was used. For each analyte, an individual sensitive analytical procedure was developed that consists of the sample's extraction, the hydrolysis, the chemical transformation of the analytes into suitable volatile derivatives, and the enantioselective capillary GC-MS separation and detection. The identities of the analytes obtained via the GC-MS technique were verified by comparing retention times (R_t) and mass spectra with literature data and external standards. Chiral analytes were resolved into enantiomers by an enantioselective GC-MS technique using Chirasil-L-Val capillary columns.

The obtained results demonstrated that carbon-to-carbon single bonds and carbon-to-nitrogen single bonds could be photochemically formed under simulated interstellar/circumstellar conditions. Even more surprisingly, 16 amino acids, six of which are among the 20 protein-constituent amino acids (i.e., glycine, alanine, valine, proline, serine, and aspartic acid) were identified on the surface of simulated interstellar dust particles (Muñoz Caro et al. 2002). Figure 7.4 shows the chromatogram of the residue obtained from the photo-processed interstellar ice analogue mixture. Most of the observed peaks were due to amino acids; three exceptions were identified as 2,5-diaminopyrrole, 2,5-diaminofuran, and 1,2,3-triaminopropane. The calculated signal-to-noise ratio was $S/N \geq 10$, even for the least abundant amino acids identified.

Glycine, the simplest amino acid, was the most abundant. For amino acids bearing only one amino group, species with higher molecular mass were less abundant. Aspartic acid was the only dicarboxylic compound found in the residues.

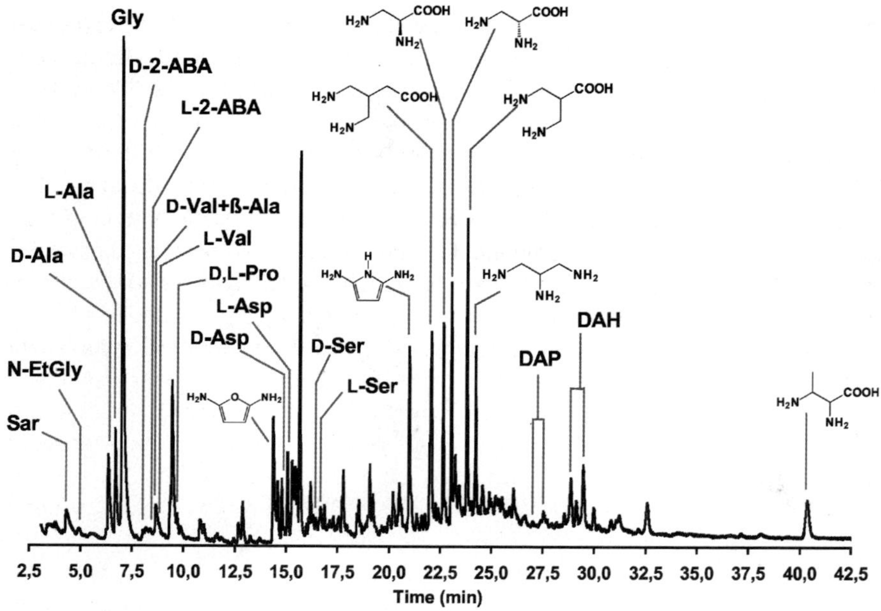

Fig. 7.4 Caught in the ice: Amino acid structures and other compounds in the gas chromatogram of an ice sample generated under simulated interstellar/circumstellar conditions. Data were obtained from enantioselective analysis of the room temperature residue of photo-processed ISM ice analogue. Chemical analysis was performed after extraction with ultra-pure water, 6 M HCl hydrolysis, and derivatization [ethoxy carbonyl ethyl ester (ECEE) derivatives taken from Huang et al. (1993) and Abe et al. (1996)]. A Varian-Chrompack Chirasil-L-Val coated capillary column 12 m × 0.25 mm inner diameter, layer thickness 0.12 μm was applied as chiral stationary phase with splitless injection, 1.5 mL min^{-1} constant flow of He carrier gas; oven temperature programmed for 3 min at 70°C, 5°C min^{-1}, and 17.5 min at 180°C. The detection of total ion current was performed with the GC-MSD system Agilent 6890/5973. DAP, diaminopentanoic acid; DAH, diaminohexanoic acid; a.m.u., atomic mass units. The first publication of the illustration by Muñoz Caro et al. (2002) is acknowledged

Decarboxylation is known as the main photo-damage of UV-exposed amino acids and probably competes with photoproduction (Johns and Seuret 1970).

Additional irradiation experiments involving a ten times higher H_2O ice concentration and/or deposition at 80 K, monitored by infrared spectroscopy, showed that the residue production proceeded similarly, although the relative abundances of the main components, hexamethylenetetramine and carboxylic acids, varied somewhat (Muñoz Caro et al. 2004; Nuevo 2005; Nuevo et al. 2007).

7.3.3.2 Isotopic Analysis

Changes of isotopic compositions of the initial gas mixture ($^{12}C/^{13}C$) combined with the mass spectroscopic isotope identification in the organic analytes provided clues of the reaction pathway. Did the amino acid structures really originate from the C-1 and N-1 units introduced in the system? Careful precaution was taken to eliminate risk of biological contamination: the reactants methanol, carbon monoxide, and carbon dioxide containing the usual ^{12}C isotopes were replaced by their ^{13}C isotopic counterparts. A similar chromatogram was obtained for the ^{13}C isotopically labelled sample (molar concentration H_2O:$^{13}CH_3OH$:NH_3:^{13}CO:$^{13}CO_2$ $= 2:1:1:1:1$). This isotope exchange resulted in increasing the mass of each amino acid's carbon atom by one. Thus, the synthesized amino acids were assumed to show a different mass, distinguishing them from biological amino acids of terrestrial origin by mass spectroscopy. The compounds corresponding to these mass increments were indeed identified as shown in Table 7.1.

To illustrate the measured isotopic mass shift, I will give an example for the enantiomers of the amino acid α-alanine: the mass spectrometric fragmentation of the ^{12}C-isotope of the α-alanine ECEE-derivative is given by $m/z = 189$, 144, **116**, and 88 atomic mass units (amu, parent peak in bold numbers). The mass spectra of the enantiomers D-Ala and L-Ala were, of course, identical. The ^{13}C-isotopic derivatives of α-alanine were $m/z = 147$, **118**, and 90 amu. The ECEE-α-alanine mass spectrometric fragment of 147 amu contains three ^{13}C atoms; the fragments of 118 and 90 amu contain two ^{13}C atoms each. According to this fragmentation pattern, one can conclude that the detected amino acids were indeed formed by the photoprocessing of the initial gas mixture. Contamination of the system by amino acids of biological origin can definitely be excluded, since these amino acids would have been composed of ^{12}C isotopes.

7.3.3.3 Hydrolysis and Meteoritic Samples

In the cases of reported analyses of meteoritic samples of carbonaceous chondrites, such as Murchison and others, the absolute abundances of the identified amino acids were enhanced by acid hydrolysis with 6 molar hydrochloric acid (Cronin 1976; Engel and Macko 2001). Similarly, in our case amino acids in the residues of simulated interstellar ices were identified in considerable amounts only after acid hydrolysis, which probably indicates that these products were originally formed as peptidic or oligomer molecules. The precise molecular structure of these oligomers and polymers is hitherto unknown; their structural elucidation is subject of numerous topical research activities: Matthews (1992, 2004), Minard et al. (1998) and Matthews and Minard (2006) expect that polymers could be major components of cometary matter; Kissel and Krueger (1987) and Kobayashi et al. (1998, 1999b) similarly suggest that complex organic macromolecules would make up the main organic ingredients. Atomic force microscopy images of similarly processed high-molecular-weight complex organic materials were recently published by Takano et al. (2007).

Free amino acids are released after hydrolysis of the polymer material. A gas chromatogram of photoprocessed simulated interstellar/circumstellar ices without hydrolysis is given in the literature (Meierhenrich et al. 2005a) showing minor quantities of the amino acid glycine among N-heterocyclic compounds. It should

Table 7.1 16 Amino acids, including 6 diamino acids, identified in simulated interstellar ices

Amino acid	Quantum yield, [100Φ/Φ(Gly)]	MS ^{12}C sample [a.m.u.]	MS ^{13}C sample [a.m.u.]	R_t of analyte [min]
Glycine	100	175, 130, **102**	177, 132, **103**	6.99
α-D-Alanine	19.29	189, 144, **116**, 88	147, **118**, 90	6.33
α-L-Alanine	20.00	189, 144, **116**, 88	147, **118**, 90	6.68
β-Alanine	4.29	189, 160, 144, 116, **115**, 102, 98	192, 147, 119, **117**, 103, 101	8.66
Sarcosine	5.71	189, 144, **116**, 88	192, 147, **118**, 90	4.29
D-2-Amino-butyric acid	0.46	130	133	8.05
L-2-Amino-butyric acid	0.48	130	133	8.37
N-Ethylglycine	1.91	**130**, 84, 58	**133**, 207	4.90
D-Valine	0.61	**144**	≤ d.l.	8.66
L-Valine	0.61	**144**	≤ d.l.	8.80
D,L-Proline	0.06	142	≤ d.l.	9.59
D-Serine	3.29	175 (McLaff.), 160, **132**, 114	116, 134	16.45
L-Serine	3.86	204, 187, 175 (McLaff.), 160, **132**, 114	116, 134	16.67
D-Aspartic acid	1.14	**188**, 142	**191**, 145	14.90

Table 7.1 (continued)

Amino acid	Quantum yield, [$100\Phi/\Phi$(Gly)]	MS ^{12}C sample [a.m.u.]	MS ^{13}C sample [a.m.u.]	R_t of analyte [min]
L-Aspartic acid	1.07	**188**, 142	**191**	15.05
D-2,3-Diaminopropanoic acid	20.0	231, 203, 188, **175**, 157, 129, 102	234, 205, 190, **177**, 159, 131, 103	22.64
L-2,3-Diaminopropanoic acid	20.0	231, 203, 188, **175**, 157, 129, 102	234, 205, 190, **177**, 159, 131, 103	23.03
2,3-Diaminobutyric acid	14.1	**203**, 157, 115, 85	**205**	40.38
3,3′-Diaminoisobutyric acid	32.9	290, 245, 238, 216, 188, 167, 156, **142**, 128, 102	249, 219, 191, **145**, 132, 103	23.79
4,4′-Diaminoisopentanoic acid	22.4	304, 259, 230, 202, 188, 182, 166, 156, 142, **116**	264, 206, 191, 187, 171, 160, 145, **118**	22.05
A-Diaminopentanoic acid *	1.72	**142** (c.i.i.), 129, 70	< d.l.	27.05
B-Diaminopentanoic acid *	1.92	**142** (c.i.i.)	< d.l.	27.54
A-Diamino-hexanoic acid †	33.1	272, **156** (c.i.i.), 128, 102, 84	272 (cyclic ion), **160**	28.86
B-Diamino-hexanoic acid †	35.4	272, **156** (c.i.i.), 128, 102, 84	272 (cyclic ion), **160**	29.48

For glycine, the quantum yield is $\Phi(\text{Gly}) = 3.6 \cdot 10^{-5}$. Numbers designated in boldface are the main mass fragments; *a.m.u.* atomic mass unit, *A* first eluting enantiomer, *B* second eluting enantiomer, *c.i.i.* cyclic immonium ion, *d.l.* detection limit, *McLaff* McLafferty rearrangement. * Exact molecular structure not yet known. Ornithine or its constitutional isomer.† Exact molecular structure not yet known. Lysine or its constitutional isomer.

be noted that in the Urey-Miller model, the intermediate products like amino and hydroxy nitriles also require a similar hydrolysis step to form amino acids and hydroxy acids.

Most of the amino acids reported in the experiment above were previously identified in meteorites called carbonaceous chondrites (Kvenvolden et al. 1979; Cronin and Pizzarello 1983, 1997; Engel and Macko 2001; Ehrenfreund et al. 2001b; Botta et al. 2002). However, no amino acids with more than one amino group were detected in these samples. The above experiment suggests that these species might also be present in meteorites. Subsequent enantioselective ultra-trace analyses of the Murchison meteorite verified indeed the presence of a variety of diamino acids (Meierhenrich et al. 2004).

7.3.3.4 Diamino Acids in the Simulated Interstellar Medium

The amino acids identified in interstellar and circumstellar ice analogues were not a random mixture of organic molecules. From the 16 generated amino acids, the diamino carboxylic acids[1] were of particular interest, since they belong to a new family of amino acids, also called diamino acids. These molecules have now been identified in residues of UV-photoprocessing of interstellar ice analogs and in the Murchison meteorite. They were proposed to provide hints for elucidating important reactions in prebiotic chemistry with respect to the further development of early genetic material.

The RNA-world scenario (Gilbert 1986), in which informational and catalytic properties of the first self-replicating system were provided by single RNA molecules, gives rise to some unanswered questions concerning the central ribose molecule. First, the ribose molecule is difficult to synthesize under prebiotic conditions. The formose reaction generates about 40 sugar compounds with a low yield of ribose. This was recently demonstrated by George Cooper and his colleagues who identified a great number of different polyols in meteoritic samples (Cooper et al. 2001). The evolutionary selection of the ribose molecule out of this random mixture of polyols remains problematic and until now unresolved. Second, ribose molecules are unstable, particularly in liquid solution.

Therefore, at present there is a growing consensus that the RNA-world did not evolve *directly* from monomeric molecules formed by prebiotic processes (Ferris 1993). It is assumed that RNA itself was preceded in evolution by a pre-RNA material of oligomer or polymer structure. In addition to numerous proposed alternative structures (mainly based on substituted ribose analogues), an attractive substituent of the ribose phosphate backbone in the RNA molecule is that of a peptide nucleic acid (PNA) (Nielsen et al. 1991; Nielsen 1993). The PNA molecules serve as DNA analogues and consist of peptide backbones made of diamino carboxylic acids to which nucleotide bases are attached via carbonyl methylene linkers (see

[1] Diamino acids identified in interstellar ice analogues are 2,3-diamino propanoic acid, 2,3-diamino butanoic acid, diamino isobutanoic acid, diamino isopentanoic acid, diamino pentanoic acid, and diamino hexanoic acid.

Chap. 8). PNA molecules were found to bind to oligo(deoxy)ribonucleotides and obeyed Watson-Crick base pairing rules, i.e., A-T and G-C base pairs are highly preferred (Egholm et al. 1993a). In template-directed reactions, information can be transferred from PNA to RNA, and vice-versa (Brack and Ferris 2002).

PNA can be easily produced from diamino carboxylic acids. The first identification of 6 molecular structures of diamino carboxylic acids in the residues made from UV-photoprocessing of interstellar ice analogues reported here might stimulate future research through the study of the reaction pathways from diamino carboxylic acids that result in PNA structures. Chapter 8 will present more details on this 'new' amino acid family including their eventual role in chemical evolution.

7.3.3.5 Topical Discussions on Amino Acids in Residues made by UV-Processing of Simulated Interstellar Ice Analogues

Based on the previous observations, how can we obtain additional experimental evidence that the identified amino acid structures in interstellar ice analogues really triggered the appearance of life on Earth? Jordan et al. (2005) compared proteins from organisms representing all three domains of life (Bacteria, Archaea, and Eukaryota) with regard to their amino acid content. Amino acids with declining amounts in proteins during biological evolution were identified, and some have increased in amounts. The amino acids with declining frequencies were proposed by Jordan et al. (2005) to be among the first incorporated into the genetic code, and most of those with increasing frequencies were probably recruited later. Based on these results, a consensus order of amino acid recruitment into the genetic code was established starting with glycine, alanine, aspartic acid, valine, then skipping over proline to serine. These six amino acids are consistent with the six proteinaceous amino acids identified in residues made from UV-photoprocessing of interstellar ice analogues.

But it is still a question of topical debate how amino acids got into the meteorites in the first place (Elsila et al. 2007). There appear to be two reaction-pathways. One suggestion is that Strecker-type chemical reactions involving liquid water took place on or in the rocky bodies that formed when the Solar System was young. The herewith-related presence of liquid water in solar system objects will be discussed in Chap. 9. The alternative model, discussed here, assumes that interstellar and circumstellar photoreactions produce radicals in interstellar ices that recombine at higher temperatures to form a complex polymer material. After hydrolysis this polymer material releases amino acids and diamino acids. At present, we are not in a position to explain the mechanism of interstellar amino acid formation in detail – neither the intermediate precursors nor the function of water in the form of solid or liquid. Since analyses of amino acids in meteoritic samples and in residues made from UV-photoprocessing of interstellar ice analogues both require a hydrolysis step, we assume for these matrices a chemical composition of peptide-like molecules that contain amino acid structures. Yet this assumptions needs to be verified in the future.

Several research groups tried to shed some light on the interstellar amino acid formation pathway after the simultaneous experimental identification of amino acid structures in residues made from UV-photoprocessing of interstellar ice analogues by NASA Ames researchers and our European group (Bernstein et al. 2002; Muñoz Caro et al. 2002). These teams focused on (*a*) the theoretic ab initio calculation of spontaneous interstellar and circumstellar amino acid synthesis by quantum chemical modelling studies (e.g. Woon 2002), (*b*) the application of different energy sources for irradiation of interstellar ices and its effect on the synthesized organic molecules (Klánová et al. 2003; Takano et al. 2004b, 2007), or (*c*) the deposition of different types of gas mixtures (Takano et al. 2003b; Takahashi et al. 2005; Nuevo 2005; Nuevo et al. 2007). The precise mechanism, however, remains unknown.

The simulated interstellar formation of amino acid structures has been pointed out by Nobel Prize laureate Christian de Duve (2003) who said that there is now clear evidence that large quantities of organic molecules continually arise at many sites in outer space by mechanisms that are beginning to be understood and even reproduced in the laboratory. However, – and this is the central conflict – amino acids have not been identified unambiguously by spectroscopic methods in interstellar environments. Why not? One argument is based on the so-called partition function problem, which is valid for spectroscopic measurements of spectroscopically "complex" molecules like amino acids, making specific electronic transitions and vibrations spectroscopically invisible at a given detection limit among a variety of other electronic transitions and vibrations.

Another argument states that amino acids in interstellar environments mainly occur embedded in polymers, which release free amino acids after acid hydrolysis only. This refractory polymer material remains in the solid state making it invisible to radio astronomy techniques. Thus, amino acids cannot be identified spectroscopically in different interstellar environments with classical methods. Recently, Martin Schwell from Paris University Denis Diderot presented experimental evidence for a third argument in this conflict: Schwell's experimental set-up was constructed in order to study photolysis reactions (i.e., photodegradation reactions) of amino acids in well-defined vacuum ultraviolet radiation fields. In the chosen irradiation fields, organic molecules and amino acids in particular exhibited limited stability against electromagnetic radiation. Schwell et al. irradiated amino acids in the gaseous phase, characterized their photochemical properties, and deduced that amino acids should undergo photodecomposition reactions if they are subjected to energetic interstellar vacuum ultraviolet irradiation (Schwell et al. 2006; Schwell 2007). Regarding photodecomposition, the same can be said about methanol. Methanol photodecomposes "easier" than amino acids, yet methanol is observed abundantly both in the solid and gas phases. Almost any molecule can photodecompose, but one has to take into account that there is an equilibrium between formation and destruction of interstellar molecules by ultraviolet irradiation.

Ehrenfreund et al. (2001a) studied the photostability of amino acids in an argon matrix at 12 K, irradiated by energetic radiation of 135 to 165 nm wavelengths. These experiments also indicated that amino acids are degraded photochemically. Nevertheless, ultraviolet destruction cross-sections were determined in an argon matrix and the extrapolation of the here determined values to other solid phases, for example to interstellar grains, or the gas phase should be handled with caution.

The tentative spectroscopic identification of the amino acid glycine in the interstellar clouds Sagittarius B2(N-LMH), Orion KL, and W51 e1/e2 was published recently (Kuan et al. 2003), but is waiting for confirmation by other research teams. At present, radio astronomers do not rely on the results of Kuan and coworkers. In summary, amino acids have not yet been identified by direct measurements in interstellar environments. This difficulty may be explained by various reasons, as outlined above.

Nevertheless, the obtained experimental data support the assumption that tiny ice grain mantles can play host to important reactions when irradiated by ultraviolet light. It is possible that such ice mantles could have become incorporated into the cloud that formed our Solar System and ended up on Earth, assisting the start of life. Several lines of evidence suggest that some of the building blocks of life were delivered to the primitive Earth via (micro-) meteorites and/or comets from space. These results suggest that interstellar chemistry may have played a significant part in supplying Earth with some of the organic materials needed to trigger life. Furthermore, since new stars and planets are formed within the same clouds in which new amino acids are being created, this probably increases the chance that life has evolved elsewhere.

Coming back to our original question on the origin of biomolecular homochirality, the chiral amino acids produced by simulated interstellar photochemical reactions showed racemic occurrence, contrary to that found in all proteins of living organisms. Nevertheless, the above experiment is assumed to be of crucial importance for the understanding of biomolecular asymmetry. It provides evidence that chiral organic molecules like amino acids spontaneously form under interstellar and circumstellar conditions (Fig. 7.5).

In the experiment outlined above, non-polarized light was used to induce photochemical reactions of the simulated interstellar/circumstellar ice mixture. However, in interstellar and circumstellar regions, electromagnetic radiation has asymmetric components and is partly circularly polarized. The transfer of asymmetry from 'chiral photons' (in the form of circularly polarized electromagnetic radiation) to prochiral or racemic organic molecules can provoke molecular enantiomer enrichments and has experimentally been proven to be feasible as described for enantioselective *photolysis* reactions in Chap. 6. Here, I will present recent experimental approaches on the enantioselective *synthesis* of amino acids under simulated interstellar conditions.

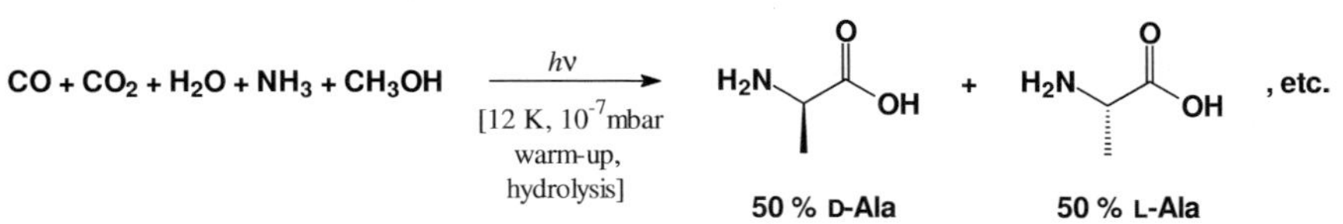

Fig. 7.5 Photochemical formation of amino acids with unpolarized light from C_1 and N_1 components, resulting in the formation of racemic mixtures mimicking the interstellar medium. Alanine and 15 other amino acids were produced under specific low temperature/low pressure conditions. The first publication of the illustration by Griesbeck and Meierhenrich (2002) is acknowledged

7.3.4 Illumination with Circularly Polarized Light

Interstellar electromagnetic radiation is known for its energy and its intensity. What about its polarization? *Circularly* polarized electromagnetic radiation was identified in the Orion star formation region by Bailey and his group (1998). Maps of circularly polarized infrared radiation from the OMC-1 region of Orion were recorded by Buschermöhle et al. (2005). The authors concluded that the observed circular polarization in the near infrared can also produce circular polarization in the range of ultraviolet wavelengths (see Chap. 6).

Consequently, in advanced experimental series, the light source described for the above experiments on the interstellar amino acid formation – emitting unpolarized light – was substituted by a circularly polarized synchrotron beam in the French synchrotron facility LURE, Paris. Interstellar ice analogues were produced in the closely located team of Louis d'Hendecourt at the Institut d'Astrophysique Spatiale (IAS) following a proposition made by J. Mayo Greenberg. With these experiments, the aim was to mimic interstellar and circumstellar conditions and to study whether one could introduce an enantiomeric excess into the produced chiral amino acid structures by absolute asymmetric photochemical *synthesis* (Griesbeck and Meierhenrich 2002; Meierhenrich and Thiemann 2004). Our common aim was to study hitherto unknown enantioselective photo*synthesis* reactions of amino acids in contrast to well-known enantioselective photo*decomposition* reactions already described by Balavoine et al. (1974), Flores et al. (1977), and Meierhenrich et al. (2005b).

Experiments were conducted with right- and left-circularly polarized light from the synchrotron in two separate runs. Each irradiation took two days, the wavelength of the synchrotron radiation was set to $\lambda = 167$ nm. After irradiation at liquid nitrogen temperature ($T = 77$ K), the ice samples were warmed to room temperature and the organic residue products remaining at room temperature were analyzed by infrared spectroscopy, residue hydrolysis (in 6 molar hydrochloric acid), and derivatization, followed by GC-MS analysis using an enantioselective column. Carbon-13 isotope labelled molecules were used in the initial gas mixture in order to carefully exclude biological contaminants. As a first result, we were able to identify eight amino acids, all of them composed of ^{13}C isotopes, in samples produced with circularly polarized light (Table 7.2) (Nuevo et al. 2006).

The amino acids have been analyzed with special emphasis for any evidence of an enantiomeric excess in either L or D. The results and discussion focused on two of the chiral amino acid products: α-alanine, the most abundant chiral proteinaceous amino acid, and 2,3-diaminopropanoic acid (DAP), a non-proteinaceous chiral diamino acid recently identified in the Murchison meteorite (Meierhenrich et al. 2004). After photochemical synthesis and hydrolysis of the interstellar ice analogues, these amino acids were detected in form of the ^{13}C-ECEE derivatives. The obtained enantiomeric excesses were compared to those measured for the same amino acids produced by unpolarized ultraviolet irradiation of the same ice mixtures (expected to be zero) in order to determine the contribution of circularly polarized

Table 7.2 Amino acids and diamino acids identified in simulated interstellar ices irradiated with circularly polarized light (Nuevo et al. 2006)

Amino acid	MS ^{13}C sample [a.m.u.]	R_t of analyte [min]
Sarcosine	103, **118**	9.81
α-D-Alanine	90, 103, **118**, 147	11.05
α-L-Alanine	90, 103, **118**, 147	11.40
Glycine	**103**, 132, 177	11.81
D-Aminobutyric acid	103, **133**	12.85
L-Aminobutyric acid	103, **133**	13.12
β-Alanine	101, 103, **117**	13.64
D,L-Aspartic acid	**191**	20.17
D-2,3-Diaminopropanoic acid	103, 131, 159, **177**	28.52
L-2,3-Diaminopropanoic acid	103, 131, 159, **177**	29.16
3,3'-Diaminoisobutyric acid	103, 117, **145**, 191	31.24

Numbers designated in boldface are the main mass fragments; *a.m.u.* atomic mass unit

light. A careful estimate of all the associated uncertainties such as statistical and systematic errors was developed.

It appeared that any enantiomer-favouring photochemical effect at this wavelength is weak, since both α-alanine and DAP enantiomeric excesses were found to be small, at most of the order of 1% as shown in Table 7.3. Here, the *e.e.* = 1% falls within the error bars of the experiment and the detection technique. The original chromatograms are depicted in Fig. 7.6.

The small amount of material was the main limiting parameter in the precise determination of an enantiomeric excess. In the near future, higher amounts of material might be produced with a larger photon flux, possible with the new synchrotron radiation beamline DESIRS at SOLEIL, that just finished construction in France. Furthermore, in future experiments, the energy of the circularly polarized synchrotron radiation might be changed from $\lambda = 167$ nm to other wavelengths, better corresponding to circular dichroism spectra of amino acids and their molecular precursors in the vacuum ultra violet region.

In light of these results, the hypothesis that circularly polarized light might be one source responsible for enantiomeric excesses of amino acids in some meteorites and, more generally, that circularly polarized light may be directly related to the origin of biomolecular homochirality on Earth remains untested.

Table 7.3 Enantiomeric excesses identified in the amino acids alanine and 2,3-diaminopropanoic acid after irradiation with circularly polarized light

Amino acid	Polarization of light	Corrected *e.e.* [%]
α-Alanine	unpolarized	0.00
α-Alanine	RCPL	0.09*
2,3-Diaminopropanoic acid	unpolarized	0.00
2,3-Diaminopropanoic acid	RCPL	1.11*

* Designated *e.e.*-values give excesses for α-L-alanine and L-DAP enantiomers

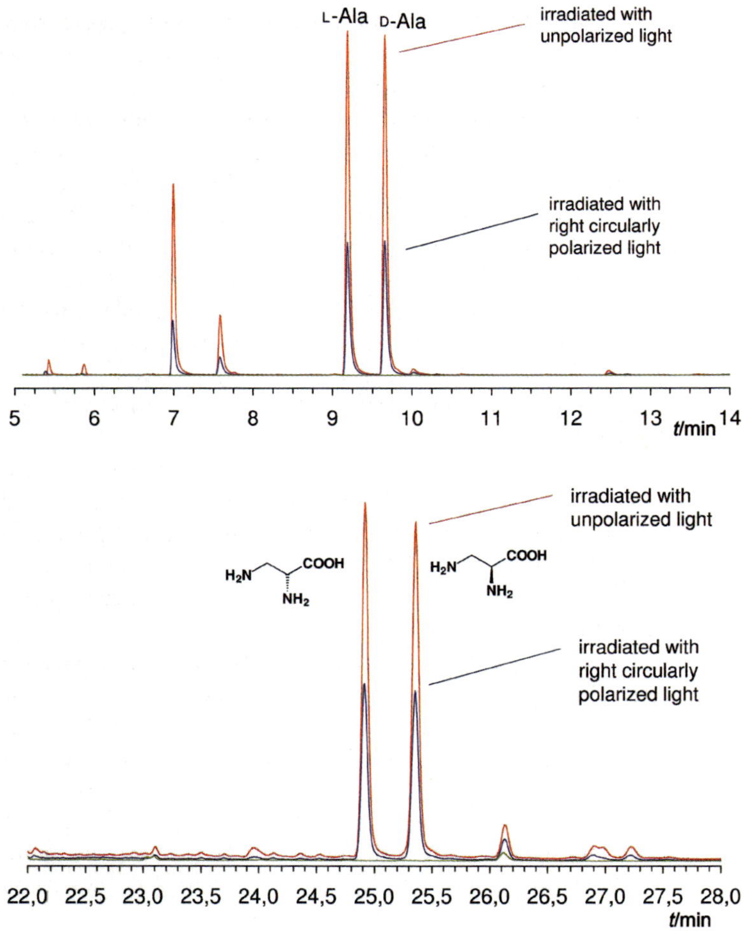

Fig. 7.6 Gas chromatograms of the chiral amino acids α-alanine (*top*) and 2,3-diaminopropanoic acid (*bottom*) identified in residues made from UV-photoprocessing of interstellar ice analogues after irradiation with unpolarized light (*red line*) and right-circularly polarized synchrotron irradiation (*blue line*). Chromatograms were taken on a 25 m Chirasil-D-Val stationary phase and recorded in single ion monitoring (SIM) mode at 118 amu for alanine and 177 amu for 2,3-diaminopropanoic acid. In the blank sample (*green line*) no amino acids and/or diamino acids were detected

In the field of absolute asymmetric photochemistry, it is hard to obtain evidence by laboratory *simulation* experiments only. Because of this, all of the experimental results based on the use of circularly polarized light, and in particular those obtained for amino acids (Flores et al. 1977; Meierhenrich et al. 2005b), should be compared with extraterrestrial material data, i.e., analyses of meteorites such as Murchison and Murray, where amino acids were detected and found to display L-enantiomeric excesses after chemical analysis. The next chapter is therefore dedicated to systematically describing the enantioselective analysis of extraterrestrial samples.

The determination of values for enantiomeric excesses in various extraterrestrial samples has become part of ambitious space scientific programs. Today, the measurement of enantioenrichments is included for the first time in one of the cornerstone missions of the European Space Agency (ESA) whose ‘chirality-part’ was constructed at the Max Planck Institute for Solar System Research in Katlenburg-Lindau, Germany. For this program, the cometary mission ROSETTA is designed – among else – to separate chiral organic molecules on the surface of a comet and to determine values of enantiomeric excesses for a wide range of chiral organic molecules. The enantioselective GC-MS analyses taken on the Agilent 6890/5973 GC-MSD system in our laboratory are indeed envisaged to be reproduced by the GC-MS system developed for the ROSETTA-Lander Cometary Sampling and Composition experiment COSAC. The COSAC gas chromatograph-module is consequently equipped with the same set of capillaries/stationary phases (Goesmann et al. 2007b) that will be explained in Chap. 9.

The US space agency NASA has independently demonstrated its interest in measurements of the enantiomeric composition of chiral molecules in extraterrestrial samples. Scientists are willing to separate and quantify enantiomers of hydrocarbons originating from Saturn’s moon Titan. A future mission, described in a special 2001 issue of the journal ‘Enantiomer’ called ‘Chirons on Titan: A Search for Extraterrestrial Enantioenrichment’, will focus on this task.

Moreover, in 2014, the joint ESA-NASA mission ExoMars envisages to set down two instruments on the surface of the planet Mars that are capable of distinguishing between enantiomers. One instrument, the Mars Organic Molecule Analyser (MOMA) is equipped with an enantioselective gas chromatograph on chiral stationary phases; the other is the UREY-instrument, which relies on enantioselective capillary electrophoresis with a liquid mobile phase. The objectives, scientific concepts, advantages (and eventual limitations) of these instruments, which are of crucial importance to the study of the origin of biomolecular asymmetry, will be introduced in Chap. 9.

Chapter 8
A New Record for Chiral Molecules in Meteorites

Recently published data gave fresh support for the assumption that interstellar processes had originally generated the chiral asymmetry of life.[1] These data were obtained in Earth-based laboratories giving access to the observation of chemical processes under *simulated* interstellar conditions. The results indicated that chiral amino acid structures form in laboratory-analogues of dense interstellar clouds and that these chiral structures were subjected to interstellar asymmetric radiation with the potential to generate a small but decisive enantiomeric excess tipping the balance in one direction towards life's homochirality. To investigate this theory for an interstellar origin of biomolecular asymmetry we will now widen our view and describe the stereochemistry of organic molecules identified in *authentic* samples of extraterrestrial origin. We will focus on samples of meteorites, comets, and Mars. The knowledge of enantiomeric ratios on an extraterrestrial body can be assumed to provide significant information to verify the above assumption based on simulated interstellar data and thus to better understand the origin of biomolecular asymmetry on Earth.

8.1 Chiral Organic Molecules in Meteorites?

In the morning of September 28th, 1969, a meteorite exploded over the small town of Murchison in Australia, spreading a total of about 100 kg extraterrestrial material in different fragments on Earth. A meteorite is by definition a solid-state object that originated in extraterrestrial space, passed Earth atmosphere, and fell down on Earth's surface without being destroyed. This particular meteorite was of a type called carbonaceous chondrite, extremely rich in carbon compounds. Chondritic meteorites, in particular the CM type carbonaceous chondrites such as Murchison, make up a unique subset of primitive meteorites, which is of particular interest in

[1] According to this assumption – just imagine its far-reaching consequences e.g. on the probability on the existence of life elsewhere in the Universe – not only the molecular building blocks of living organisms such as amino acids and other organic molecules had been generated in interstellar space ("we are made of stardust" – Cliff Matthews), but also the preselection of life's asymmetrical use of mirror image isomers had been performed by extraterrestrial processes.

the context of origin-of-life research. Not only because of their relatively high carbon content but also since most of this carbon is present as organic matter. This organic matter is a diverse mixture of compounds that in particular includes carboxylic acids, dicarboxylic acids, hydroxy acids, sulfonic acids, phosphonic acids, and amino acids in the form of monoamino alkanoic acids and monoamino alkandioic acids (Cronin and Chang 1993) but also in the form of diamino acids. Some of these species are chiral.

Among these classes of compounds, a small fraction of the isomers is indeed believed to support prebiotic evolutionary processes. The molecular building blocks of both protein enzymes and genetic material might have been originated in interstellar and circumstellar environments and subsequently been delivered by meteorites, interplanetary dust particles, and/or by comets (Oró 1961; Kissel et al. 1986a, 1986b; Anders 1989; Jessberger 1999) to the early Earth where they have triggered the appearance of life (Chyba and Sagan 1992; Ehrenfreund 1999). Based on persisting advances in chemical analysis techniques, amino acids (Cronin 1976; Cronin and Chang 1993; Cronin and Pizzarello 1997; Cronin 1998; Engel and Macko 2001; Botta et al. 2002; Meierhenrich et al. 2004), and molecular DNA constituents such as purine- (Van der Velden and Schwartz 1977) and pyrimidine bases (Stoks and Schwartz 1979), as well as sugar-related organic compounds (Cooper et al. 2001) have already been identified in various carbonaceous meteorites.

Which conclusions can be drawn today from the meteoritic occurrence of different amino acids? Is any chiral configuration preferred by meteoritic amino acids and if "yes", is there any chiral fingerprint written in these amino acids informing us on their eventual contribution to the evolution of life on Earth?

8.1.1 Amino Acids in Meteorites

Until today, 89 different amino acids have been detected in the Murchison meteorite; a list of which can be found in the Annex. Other carbonaceous chondrites also include a high content of a variety of amino acids. Most of them show slightly different constitutions of the side chains located at the α-carbon atom. Often, the presence of amino acids in carbonaceous meteorites was confirmed by their isotopic composition (Pizzarello et al. 2003), clearly indicating their extraterrestrial origin.

What about their enantiomeric enhancements? Experimental investigations on meteoritic amino acids resulted in the detection of a rather small, yet significant, enantiomeric excess of a special family of amino acids called α-methyl-α-amino acids in the Murray and Murchison meteorite. The well-known research team of John Cronin and Sandra Pizzarello at Arizona State University detected the highest enantiomeric excesses of an amino acid in a meteorite. For isovaline, a non-proteinaceous amino acid, an enantiomeric excess of 8.4% was determined for its L-configuration, and the 2*S*, 3*R*-enantiomer of 2-amino-2,3-dimethylpentanoic acid DMPA was detected in the Murchison meteorite with an *e.e.* = 9.1% (Cronin and Pizzarello 1997, 1999; Pizzarello and Cronin 2000).

More recent measurements of the Cronin-Pizzarello research team on several Murchison meteorite fragments revealed that L-isovaline displays enantiomeric excesses ranging up to 15.2% (Pizzarello et al. 2003). Some proteinaceous amino acids were also found to be non-racemic, but with smaller enantiomeric excesses: An excess of 2.2% was detected for L-valine, and 1.2% for L-alanine (Cronin and Pizzarello 1999), which is actually a value closely matching the above outlined experimental data obtained by photolytic irradiation of amino acids with circularly polarized light under simulated interstellar/circumstellar conditions.

In general, meteoritic α-methyl (α-Me) amino acids, i.e., amino acids with a methyl group bound to the α-carbon atom just next to the carboxyl group, display higher enantiomeric excesses than their α-hydrogenated (α-H) analogues. Actually, α-hydrogen amino acids are known to racemize more efficiently than α-Me amino acids (Pollock et al. 1975), because a hydrogen atom bound to the α-carbon close to a carbonyl can be removed by deprotonation. When an α-hydrogen amino acid lose its α-hydrogen, the asymmetric carbon loses its chirality. The resulting plane geometry around this atom allows protonation from either side with roughly the same probability, leading to a racemization. This racemization process cannot occur by this mechanism with α-Me amino acids. In this context it is noteworthy to remark that other different racemization mechanisms are known, for instance, the tunnelling racemization of enantiomers (Cattani and Bassalo 1998).

A racemization procedure could explain – at least in part – why, after interstellar thermal, photochemical, and radiolytic adulteration of meteoritic samples, proteinaceous amino acids, which are all α-hydrogen compounds, were found to display excesses with factors up to five times smaller than α-Me amino acids (Cronin and Pizzarello 1999). The study of α-Me amino acids produced by vacuum ultraviolet circularly polarized light irradiation of simulated interstellar ices would thus be really interesting in the future. In this kind of studies, particular attention should be paid to the asymmetric photolytic behaviour of α-Me amino acids, checking if such compounds display high enantiomeric excesses.

α-Methyl amino acids such as isovaline were recently shown to be more effective asymmetric catalysts than α-H amino acids, i.e., they can efficiently influence chemical enantioselective reactions to the production of one type of enantiomer (Pizzarello and Weber 2004), and thus they appear to be key compounds to investigate in order to elucidate their role in the origin of homochirality. However, until today α-Me amino acids have remained difficult to study in simulated interstellar ices since they have been produced in much lower abundances than α-H compounds in the residues. It was very recently that isovaline[2] was identified in an organic residue formed by unpolarized ultraviolet irradiation of a $CH_3OH : NH_3 = 1 : 1$ ice mixture (Nuevo et al. 2007). In simulating experiments of our laboratories involving *circularly* polarized light, no α-Me amino acids were detected yet, confirming their relatively low production yields (Nuevo et al. 2006).

[2] The total amount or relative quantity of isovaline in simulated interstellar ices is not precisely given.

We have seen that proteinaceous amino acids such as α-alanine and α-methyl amino acids such as isovaline have been unambiguously identified in the Murchison meteorite. What about the identification of other families of chiral organic molecules in this meteorite and their possible contribution to the origin and evolution of life on Earth? Very recently, diamino acids moved into the center of scientific interest. Why is this 'new' class of particular interest?

8.1.2 The Untold Story of Diamino Acids in Meteorites

Ever since the identification of deoxyribonucleic acid DNA by James Watson, Francis Crick, and Rosalind Franklin as the genetic material and the elucidation of its unique information bearing double-helical structure there have been speculations as to the evolutionary origin of this material and to life as we know it. Did DNA evolve in evolutionary timescales before proteins or, alternatively, did proteins precede the evolution of DNA double helices? Until the 1980s this question remained difficult to answer, since DNA requires proteins for molecular self-replication and the information how to construct proteins is written in the genetic code of the information-bearing DNA. Fig. 8.1 tries to illustrate this evolutionary paradox by referring to the old hen-or-egg enigma.

So far, based on abundant experimental data we have de facto information on the early development of the DNA genome. Several independent lines of evidence suggest that DNA- and protein-based life was preceded by a simpler form of life based

Fig. 8.1 The picture refers to the older question on the origin of chicken and egg. What was first? Genes like DNA (illustrated by the chicken/hen) or proteins (symbolized by the egg)? In order to answer this longstanding question, data of new meteorite analysis include hints that – in this analogy – "self-replicating eggs" may exist that do not necessarily require a hen for originating (Bredehöft 2007). Image of [SH]^2 Kommunikations-Design, Bremen

primarily on ribonucleic acid (RNA). This earlier era is referred to as the 'RNA-world' (Gilbert 1986; Gesteland et al. 1999; Joyce 2002), during which the genetic information resided in the sequence of RNA molecules and the phenotype derived from the diverse catalytic properties of RNA (Joyce 2002). In general, the approach of the RNA-world for the understanding of the origin of life seems to be widely accepted, even if some variations exist.[3]

The instability of the ribose molecule and considerable difficulties in finding a selective prebiotic reaction yielding ribose (Cooper et al. 2001) remain unresolved features of the RNA-world scenario. Therefore, at present there is a growing consensus that the RNA-world could not have evolved directly from monomer molecules formed by prebiotic processes (Ferris 1993). Pre-RNA genetic material is assumed to be required and this fundamental question provoked that numerous world-leading research teams today try to elucidate its molecular structure.

The main obstacle to understanding the origin of pre-RNA oligonucleotides is identifying a plausible mechanism for overcoming the gap between prebiotic chemistry and primitive biology. It is generally assumed that chemical processes have led to a substantial level of complexity before the pre-RNA world came into play. The precise knowledge of the chemical constitution of the potential molecular building blocks of pre-RNA oligonucleotides can be assumed to fill the substantial gap in the scientific understanding concerning how the RNA world arose. The data given below might contribute to fill this gap:

Due to astrophysical research today, we have numerous and undebated data, where exactly in the universe organic molecules can be synthesized spontaneously: In samples coming from interplanetary, interstellar, and circumstellar regions potential molecular building blocks of biopolymers were detected. This is particularly true for a wide variety of amino acids identified in carbonaceous meteorites (Cronin, 1976; Cronin and Chang, 1993; Cronin and Pizzarello, 1997; Cronin, 1998; Engel and Macko, 2001; Botta et al., 2002) and interstellar ice analogues simulated in the laboratory (Bernstein et al., 2002; Muñoz Caro et al., 2002). However, the molecular constituents of pre-RNA oligonucleotides have never been identified, – neither in simulated interstellar and circumstellar ices nor in situ.

In 2004, chemical analyses of the Murchison meteorite led to the identification of diamino acids, a 'new' class of organic compounds: Analytical studies were performed with a sample of the carbonaceous Murchison chondrite in the Department of Physical Chemistry at the University of Bremen. The sample itself was obtained from a collection of the Max-Planck-Society in Mainz, where it was specially taken from the protected interior of the Murchison meteorite. 1 g of the Murchison meteorite was powdered by a planetary micro mill, imaged by a raster electron microscope (Fig. 8.2), and extracted with water in a positive pressure "Class 100"

[3] Christian de Duve, for example, favors a specific "weak version" of the RNA world, in which most RNA acting as catalyst was relatively short. The catalytic activity of this RNA is supposed to be supported by short peptides in combination with other compounds that de Duve calls multimers (de Duve 2006).

Fig. 8.2 A meteorite is an object originating from outer space impacting on Earth's surface. (*Top*) As an example, a 110 kg fragment of a typical iron meteorite found in the Canyon Diablo, Arizona, is given. (*Bottom*) Raster electron microscope image (CS44, CamScan, Cranberry Township, PA) of the powdered sample of the Murchison meteorite showing a rather homogenous particle size distribution in the low micrometer range. An aliquot of this sample was subjected to diamino acid analysis

clean room. The obtained extracts were hydrolysed,[4] derivatized,[5] and analysed subsequently by enantioselective GC-MS technique using a Chirasil-L-Val stationary phase of 12 m length. With a new efficient derivatization method and an improved detection technique we focused on chiral compounds with three and more functional groups.

[4] Hydrolysis for 24 h at 110 °C in 6 molar HCl for amino acid analysis. Such conditions are typical for the hydrolysis of proteins to release "free" amino acids.

[5] Due to a method described by Abe (1996), diamino acids were transformed into *N*,*N*'-diethoxycarbonyl diamino acid ethyl ester (ECEE) derivatives.

Surprisingly, the application of this technique enabled the identification of diamino acids in the meteorite, never seen before (Meierhenrich et al. 2004).[6] Diamino acids are amino acids with an additional amino group.

The presence of diamino acids in Murchison was confirmed by comparison with external standards showing identical mass spectra and retention times. Figure 8.3 combines chromatograms of the Murchison carbonaceous chondrite sample and a simulated interstellar ice sample in which the diamino acids 4,4′-diaminoisopentanoic

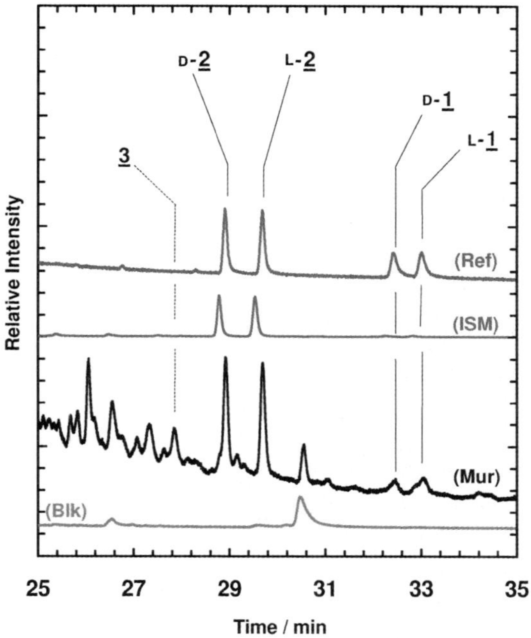

Fig. 8.3 Treasure trove of amino acids: Gas chromatogram of a sample of the Murchison meteorite (Mur) showing the diamino acids 4,4′-diaminoisopentanoic acid 3, D-2,3-diaminopropanoic acid D-2, L-2,3-diaminopropanoic acid L-2, D-2,4-diaminobutanoic acid D-1, and L-2,4-diaminobutanoic acid **L-1**. Detection in the single-ion monitoring mode (SIM) of $m/z =$ 175 amu, typical for amino acid ethoxy carbonyl ethyl ester derivatives (see Table 8.1). The insets show the external standard of the enantiomer separated diamino acids D,L-2,3-diaminopropanoic acid and D,L-2,4-diaminobutanoic acid (Ref) purchased from Fluka and detected in the total ion current (TIC); a laboratory sample produced by ultraviolet irradiation of circumstellar/interstellar ice analogues (ISM) detected in SIM of mass trace $m/z = 175$ amu and a serpentine blank (Blk) at mass trace $m/z = 175$ amu that had been heated before extraction for 4 h at 500°C and passed through the analytical protocol

[6] Until 2004, diamino acids had not been identified in samples of meteorites. The reason for this is probably to be found in the applied analytical procedures. Until 2004, for the analyses of meteorites mostly capillary columns of 30–50 m length were used to obtain high resolution for potential enantiomers. Capillary columns with such a performance are generally too long to elute diamino acids. Here, capillary columns of 12 m length were applied (as they are installed in the COSAC-experiment onboard the cometary ROSETTA mission), which were suitable for the elution and enantioselective resolution of diamino acids.

Table 8.1 Diamino acids identified in the Murchison meteorite

Diamino acid	MS [a.m.u.]	R_t of analyte [min]	Quantity [ppb]
2,3-Diaminobutanoic acid[a]	**203**, 157, 85	38.62	89.9 ± 6.2
D-2,3-Diaminopropanoic acid	**175**, 129, 102	28.92	49.9 ± 4.2
L-2,3-Diaminopropanoic acid	**175**, 129, 102	29.69	49.8 ± 4.2
3,3′-Diaminoisobutanoic acid	188, **142**, 115	32.58	48.6 ± 4.1
4,4′-Diaminoisopentanoic acid	**142**, 116	27.79	32.2 ± 3.3
D-2,4-Diaminobutanoic acid	**175**, 116	32.47	31.6 ± 3.3
L-2,4-Diaminobutanoic acid	**175**, 116	33.06	29.9 ± 3.2
D-Ornithine	**142**, 129	43.26	<5
L-Ornithine	**142**, 129	44.54	<5

MS mass spectrum, *a.m.u.* atomic mass units, R_t retention time, *ppb* parts per billion (ng analyte per g Murchison), numbers shown in boldface are the main mass fragments.
[a]Gas chromatogram taken in deviation from the standard protocol with a specifically optimized temperature program of the chiral capillary column

acid, D,L-2,3-diaminopropanoic acid, and D,L-2,4-diaminobutanoic acid had been identified previously (Muñoz Caro et al. 2002). Achiral 2,3-diaminobutanoic acid and 3,3′-diaminoisobutanoic acid were also detected in the carbonaceous chondrite.

Chiral diamino acids in the Murchison meteorite showed a racemic ratio as it is indicated in Table 8.1, accompanied by corresponding confidence intervals.

Attention was paid to exclude contamination with biological products. A serpentine sample, which was heated for 4 h at 500°C to entirely destroy thermolytically all eventually present amino acids and diamino acids, was taken as a blank, and subjected to the entire analytical protocol. In this context, serpentine has quite often been used as a blank sample since it possesses a sheet silicate structure with the potential to resorb amino acids and other organic compounds, like a sponge. As depicted in Fig. 8.3, no diamino acids were detected in the heated serpentine blank sample. These data show that the identified meteoritic diamino acids were not the result of contamination in the laboratory.

Diamino acids were quantified by comparison of the peak areas with external standards. The concentrations of the observed diamino acids were lower than those of monoamino acids in the Murchison carbonaceous chondrite (Cronin and Chang 1993; Cronin 1998). In general, the abundances of amino acids in carbonaceous chondrites decrease logarithmically with increasing carbon number. In the reported analyses of meteoritic material, the absolute abundance of amino acids generally is enhanced by previous 6 molar hydrochloric acid hydrolysis (see also Chap. 7; Cronin 1976). The concentration of free diamino acids in the Murchison meteorite was also observed to increase after hydrolysis by factors between 1.7 and 6.0 relative to the extraction with hot water at 100°C. After cold-water extraction, no diamino acids were detected. These data indicate that these products must have been originally formed as molecules of higher molecular mass, that release free diamino acids after hydrolysis.

For the past 40 years, the hydrolysis has been a reasonable and required step in any meteorite analysis protocol in order to make the incorporated amino acids visible. The amino acid structures including their molecular skeleton composed of carbon-carbon and carbon-nitrogen bonds are preformed in original molecules of higher molecular mass. The hydrolysis step is neither capable of forming a carbon-carbon bond nor a carbon-nitrogen bond and this step can thus not be interpreted as a chemical reaction that "produces" amino acids. However, we may seriously ask the question concerning the chemical evolution of amino acids "where and under which environmental conditions do hydrolysis steps take place during chemical evolution releasing free (and chiral) amino acids?" Today, we assume that this hydrolysis step might have occurred over long time spans either in terrestrial liquid water, or – partly – in liquid water present in meteorite parent bodies. It is estimated that the parent body of the Murchison meteorite has reached temperatures of up to approximately 300 K due to radioactive heating.[7] So do various meteorites in which fluid inclusions in salt crystals (Shock 2002) and specific mineralogical compositions show evidence of a water-rich past on their parent bodies indicating aqueous alteration at even much higher temperatures.

It is appropriate to make a comparison between the identified organic molecules such as amino acids and diamino acids in meteorites, and amino acids and diamino acids obtained from UV-irradiation of circumstellar/interstellar ice analogues (Chap. 7). Such a comparison, however, is not necessarily straightforward, because interstellar ices are believed to be pristine while carbonaceous chondrites experienced a different history and underwent – possibly frequent – alterations through aqueous processing, thermal effects, or shock, which all may have affected their original composition (Kerridge 1999; Schwartz and Chang 2002). The Murchison meteorite is classified as a CM2-type chondrite, a type that is characterised by minimal alteration relative to other chondrites studied for organic components (Cronin and Chang 1993). The diamino acids described above have previously been identified as refractory products of photo- and thermal processing of circumstellar/interstellar ice analogues in the laboratory (Muñoz Caro et al. 2002), and thus point to a correlation between the organic matter that might be present on icy grain mantles residing in star-forming regions (and hence likely also in circumstellar disks) and carbonaceous chondrites. The exact pathways of formation of the amino acids and diamino acids in carbonaceous chondrites are hitherto unknown. Besides the well studied mono-α-amino acids, β-amino acids (Engel and Macko 2001; Ehrenfreund et al. 2001b), and diamino acids have been identified in both laboratory simulations of circumstellar/interstellar ice that was photolytically and thermally processed (Chap. 7), and – as presented here – in carbonaceous chondrites. The latter organic compounds could not have been formed by the known Strecker-cyanohydrin synthesis (Cronin 1998). The results suggest an alternative formation mechanism, namely, that amino acids and diamino acids as refractory products of experimental simulations of ice photoprocessing in circumstellar regions and in carbonaceous chondrites are synthesized photochemically via complex recombination processes of radicals.

[7] Personal communication with Rainer Merk from Tel Aviv University in Mai 2007.

Why are diamino acids of immense interest for studies on the origin and the evolution of life on Earth? The function of diamino acids in the prebiotic development of polypeptide β-structures was intensively studied by Brack and Orgel (1975) and Brack (1987, 1993). More recently, it became apparent that diamino acids might have contributed to the origin of life's genetic material as well.

8.2 Diamino Acids as Gene Trigger

The delivery of organic compounds by meteorites, interplanetary dust particles, or by comets (Oró 1961, Jessberger 1999) to the early Earth is one mechanism thought to have triggered the appearance of life on Earth (Chyba and Sagan 1992; Ehrenfreund 1999). Particularly the prebiotic formation of peptides and proteins on Earth by the use of amino acid monomers produced asymmetrically by interstellar photochemical processes is considered as an elegant model. However, to refer to one of the initial questions of this chapter, there is so far much less conclusive evidence on the origin of genes and the early development of genetic DNA material and its analogues.

In the course of the evolution of genetic material, it is widely accepted that today's "DNA-RNA-protein world" was preceded by a prebiotic system in which RNA oligomers functioned both as genetic materials and as enzyme-like catalysts (Gilbert 1986). RNA itself remains difficult to synthesize under prebiotic conditions. Furthermore, it is unstable in aqueous solution and was thus proposed to have been preceded by pre-RNA genetic material. Hitherto, the molecular structure of life's first genetic material is not understood and hence various world-leading research groups put much effort in elucidating its structure. This remains a rather challenging endeavour since life's first genetic material is probably no more present on Earth today. Even in biological organisms it might remain difficult – if not impossible – to detect its molecular relicts. Nevertheless, structural analogues of DNA and RNA can be synthesized in the laboratory and their solubility and stability as well as the capability of selected sequences to combine with DNA and RNA strands to form duplex structures can be studied including investigations of the physico-chemical properties of the duplexes.

After years of intensive international research, numerous molecular candidates have been proposed to serve as pre-RNA oligonucleotides. We will start our guided tour through all these candidates by having a closer look at one of the most intriguing structures, namely the peptide nucleic acid PNA. Peter Nielsen from the University of Copenhagen introduced PNA structures (Nielsen et al. 1991; Nielsen 1993). PNA is composed – as common oligonucleotides are as well – of a backbone of a regular pattern to which a sequence of the nucleotide bases adenine, guanine, cytosine, thymine, and uracil can be attached via appropriate spacer groups, eventually carrying life's first genetic code (Fig. 8.4). Two principal alternatives exist for the molecular architecture of the PNA backbone, letting us a priori distinguish between aegPNA and daPNA:

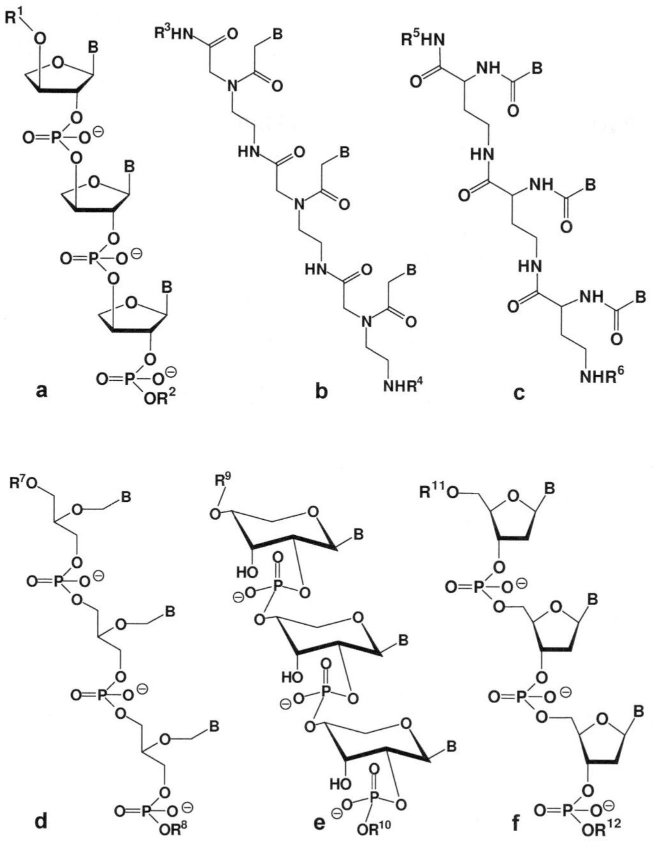

Fig. 8.4 Chemical structures of candidate precursors to RNA during the early history of life on Earth. **a**, Threose nucleic acid (TNA); **b**, peptide nucleic acid based on N-(2-aminoethyl)glycine monomers (aegPNA); **c**, peptide nucleic acid based on 2,4-diaminobutanoic acid monomers (daPNA); **d**, glycerol-derived nucleic acid analogue; **e**, D-β-ribopyranosyl-($2' \rightarrow 4'$) oligonucleotide (p-RNA); **f**, ribonucleic acid (RNA). **B**, Nucleotide base

1. The PNA backbone is composed of N-(2-aminoethyl)glycine (aeg) monomers leading to aegPNA molecules. The monomer aeg was shown to be produced directly in electric discharge reactions from CH_4, N_2, NH_3, and H_2O (Nelson et al. 2000).
2. The PNA backbone is composed of diamino acids (da) as monomers leading to daPNA structures. Various structural analogues of diamino acids are known to form daPNA.

Different daPNA molecular structures are known to serve as DNA/RNA analogues. They consist of peptide backbones made of diamino carboxylic acids to which nucleotide bases are attached for example via carbonyl methylene linkers. Interestingly, the PNA backbone can be produced from diamino carboxylic acids by simple condensation reactions (Nielsen 1993)! The diamino acids 2,4-diaminobutanoic acid, and ornithine, which we detected in the Murchison meteorite as racemic pairs

(Meierhenrich et al. 2004), had been suggested to be essential constituents of PNA monomers (Nielsen 1993). The first identification of 6 molecular structures of diamino carboxylic acids in simulated interstellar ice analogues is reported in Chap. 7 and Muñoz Caro et al. (2002). A racemic mixture of 2,3-diaminopropanoic acid was synthesized by spark discharge experiments (Engel et al. 1995). Taken together, these results on the quasi omnipresence of diamino acids in relevant samples for the study of the origin of life might stimulate future research activities on appropriate reaction pathways from diamino carboxylic acids resulting in a variety of different PNA structures (Meierhenrich et al. 2002a). These data might then hold the key to fill the gap in our understanding between chemical evolution and biological evolution.

Besides peptide nucleic acids, other systems of oligonucleotides have been investigated and were proposed to have preceded RNA, just as RNA itself preceded DNA and protein. A systematic investigation of potentially natural nucleic acid analogues has led to the recognition of four candidate precursors to RNA.

1. The molecular structure of L-α-threo-furanosyl-(3′→2′) nucleic acid (TNA) is based on L-α-threo-furanosyl units joined by 3′,2′-phosphodiester linkages (Fig. 8.4 a) and able to form stable Watson-Crick base pairs with itself and with RNA (Fig. 8.4 f). In chemical evolution, TNA molecules are assumed to be more advantageous than RNA because of their relative chemical simplicity (Schöning et al. 2000).
2. In the uncharged peptide nucleic acid (PNA) the sugar-phosphate backbone of the RNA genome is replaced by a backbone composed of N-(2-aminoethyl)glycine units held together by amide bonds (Fig. 8.4 b). Nucleic acid bases are attached via methylenecarbonyl spacers to the PNA structure. PNA polymers of N-(2-aminoethyl)glycine form stable double helices with complementary molecules of RNA (Egholm et al. 1992, 1993b; Nelson et al. 2000). In template-directed reactions (Schmidt et al. 1997a, 1997b) information can be transferred from aegPNA to RNA, and vice versa, and aegPNA-DNA "chimeras" form readily on either DNA or aegPNA templates (Koppitz et al. 1998). The polycondensation reaction of diamino acids like 2,4-diaminobutanoic acid or ornithine yields daPNA structures of a corresponding constitution (Fig. 8.4 c), which also serve as potential pre-RNA oligonucleotides. The structural monomer 2,4-diaminobutanoic acid is clearly visible in the molecular daPNA structure given in Fig. 8.4.
3. Glycerol-derived nucleic acid analogues (Fig. 8.4 d) were proposed for pre-RNA self-replicating systems (Spach 1984; Joyce et al. 1987; Chaput and Switzer, 2000), although sufficient experimental support is missing to consider them as a strong candidate (Joyce 2002).
4. Members of the family of pentapyranosyl-(2′→4′) nucleic acids (p-RNA, Fig. 8.4 e) were systematically synthesized in the laboratories of Albert Eschenmoser for the comparison with RNA with respect to those chemical properties that are fundamental to RNA's biological function (Eschenmoser 1997; Bolli et al. 1997). In these studies, the internal base-pairing-strength of the pentapyranosyl oligonucleotide systems was found to be higher than that of the corresponding pentafuranosyl RNA system (Beier et al. 1999).

The detection of diamino acids in samples of the Murchison meteorite enabled to explore new routes to interpreting chemical evolution. The results stated that (a) amino acid structures, the molecular building blocks of proteins and (b) molecular building blocks of pre-RNA oligonucleotides are present in interplanetary, interstellar, and circumstellar samples. The results are new clues for the assumption that organic ingredients of living systems have been delivered via (micro-) meteorites and comets to the early Earth from regions of the interstellar medium. After the transport these molecules were potentially participating in initial prebiotic reactions, which turned out to be of central importance for the origin of life on Earth.

The results would suggest that both early proteins and early genetic material were synthesized from molecular building blocks that had been delivered from interstellar/circumstellar space to the early Earth (see also Strasdeit 2005). We have outlined some guesses that life's genetic material might have arisen originally starting with the formation of diamino acids under interstellar condition followed by the formation of peptide nucleic acid structures by polycondensation reaction. The subsequent formation of RNA might have triggered the origin of today's DNA/protein world. The reader should be aware that this proposal on the evolutionary origin of DNA is only one hypothesis amongst others. It is not the standard model of the scientific community for the origin of the genetic material on Earth. Until now, such a standard model does not exist.

8.3 Survival of Organic Molecules After Impact on Earth

Often, the question arises whether organic molecules and particularly amino acids embedded in comets and/or meteoroids would survive an impact on Earth from outer space. For comets, the Greenberg-model suggested that organic molecules would at least partly survive such an impact (Greenberg et al. 1994). Greenberg expected that the survivability of cometary organic molecules depends on both the size of the comet and the scale height of the atmosphere as well as the comet density and morphological structure. Since comets are made up of aggregated interstellar dust particles (see next chapter) the heating of an impact would be concentrated at the surface because of their exceedingly low thermal conductivity. This would lead to rapid water (ice) evaporation and further fragmentation. This sequence would lead to smaller and smaller fragments whose deceleration in the atmosphere is progressively less and whose heating is consequently smaller. The general result is then, that a comet, rather than impacting as a single body with high kinetic energy, breaks up into smaller and smaller components, each of which is subjected to lower degrees of heating. The dissipation of the energy by impulsive evaporation of the volatiles can consequently lead to the survival of the refractory organic molecules during atmospheric entry (Greenberg 1993).

More recently, the hypothesis of the survival of organic molecules during atmospheric entry was strengthened for the case of meteorites by the STONE experiment. This experiment deserves attention, however at present it is not very well known:

An "artificial meteorite" was designed in order to investigate physical modifications to sedimentary rocks and chemical modifications to their organic ingredients during terrestrial atmospheric entry. Three different samples (basalt, dolostone, and an artificial rock simulating Martian soil) were doted with amino acids and fixed onto the heat shield of the recoverable Foton-12 spacecraft. The spacecraft was launched, taken in low Earth orbit followed by the atmospheric re-entry and recovery of the capsule. Since the basalt sample was lost during entry, mineralogical and isotopic studies were performed with the partly recovered dolostone and the simulated Martian regolith (Brack et al. 2002). The results indicated that intact organic molecules such as amino acids could be recovered from the artificial meteorite.

The delivery of organic molecules from interplanetary space to Earth might also have occurred via micrometeorites (Maurette 2006). These micrometeorites have a typical size of 20–500 μm and can be collected from deep-sea sediments, terrestrial sand, sedimentary rocks, Greenland lake sediments, and by melting large amounts of Antarctic ice and snow (Duprat et al. 2007). Micrometeorites are even assumed to be the dominant mass fraction of extraterrestrial material delivered to Earth today bringing to the Earth about 100 times more material than objects found outside this size range, including the much larger meteorites (Brinton et al. 1998). Enantioselective analysis of a collection of micrometeorites revealed that they include terrestrial L-amino acid components making conclusions on eventual original enantiomeric excesses difficult. In one sample, the amino acid α-aminoisobutyric acid AIB was found to be present at a level significantly above the background blanks (Brinton et al. 1998). These data let us conclude that amino acids – at least in part – would survive an impact on Earth via micrometeorites.

8.4 Space Exploration and Chirality: What Next?

In future investigations the stable isotope compositions of hydrogen, nitrogen, and carbon in the carbonaceous chondrite's diamino acids need to be determined. The $\delta^{2}H$, $\delta^{15}N$, and $\delta^{13}C$ values of amino acids in Murchison generally lie outside the ranges of organic matter on Earth (Engel and Macko 2001; Pillinger 1984). A similar shift of the isotopic composition is expected for the diamino acids in the Murchison meteorite. Moreover, different constitutional isomers of diamino acids might be determined in other carbonaceous chondrites in the laboratory, in cometary samples by the GC-MS instrumentation onboard the ROSETTA mission and possibly by measurements to be performed by Mars mission ExoMars in the coming years (see Chap. 9). In addition, the synthetic, bio-organic chemical pathways of PNA structures, that start with the diamino acids like those reported here as present in meteorites for the first time, will be studied more intensely in order to elucidate possible origins of life's genetic material.

Chemical and analytical investigations of meteorites will most likely continue to attract the scientific interest in future times, since meteorites that originate from extraterrestrial environments contain molecular ingredients that can be assumed to have been present when life formed on the early Earth. On terrestrial samples, however, no such molecular relicts can be identified today since here the environment is 'contaminated' with life, throughout effacing information on its origin. One can thus hope for future experiments that the advancement in the development of analytical techniques might allow the identification of new families of organic compounds in carbonaceous chondrites; compounds that might be interesting in the context of the origin of life and its genome. Separation of organic molecules based on multidimensional GC×GC techniques and their identification with powerful time-of-flight mass spectrometers or modern UPLC instruments in liquid chromatography will be applied to the analysis of meteorites in the future. These and other techniques might provide us with structural elucidation of the 'organic polymer' present in meteorites and in the end enables us to better understand the origin of life on Earth including its chiral asymmetry.

Meanwhile, other research teams might trace back biological evolution by, e.g., comparing RNA-sequences and advance in the understanding of the molecular structure of first self-replicating systems on Earth. Thereby, the gap in our understanding between chemical and biological evolution should become ever smaller.

Chapter 9
The New Space Race: Chiral Molecules on Comets and on Mars

9.1 In Search of Chiral Molecules in Comets

Comets are fascinating celestial objects. But they are much more than that. It is probable that comets, to a large extent, consist of pristine interstellar material (Greenberg 1982; Irvine et al. 2000).[1] This material could have been delivered to the early Earth by comet dust, as well as asteroids, and interplanetary dust particles, during the epoch of heavy bombardment (Oró 1961; Chyba and Sagan 1992). Comets were thus discussed to hold the key of the origin of life on Earth and they may have played a crucial role in the origin of biomolecular asymmetry. It is therefore of importance to understand the chemical composition of comets and particularly the stereochemistry of cometary organic ingredients more in detail. The determination of enantiomeric ratios in the matter of such extraterrestrial bodies in situ is long overdue (Thiemann 1975; Brack and Spach 1986). The cometary sampling and composition experiment (COSAC) is part of the payload of the Lander Philae in ESA's cometary mission ROSETTA (Biele and Ulamec 2008). It will provide the first opportunity of this kind when it begins its measurements on the nucleus of comet 67P/Churyumov-Gerasimenko after its touchdown in November 2014 (Glassmeier et al. 2007). Previous to the discussion of the 'Chirality-Module' of this mission, we will have to understand: *What is a comet*? and *how does it form itself*?

[1] As we will see later in this chapter, at least some comets are not so pristine as the Greenberg-model proposed. This is what recent results from cometary missions tell us. The presence of crystalline silicates, for example, inferred from astronomical observations (Bregman et al. 1987; Hanner 1999), and minerals formed at high temperatures present in the dust of comet Wild 2 (Brownlee et al. 2006) suggest that at least some comets are not that pristine as previously thought. Such minerals probably formed in the hot inner regions of the solar nebula. Comets likely contain a mixture of interstellar and solar nebular materials.

9.1.1 Comet Formation: Theory and Experimental Results

Comets show a visible coma (a dust hull) and a nucleus, invisible from Earth and Earth based telescopes. In 1986, the cometary mission GIOTTO took the first detailed image of a cometary nucleus (Fig. 9.1). Our knowledge of the composition of cometary nuclei is predominantly based on the evaporation of volatile species and thermal emissions from siliceous and carbonaceous dust when bright comets approach the sun. Sometimes comets such as 73P/Schwassmann-Wachmann 3 split into fragments and offer a look onto fresh material from the comet's interior (Dello Russo et al. 2007). From photospectrometric measurements of this kind, we know today that comets are made of silicates ($\sim$25%), organic refractory material ($\sim$25%), and $\sim$50% of water ice with small admixtures of other ice species. Comets are assumed to include 33 to 50 % of carbonaceous material (Lisse et al. 2006). Moreover, comets can be observed from Earth when they are far away from the sun and therefore inactive. Spectroscopic methods already allowed the identification of many elements and specific compounds in comets' comae, including hydrogen, carbon, nitrogen, oxygen, sulphur, water, carbon monoxide, carbon dioxide, and formaldehyde (A'Hearn and Festou 1990), methanol (Greenberg et al. 1994), and furthermore hydrogen sulphide, sulphur dioxide, methane, ethane, formic acid, ammonia, and acetonitrile. Polycyclic aromatic hydrocarbons, water vapour and ice, and sulphides were found by spectral observations after NASA's Discovery mission Deep Impact had sent an impactor into the nucleus of comet 9P/Tempel 1 (Lisse et al. 2006)

The nucleus of a comet is thought to be built up of aggregates of interstellar dust particles (Greenberg 1993). Interstellar dust particles are silicate grains surrounded by a mantle of H_2O, CO_2, CO, and other ices, which may serve as a matrix for many kinds of atoms and molecules (Fig. 9.2). Due to the results of laboratory simulation experiments, we have reason to assume that specific chiral organic molecules like amino acids can be found in cometary matter too (Muñoz Caro et al. 2002; Bernstein et al. 2002).

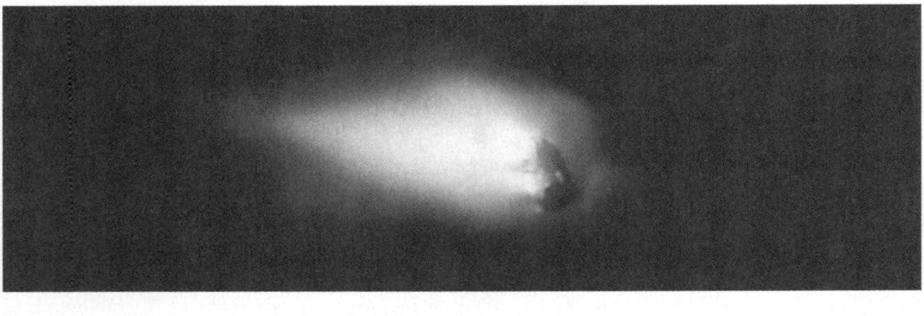

Fig. 9.1 Comet Halley pictured by the Halley Multicolour Camera onboard ESA's GIOTTO spacecraft. Coma and nucleus are visible. The ROSETTA mission's target is the nucleus of comet Churyumov-Gerasimenko. The ROSETTA mission was launched on 2 March 2004 by the Ariane 5 launcher. "Chury" will be reached by May 2014, the robotical lander Philae will then detach from the orbiter and start the landing manoeuvre. Credit: Max Planck Institute for Solar System Research, Katlenburg-Lindau, Germany. Image courtesy: Dr. H. U. Keller

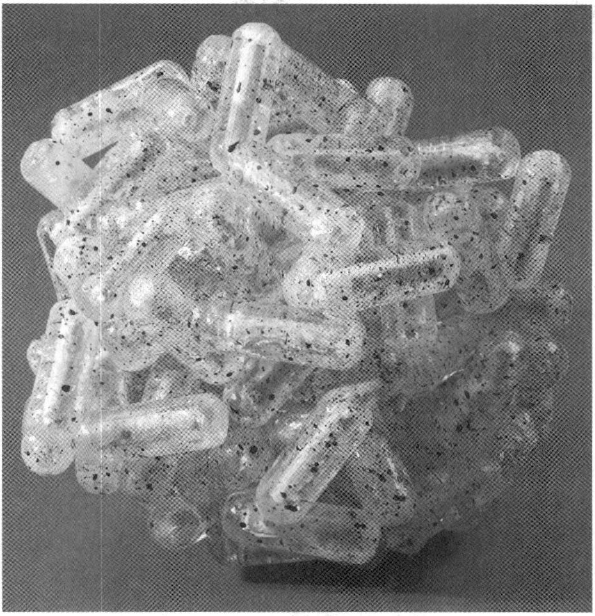

Fig. 9.2 A piece of a fluffy comet manufactured by the Greenberg research team at the Leiden Observatory for Astrophysics, the Netherlands: Model of an aggregate of 100 average interstellar dust particles. Each particle consists of a silicate core, an organic refractory inner mantle, and an outer mantle of predominantly water ice in which are embedded the numerous very small ($\leq 0.01\,\mu m$) particles responsible for the ultraviolet 216 nm absorption and the far ultraviolet extinction. Each particle as represented corresponds, in reality, to a size distribution of thicknesses starting from zero. The packing factor of the particles is about 0.2 (80% empty space) and leads to a mean comet mass density of $0.28\,g\,cm^{-3}$. Once the ices removed, the dust mass density is $\approx 0.1\,g\,cm^{-3}$ and the porosity is ≈ 0.96 (Greenberg 1996). The overall dimension is scaled to about $4\,\mu m$ (Greenberg 1986)

9.1.2 ROSETTA: A Cometary Mission Full of Answers

After the remarkable success of the cometary GIOTTO mission in 1986, the European Space Agency (ESA) directed main parts of its activities on cometary research and aimed – for the very first time – to create a mission capable of landing on a cometary nucleus. Such landing – so the assumption – would open possibilities for picturing in previously unknown detail, as well as doing chemical and physical analysis of cometary matter in hitherto unknown precision. Consequently, the cometary mission ROSETTA was designed and constructed and became ESA's cornerstone mission with an important budget of about 1 billion Euros.

Originally, 600 m radius comet 46P/Wirtanen (Jorda and Rickman 1995) was selected as a target comet of mission ROSETTA. Due to substantial problems with the Ariane 5 launcher at the beginning of 2003, resulting in a one-year launch delay, another comet had to be selected. Finally, the ROSETTA mission was successfully launched in March 2004 targeting the considerably larger comet 67P/

Churyumov-Gerasimenko. The radius of this comet's nucleus was estimated with the help of data from Hubble Space Telescope to be between 2 000 m (Lamy et al. 2003) and 4 740 m (Lamy et al. 2007).

The ROSETTA spacecraft carries its small subsatellite, the ROSETTA Lander Philae (Ulamec et al. 1997; Bibring et al. 2007), to be detached from the orbiter and set down on the surface of the comet's nucleus at about 3 astronomical units (AU) from the sun with near-zero gravity and negligible gaseous atmosphere (Fig. 9.3). Philae was constructed under scientific guidance of Helmut Rosenbauer from the Max Planck Institute for Solar System Research in Katlenburg-Lindau, Germany.

One of the scientific interests of the ROSETTA Mission is to find out whether chiral organic compounds, such as amino acids in cometary matter brought to the Earth by cometary impacts might have had, due to corresponding enantiomeric excesses, a seed function in determining the handedness which is characteristic of homochiral compounds employed by life on Earth. For this reason, we have developed an experiment for the ROSETTA mission, named Cometary Sampling and Composition Experiment (COSAC). The COSAC experiment was installed on the ROSETTA Lander Philae and its Chirality-Module is dedicated to the enantioselective gas chromatographic and mass spectrometric (GC-MS) analysis of chiral organic constituents of the cometary nucleus (Goesmann et al. 2007b).

Fig. 9.3 This artist's impression shows the ROSETTA Lander Philae anchored to the comet's surface. Philae will work for a minimum mission target of 65 hours, but its operations may continue for many months. The Lander's structure consists of a baseplate, an instrument platform, and a polygonal sandwich construction, all made of carbon fibre. Some of the instruments and subsystems are beneath a hood, which is covered with solar cells. An antenna transmits data from the surface to Earth via the orbiter. The lander carries nine experiments, with a total mass of 21 kilograms. It also carries a drilling system to take samples of subsurface material. Credits: ESA/AOES Medialab

9.1.3 The Chirality-Module of Mission ROSETTA

The simulation experiments of interstellar environments presented in Chap. 7 strongly suggest that chiral amino acid structures are present in both the surface of interstellar dust particles and in comets, because comets are – in a first approximation – aggregates of interstellar dust particles. In order to resolve and quantify chiral organic molecules in cometary matter in situ, dedicated enantiomer-separating chromatography will be used.

The cometary sample will be filled by the Lander's Sample Drill and Distribution Subsystem (Ercoli Finzi et al. 2007) in small ovens, mounted on a "carrousel", and moved by rotation, from a position where the sample will be filled in, to the so-called "tapping station", where the ovens are going to be closed and heated for obtaining material in gaseous phase. The ovens will be heated stepwise to levels programmable by ground commands. The pressure developed during each heating step will be measured and recorded, too, because it is indicative of the total amount of gas released (Rosenbauer et al. 1999, Goesmann et al. 2007b). To volatilize the organic compounds sampled on the cometary nucleus, where macromolecules and complex organic polymers of low volatility are expected to make up a major part of organic matter, the combination of two pre-processing techniques will be applied:

1. Evaporation and pyrolysis by stepwise heating of the cometary samples in the ovens of the pyrolysis section, and
2. Chemical derivatization, in which a reagent transforms the parent compound by esterification into a volatile form to be analyzed by gas chromatography.

The analytes in the gaseous phase will be injected into capillary columns coated with chiral as well as non-chiral stationary phases. The chromatographically separated analytes are going to be detected, identified, and quantified on miniature thermal-conductivity detectors, which are coupled with a multi-reflectron TOF mass-spectrometer as schematically illustrated in Fig. 9.4.

Further technical descriptions of the instrument including a test gas chromatogram of an argon, butane, methanol, isopropanol mixture in helium, and mass spectra of a H_2O/D_2O mixture and perfluorotributylamine can be found in Goesmann et al. (2007b). Here, the interested reader will also find a description of the international COSAC project team including individual contributions under scientific guidance of the Max Planck Institute for Solar System Research in Katlenburg-Lindau, Germany.

ROSETTA's capillary columns coated with chirally active liquid films received considerable attention, because they were designed for the resolution of non-complex enantiomers to allow the determination of enantiomeric ratios of cometary chiral organic compounds (Fig. 9.5). Consequently, they might provide crucial information about the origin of molecular parity violation in biomolecules (Meierhenrich et al. 1999, 2001b, 2001d, 2003a; Thiemann and Meierhenrich 2001; Szopa et al. 2002; Meierhenrich 2002; Goesmann et al. 2007b).

I would like to point out, that ROSETTA's Chirality-Module is not designed to resolve one or two specific families of chiral organic molecules. It is rather planned to

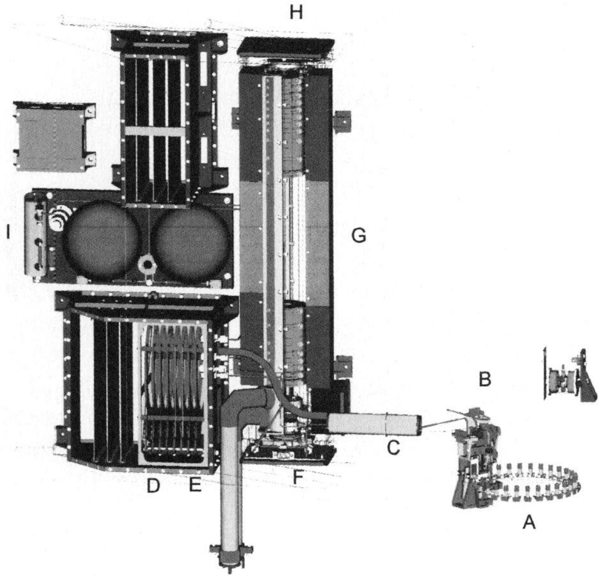

Fig. 9.4 Schematic view of the newly developed COSAC-instrument onboard ROSETTA Lander Philae. The cometary sample will be filled into pyrolysis ovens mounted on carrousel **A**. After measurement of an infrared spectrum of the cometary sample, the selected oven will rotate into tapping station **B** where the oven is closed and heated stepwise. Volatile organic compounds will evaporate and be pressure-controlled transferred via transfer line **C** into the gas chromatograph **D**, which consists of eight capillary columns working in parallel. The gas chromatographically separated organic compounds and enantiomers will be detected by thermo-conductivity detectors TCDs **E** and given into the inlet system **F** of the linear reflectron time-of-flight mass spectrometer rTOF-MS **G**, ionized by electron impact, and analyzed after their time-of-flight with a multi-spherical-plate detector **H**. The obtained data will be sent to Earth. Carrier gas and calibration gas are stored in **I**. With the help of the COSAC-instrument, enantiomers of organic molecules can be identified, because they will be separated with chiral stationary phases and show identical mass spectra in the TOF mass spectrometer. Credit: Max Planck Institute for Solar System Research, Katlenburg-Lindau

resolve a wide variety of different classes of chiral compounds. This can be done by gas chromatography, using different stationary phases in which chiral molecules like cyclodextrines are embedded. Chiral stationary phases suitable to perform enantioselective gas chromatography and applicable to a wide range of chiral organic compounds including alcohols, dioles, amines, amino acids, hydroxy carboxylic acids, and even hydrocarbons were therefore developed and investigated in the laboratory before sending them incorporated in ROSETTA's COSAC instrument toward the comet.

After thorough tests in practice, three types of chiral stationary phases with different polarity and ring size (β- and γ-cyclodextrins) seem to give an optimum performance for this kind of problem and were selected to be send to comet 67P/Churyumov-Gerasimenko: A Chirasil-L-Val phase (for the chemical structure confer Chap. 2), a cyclodextrin G-TA phase (Fig. 9.6), and a Chirasil-Dex CB phase (chemical structure in Chap. 2).

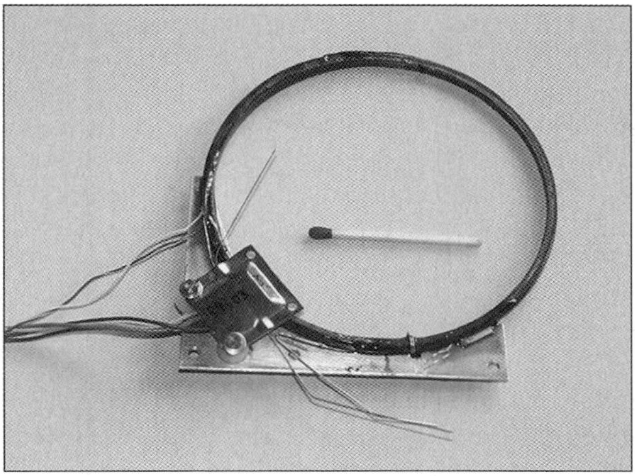

Fig. 9.5 COSAC's capillary column with its chiral stationary phase including heater, temperature sensor, and thermo-conductivity detector. Eight such subunits form the set of COSAC's chromatographic columns. Credit: Fred Goesmann, Max Planck Institute for Solar System Research, Katlenburg-Lindau

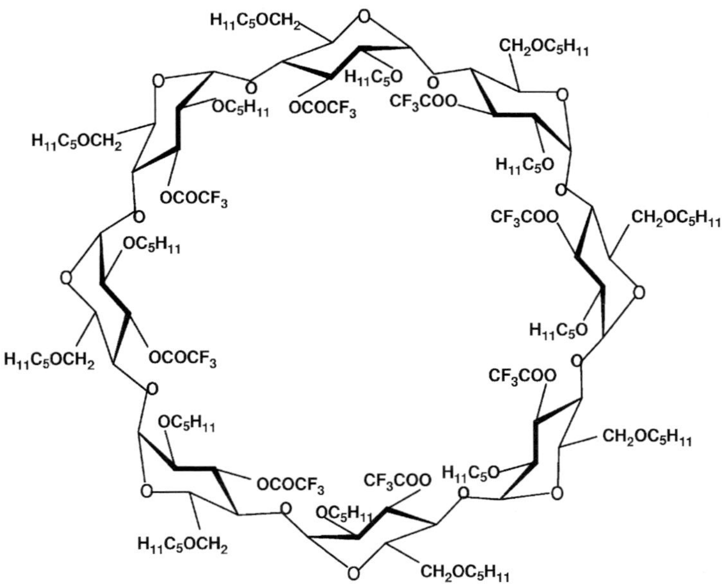

Fig. 9.6 Chemical structure of the cyclodextrin G-TA stationary GC phase, selected for the COSAC instrument onboard the ROSETTA Lander Philae

Why were these columns chosen for the ROSETTA mission among a wide variety of rival chiral selectors? The Chirasil-L-Val phase was selected for the resolution of chiral amino acids and carboxylic acids, compounds able to build two hydrogen bridge bonds to the stationary polymer phase. Enantiomers with only one functional group cannot be resolved by this phase. Therefore, the Cyclodextrin G-TA phase will be used for the resolution of chiral alcohols and diols. The G-TA phase is less stable, since the γ-cyclodextrin molecules are not chemically bonded to the stationary polymer. Finally, the Chirasil-Dex CB phase will be applied for the separation of chiral hydrocarbons. Characteristics of the chosen chiral GC selectors for the ROSETTA mission are summarized in Table 9.1. More detailed information on the resolution of enantiomers including test data can be found in the Annex.

After ROSETTA's Lander Philae touchdown on the cometary surface in November 2014, the gas chromatographic in situ resolution of enantiomers is designed to allow the quantification of specific enantiomers of cometary material and thus the determination of enantiomeric excesses which, should they be found, might be helpful to understand the phenomenon of enantiomeric asymmetry of biomolecules on Earth.

Several enantiomers were selected to be investigated as potential candidates for cometary organic constituents. Chemical compounds in general contain asymmetric centres increasing with the number of carbon atoms. Below a certain carbon number, there are no isomers with stereogenic carbon atoms, i.e. no enantiomers. As for the case of (a) meteorites and (b) laboratory experiments generating hydrocarbons, the absolute (and relative) quantity of each specific isomer decreases exponentially with growing number of carbon atoms. The ability to analyse the individual enantiomer vanishes accordingly above a critical number of carbon atoms, because the quantity of the individual compound would ultimately drop below the detection limit. Consequently a "detection optimum" for chiral cometary molecules is imposed by the intersection of a minimal number of carbon atoms providing a sufficient abundance and a sufficient number of carbon atoms to generate a stereogenic center.

In accordance with the above limitations, the following families of chiral organic molecules, as they are: chiral amino acids, chiral carboxylic acids, chiral macromolecules, chiral hydrocarbons, and chiral amines, alcohols, and diols (see Annex) were selected as candidates for their gas chromatographic resolution. In a first attempt, we tried to achieve this goal without a derivatization step, because performing such a chemical procedure on a cometary nucleus would pose considerable additional problems. Later on, non-volatile carboxylic acids and amino

Table 9.1 Chiral stationary GC phases onboard the ROSETTA Lander Philae. The total set of columns used in the COSAC gas chromatograph including the achiral stationary phases is given in Goesmann et al. (2007b)

Chiral stationary phase	*L* [m]	*ID* [mm]	*LT* [μm]	Targeted chiral analytes
Chirasil-L-Val	12.5	0.25	0.120	Amino acids, carboxylic acids, amides
Cyclodextrin G-TA	10.0	0.25	0.125	Alcohols, diols, amines
Chirasil-Dex CB	10.0	0.25	0.250	Hydrocarbons

L length, *ID* inner diameter, *LT* layer thickness

acids were investigated with a derivatization step prior to gas chromatographic analysis.

9.1.3.1 Resolution of Chiral Amino Acids

The identification and resolution of chiral amino acids are probably the most fascinating targets for ROSETTA's chirality-module. However, we have to consider that D,L-amino acids are too polar to get direct access to their analysis by gas chromatography. D,L-Amino acids are affected with insufficient gas chromatographic properties. The volatility of these zwitterionic compounds is too low.

Compared to amino acids, esters are non-polar, much more volatile, and do gas chromatograph well. Esterification is therefore a good choice for a derivatization of amino acids, particularly for the intended gas chromatography on the comet's nucleus. As requested by ESA, solvent chemistry had to be avoided, because the transformation of polar organic compounds into derivatives suited for chromatographic resolution has to be performed right on the surface of a cometary nucleus under near zero-gravity. Conventional methods of esterification include reaction with diazomethane or higher diazoalkanes or esterification with methanol-acid mixtures. Diazomethane is unstable, toxic, and explosive, and the transesterification step requires solution chemistry. These properties excluded conventional methods from the ROSETTA-COSAC project.

There was thus a need for a reliable working "dry" method for the preparation of amino acid esters on COSAC's gas chromatograph. Dimethylformamide dimethylacetal DMF-DMA was selected being a suitable gas phase derivatization reagent that is also assisting the pyrolysis procedure. Sufficient DMF-DMA was inserted into special alloy capsules that melt at 100°C. These capsules were inserted into half of the COSAC ovens in order to release DMF-DMA when the required gas phase temperature for the derivatization reaction is reached. Amino acids will be transformed into their dimethylamino methyl esters as it is given in Fig. 9.7 for the example of D,L-valine. The elegance of this technique is, that the whole derivatization procedure will be automatically achieved within the oven/injector system of the COSAC gas chromatograph and sweep the volatile derivatives directly into the capillary column. This fast derivatization technique can be assumed to provide advantages also for Earth-based analytical laboratories, where it may find an increasing number of useful applications.

9.1.3.2 Resolution of Chiral Carboxylic Acids and Macromolecules

Similar to D,L-amino acids, D,L-hydroxycarboxylic acids exhibit a low vapour pressure, because of their high dipole moment and polar character. In a GC injector, they are slow-evaporating and sweep rarely into the analytical column. Apart from this, underivatized D,L-hydroxycarboxylic acids tend to 'tail' because of associative effects with the stationary phase. Compared with the parent D,L-hydroxycarboxylic acids, their alkyl esters are less polar, much more volatile, and well resolved by gas

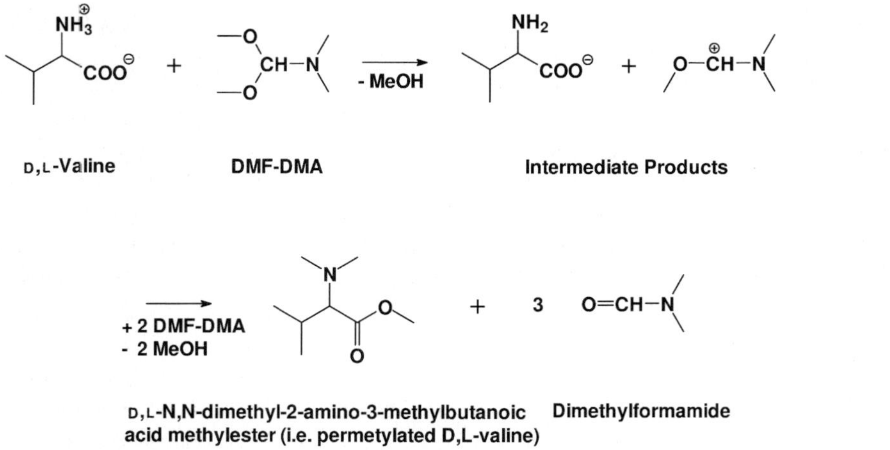

Fig. 9.7 Chemical equation depicting the pyrolysis assisting reagent dimethylformamide dimethylacetal DMF-DMA permethylating non-volatile amino acids like D,L-valine. Permethylated amino acids are volatile and thus "visible" for the COSAC GC-MS instrument

chromatography. Esterification was therefore the first choice for derivatization also in the case of D,L-hydroxycarboxylic acids.

Just like for amino acids, DMF-DMA will be used as a methylating reagent in the heated insert zone directly before their gas chromatographic enantioseparation. Results show that the D,L-lactic, D,L-malic, D,L-mandelic, and D,L-tartaric acids or mixtures thereof were transformed into their methyl esters very fast. These D,L-methylesters could easily be resolved into their enantiomers by gas chromatography employing chiral selectors embedded in stationary phases (Meierhenrich et al. 2001c).

Besides the derivatization of highly polar analytes like D,L-amino acids and D,L-hydroxycarboxylic acids, we have to consider that the organic compounds expected to be abundant on comets might not occur as monomer units. Matthews (1992) expected that macromolecules could be major components of cometary matter, Krueger and Kissel (1989), Kobayashi et al. (1998), and Takano et al. (2007) similarly suggested that complex organic oligomers would make up the main cometary ingredients. We assume that these macromolecules and oligomers would break up into their subunits during pyrolysis with DMF-DMA, because McKinney et al. (1995) pointed out that similar working reagents do not only methylate polar pyrolysis products, but could assist in bond cleavage reactions, too. In lack of samples of cometary matter, the described derivatization experiments were successfully tested with external standards.

9.1.3.3 Resolution of Chiral Hydrocarbons

It is important to take into consideration the chemical and thermal stability of chiral molecular candidates to be resolved with the ROSETTA mission. Amino acids, carboxylic acids, and carbohydrates disappear comparatively rapidly in geological timescales. They do not survive as stable compounds for very long geological periods of time, in any appreciable amounts on Earth. The relative chemical and thermal stability of carbohydrates, amino acids, carotenoids, porphyrins, and hydrocarbons increases in that order. For hydrocarbons, we have experimental data ($\Delta H_{C} = 66.5$ kJ/mol and the pre-exponential factor of the Arrhenius equation $A = 10^{14}$) for the breaking of a carbon-carbon bond. The lifetime required to decrease the amount to 1/e fraction of its original value is $10^{14.5}$ years at 400 K and 10^{27} years at 300 K. For the destruction by breaking off a pair of hydrogen atoms, using $\Delta H_{H} = 63.0$ kJ/mol and $A = 10^{13}$ we can determine the lifetime as $10^{15.5}$ years at 400 K and 10^{25} years at 300 K (Calvin 1969).

Branched saturated hydrocarbon structures can be chiral. The above calculations are also valid for the stability of chiral hydrocarbons against racemization reactions, which are expected to occur mainly via hydrogen subtractions[2]. Hydrocarbons can therefore be considered to be the most stable group of cometary compounds and may be expected to retain a significant part of their original molecular structure. The detection of a number of aliphatic hydrocarbons in samples of meteorites such as CI1 and CM2 chondrites has been demonstrated (Cronin and Chang 1993) and their abundance in the interstellar medium is established. It was thus plausible to assume that hydrocarbons occur as organic ingredients in cometary matter.

The preferred method for separating racemic pairs of non-derivatized alcohols, diols, and amines has been a classical gas chromatography on cyclodextrin phases. However, enantiomers of saturated chiral hydrocarbons like 3-methylhexane and its higher homologues could not be resolved in this classical way, because they lack the functional groups necessary to undergo "intensive" diastereomeric guest interactions with the cyclodextrin host molecules.

The first quantitative resolution of a series of branched hydrocarbon enantiomers was achieved recently in our group at the Department of Physical Chemistry at the University of Bremen. Racemates of the aliphatic hydrocarbons *R,S*-3-methylheptane, *R,S*-3-methyloctane, *R,S*-4-methyloctane, *R,S*-3-methylnonane, and *R,S*-4-methylnonane were resolved into their enantiomers by enantioselective gas chromatography with new chiral stationary phases (Meierhenrich et al. 2003a). The enantiomers of *R,S*-3-methylhexane had been separated previously (König et al. 1990). Enantiomer resolution has been increased by the use of (a) a special cryostat controlled GC instrumentation[3] delivering liquid nitrogen directly into the GC-oven for low-temperature gas chromatography and (b) the Chirasil-Dex CB phase, in

[2] Further investigations are required to understand pathways of racemization reactions of chiral hydrocarbons in detail. At present, we have reason to assume that racemizations of chiral hydrocarbons are negligible for the timescale of Gyr on Earth and particularly on comets.

[3] For the low gas chromatographic temperatures we often use an Agilent cryostat, which delivers liquid nitrogen directly into the GC oven via a monitored electromagnetic valve. Under such conditions, the stationary phase in the GC capillary column stays liquid down to minus 60°C and allows chromatographic resolution particularly for highly volatile compounds. Consider that the polymer of the stationary phase in gas chromatography with wall-coated open tubular (WCOT) columns is liquid (and not solid!) interacting with analytes via partition processes and not by adsorption of the gas phase analytes on the inner surface of capillary columns.

which the cyclodextrin molecules are chemically bonded to the polysiloxane matrix. This stationary phase was used – as for ROSETTA's COSAC instrumentation – in a length of 10 m only, allowing to focus on highly volatile hydrocarbons by applying very low temperatures. With a combination of these techniques, sufficient separation factors α and resolutions R_S were obtained for the enantiomer separation of chiral hydrocarbon molecules like trialkylalkanes resulting in a base-line separation of the enantiomers.

A gas chromatogram of a selected mixture of six chiral aliphatic hydrocarbon pairs is depicted in Fig. 9.8. Racemic mixtures were injected. Peaks of equal areas were obtained for each pair of enantiomers. At present, the absolute configuration of the separated enantiomers is unknown. The enantioselective synthesis of branched hydrocarbon structures started recently in our laboratory. Furthermore, we currently try to resolve enantiomers of the deuterated R,S-[$^2H_1$, $^2H_2$, $^2H_3$]-neopentane hydrocarbon presented in Chap. 2 by enantioselective cryogenic gas chromatography.

The mechanism of formation of aliphatic hydrocarbons in the interstellar medium and in meteorites is still subject of a controversial debate. A Fischer-Tropsch type (FTT) process was suggested as well as interstellar photoreactions via alkyl radicals, and the decomposition of hydrogen cyanide polymer structures (Minard et al. 1998).

So what about the chirality of hydrocarbon enantiomers? Today, nothing is known on the chirality of interstellar or interplanetary hydrocarbon enantiomers. In this context, however, it is useful to refer the reader to the occurrence of chiral hydrocarbons on Earth, where their origin in geochemical samples gave rise to a remarkable scientific debate: Pristane and phytane are known as the archetypes of

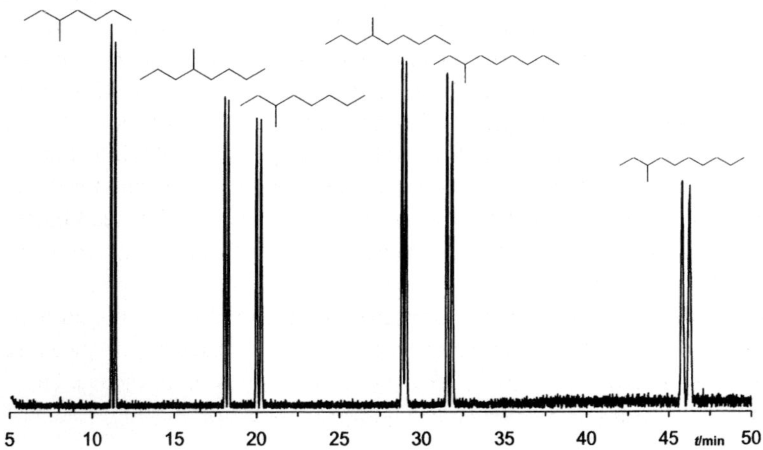

Fig. 9.8 Gas chromatogram of resolved hydrocarbon enantiomers separated on a capillary column coated with permethylated β-cyclodextrin (Chirasil-Dex CB, 25 m, 0.25 mm inner diameter, layer thickness 0.25 μm, Varian-Chrompack). Split injection 1:50 at injector temperature of 230°C, 1.3 mL/min constant flow of He carrier gas, oven temperature programmed 3 min at 35°C, 1°C/min, to 15 min at 70°C. The transfer line temperature was constant and set to 200°C. After solvent delay of 5 min detection of total ion current in mass range 50–350 amu with GC-MSD system Agilent 6890/5973

chiral hydrocarbons. They show branched aliphatic structures and belong to a group called isoprenoids. In these molecules, most carbon atoms at branching positions are stereogenic centers provoking that pristane, phytane, as well as other isoprenoids are chiral. Isoprenoids are usually quite abundant in geological matter. Pristane for example occurs in ancient sediments, rocks, shales, crude oils, coals, and bitumens (Nip 1987).

Despite the abundance of isoprenoid compounds in geochemical samples, the pathway to the formation of the enantiopure pristane and phytane molecules is not completely understood. Biological and non-biological formation processes have been discussed. Pristane-formation precursors like kerogens (Larter et al. 1979; van de Meent et al. 1980; van Graas et al. 1981), tocopherols (Goossens et al. 1984, 1988; Sinninghe Damsté and de Leeuw 1995), and methylated 2-methyl-2-(4,8,12-trimethyltridecyl)chromans (MTTCs) (Li et al. 1995) were suggested (Tang and Stauffer 1995). One particular discussion refers to the question whether enantiopure isoprenoids in geochemical samples are primarily degradation products from plants, formed e.g. by chlorophyll decomposition. Indeed, chlorophyll's side chain is a branched hydrocarbon structure including two stereogenic centers in *R*-configuration. Alternatively, an animal origin was proposed due to the here relevant tocopherol (vitamin E) side chain that is equally composed of a branched alkane with two homochiral branching points in *R*-configuration (Meierhenrich et al. 2001e).

Branched alkanes have been intensively used as biomarkers for a variety of geological compounds. Special ratios of the isoprenoid alkanes pristane, phytane, and pristene have often been used as palaeoenvironmental indicators of thermal maturity (Goossens et al. 1988) and as depositional environment indicators (ten Haven et al. 1987; Hughes et al. 1995; Large and Gize 1996). The *chiral* composition of branched hydrocarbons in crude oil or sedimentary samples may furthermore lead to information on deposition time, oil maturity, etc., since chiral hydrocarbons are assumed to show tremendously long racemization times. Stereochemical studies of isoprenoid compounds were therefore assumed to provide fundamental information on their pathways of formation. The above-presented separation technique might open a door to the resolution of enantiomers for a wide range of chiral alkanes like the branched and chiral isoprenoid hydrocarbons pristane and phytane in petrochemistry and related fields.

Summarizing, the envisaged resolution of enantiomers of chemically and thermally robust chiral hydrocarbons including isoprenoid-type structures in cometary samples is an integral part of the ROSETTA mission and might contribute to a better understanding of the origin of biomolecular asymmetry on Earth.

9.1.3.4 Thoughts on Possible Results of ROSETTA's 'Chirality-Module'

The recent identification of amino acid and diamino acid structures in interstellar ice analogues (see Chap. 7) strengthened the scientific and public interest in the ROSETTA cometary mission (cf. ESA homepage). In order to perform a more reliable estimation on the presence of amino acid structures in comets, their delivery to the early Earth, and their role in life's molecular origins including life's selection for one-handed molecules, supplementary information is required. We need to understand more precisely parameters such as UV exposure, particle fluxes, and number of collisions under interstellar and circumstellar conditions. The in situ analysis of cometary material by space probe ROSETTA is necessary for this understanding. In 2014, ROSETTA's COSAC experiment might find molecular evidence of the presence of amino acid structures in an authentic sample of comet 67P/Churyumov-Gerasimenko.

ROSETTA's 'Chirality-Module' was moreover designed to provide information on the distribution of enantiomers in cometary matter. An observable enantiomeric excess of the 'correct' left-handed amino acids will support the assumption that chiral organic molecules have indeed been delivered onto the Early Earth during the Heavy Bombardment phase via comets and/or other interstellar bodies that are thought to carry along a considerable amount of organic matter. These molecules of an interstellar/cometary origin could have had a seed function in determining the handedness of chiral biotic matter in the Earth's initial biosphere (Thiemann 1998; Thiemann and Meierhenrich 2000; Meierhenrich et al. 2003b)[4].

In this case, the enantiomeric excess measured by ROSETTA's 'Chirality-Module' is to be compared with the molecules' circular dichroism reference spectra in order to investigate whether interstellar photochemical processes might have induced the molecular asymmetry. One might be able to deduce from these data wavelength and polarization of the required extraterrestrial circularly polarized radiation. A number of non-racemic organic molecules such as amino acids, diamino acids, carboxylic acids but also alcohols, diols, amines, and hydrocarbons would provide detailed information about the chiral field that would have been necessary to introduce the molecular parity violation.

If the observed biomolecular parity violation on Earth were mainly caused by random mechanisms, one would naturally expect to find enantiomers of different handedness on different planetary or interstellar bodies. If homochirality in biomolecules were however originated by multiple (Mie) scattering or dichroic scattering (Lucas et al. 2005) in star formation regions, these scattering processes would presumably produce antipodic circularly polarized UV-photons as a function of time and location. All planetary bodies within a given solar system, including comets, would have been affected by the same light's handedness and their biomolecules would be produced stereospecifically in the same configuration. In different planetary systems or on different interstellar bodies, one would expect to detect variable enantiomeric excesses. And if biology's optical activity were determined by the universal chiral influence of the weak force, one would expect to find identical handedness of enantiomers throughout the entire universe. The results of ROSETTA's 'Chirality-Module' might give hints to the validity of these various hypotheses.

---

[4] Cometary matter showing right-handed amino acids would also be very exciting, maybe as much as left-handed cometary amino acids; it would be only more unexpected from the origin of life on Earth point of view.

Although – at this stage of investigating cometary surfaces – we do not anticipate the existence of large enantiomeric excesses of organic molecules, not to speak of homochiral samples on comets with their harsh and life excluding climatic conditions as there are low temperatures, near-zero atmosphere, probably no liquid water, and others. Yet we believe in a strong chance to find *partial* enantiomeric excesses of small molecules due to the treatment of the comet's surface with extraterrestrial chiral fields. We assume that once the comet is formed, processing occurs at the very surface since vacuum ultraviolet light penetrates less than one micrometer and cosmic rays of the order of a few meters. If a small enantiomeric excess is induced on the grain mantles in this decisive comet-forming step, such *e.e.* may be preserved in the comet nucleus. The long awaited results of ROSETTA's 'chirality-module' on the resolution of different kinds of enantiomers would help understanding at least one important aspect of the development of life on Earth and point out the importance of synthesis of chiral compounds in interstellar space, regardless of the process by which the required enantiomeric excess may have come about.

Besides this, the identification of chiral diamino acid structures, aminoethyl glycine, sugar molecules or their characteristic molecular fragments by ROSETTA's COSAC instrument on comet 67P/Churyumov-Gerasimenko might promote research activities on the eventual contribution of these compounds to the evolution of the first genetic material on Earth, which might be described as a pre-RNA oligonucleotide.

9.1.3.5 Liquid Water in Comets?

In the above chapters, we have assumed that comets and small icy bodies of the Solar System are – given the low ambient temperature of interplanetary and interstellar regions – composed of pristine, original, and unaltered material. This was one of the reasons to consider the chemical and enantioselective analysis of comets as important for obtaining clues on the origin of life and particular on the origin of biomolecular asymmetry; one of the reasons to envisage cometary in situ measurements with the help of the ROSETTA mission.

However, if we have a closer and more precise look into the interior of comet-forming small icy bodies of the Solar System, we should notice that these objects could alter due to impacts, insolation, and particle radiation. But more importantly, radioactive decay of unstable isotopes incorporated in the inorganic dust might have contributed to a non-negligible thermal evolution of comets via radioactive heating. Particularly the ^{26}Al radioactive decay with an ^{26}Al half-life of 0.72 million years could have transformed the amorphous cometary ice into crystalline ice. Even the melting of water ice was calculated to be possible – if not widespread – for cometary-like bodies of radii between 2 and 32 km. Numerical modelling of cometary thermal evolution led to a broad variety of thermally processed icy bodies and to the knowledge that the early occurrence of liquid water and extended crystalline ice interiors may be a common phenomenon (Merk and Prialnik 2006).

The calculated duration of the liquid water stage in comets and small icy bodies is illustrated in Fig. 9.9. It is clearly shown that liquid water could have been present for a duration of up to 4.5 million years (Ma), which is a function of the heliocentric distance, the radius of the object, and its ice to dust ratio. The bigger the object and the higher the ice content, the higher the temperatures.[5] According to Rainer Merk from Tel Aviv University, we can assume for the indicated liquid time spans in Fig. 9.9 that 40–60% of the radius (given at the abscissa) had been liquid. This liquid may be considered as a bygone "three-dimensional internal lake" shaded by a thick pristine crust to the exterior. The pristine nature of comets and small icy bodies seems thus to be restricted to a particular set of initial conditions and should not be taken for granted.

It is evident that a more profound knowledge on the presence of liquid water including its temporal extent, temperature, volume, pressure, concentration of polar organic molecules therein, but also information on the physical, eventual nebular-like, state of liquid water under near zero gravity in a highly porous body is crucial for 'astrobiological' questions and affected with far-reaching consequences. In particular, we assume that organic molecules of high molecular weight are present in interstellar ices and comets releasing amino acids after hydrolysis. We may now figure out if this hydrolysis may have occurred not only after the delivery of cometary organic molecules on the surface of Earth, or other planets, moons and asteroids bearing liquid water, but also in small icy bodies and comets itself in well defined interstellar regions. Nevertheless, experimental data show that no hydrated minerals were found so far in the Wild 2 grains collected by Stardust.

Comet 67P/Churyumov-Gerasimenko, target of the ROSETTA Mission, is a periodical Jupiter family comet classified as short period. This comet has probably originated from the scattered disk, located between Kuiper belt and Oort cloud. The above-mentioned numerical simulation of the early internal thermodynamic by Merk and Prialnik (2006) are in particular valid for objects from this region.

The nucleus of comet 67P/Churyumov-Gerasimenko was estimated to have a radius between 2 000 m and 4 740 m. Assuming a dust-to-ice mass ratio of about 0.8 for 67P/Churyumov-Gerasimenko (Lamy et al. 2007), this comet is to be settled on the left side of the upper Fig. 9.9. At the present state of our knowledge, it is therefore improbable that the *surface* of the nucleus of this comet, where the ROSETTA space probe is going to touch-down its robot lander Philae, has ever been transformed to liquid water in its history. But with this prediction we have to be careful, since comets might be fragments of bigger 'parent bodies'. If this is the case, comet 67P/Churyumov-Gerasimenko might have seen liquid water in a temperature range of 290 up to 350 K.[6] For this estimation, error bars are difficult to number, since

[5] The upper image in Fig. 9.9 was calculated for low dust content and consequently a low amount of radioactive ^{26}Al heating sources. Nevertheless, it shows longer time spans for liquid water (4.5 Ma compared to 2.7 Ma) compared to the lower image that was calculated for a higher dust content. This is not surprising, since dust-rich particles show a higher density, accrete more slowly, and cool down faster.

[6] Personal communication with Rainer Merk, Tel Aviv University, in May 2007.

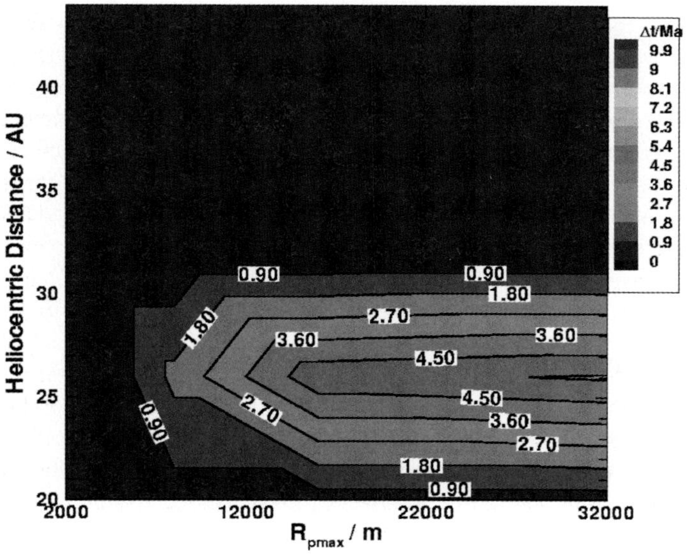

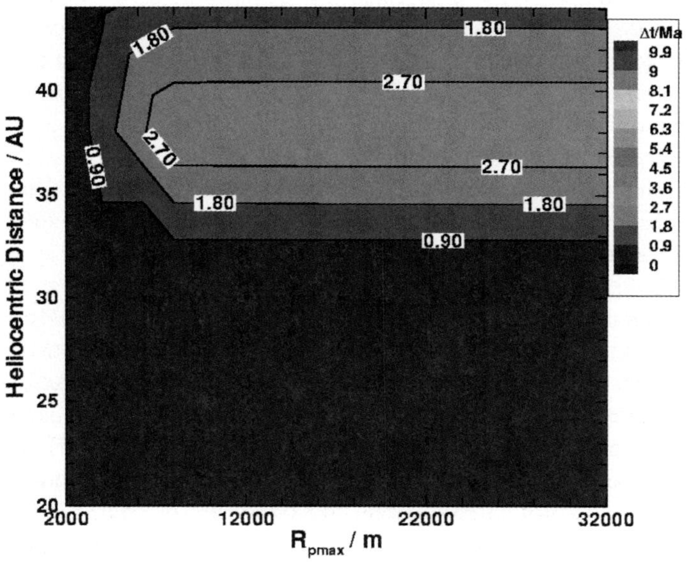

Fig. 9.9 Time spans in million years (Ma) during which comets were calculated to contain liquid water. The heliocentric distance in AU is printed as a function of the object's radius in m. (*Top*) 1:1 composition of ice and dust; (*Bottom*) ice-poor bodies. The illustration was first published by Merk and Prialnik (2006). Reproduced with permission of the authors

numerical calculations do not give random errors; systematic errors on the other hand are improbable because of the use of sophisticated standard models.

At present, we should accept that the substantial question on the ancient presence or absence of liquid water in comets cannot be answered for individual comets with sufficient precision and will remain subject to ongoing research. Direct evidence for liquid water in comets has never been seen and today we should not expect it for comet 67P/Churyumov-Gerasimenko due to the limited half-life of the radioactive heating sources.

9.2 Chiroptical Techniques and the SETH Project

Information on the eventual homochirality of organic enantiomers in interplanetary, interstellar, and extraterrestrial samples in general can be used as an elegant way to finding life traces somewhere in the universe. Focussing on this ambitious endeavour, a complementary option to the above described enantiomer-separating chromatography was proposed: Space missions were suggested to implement an optical device able to determine any optical activity, optical rotatory dispersion, resp. circular dichroism of a condensed sample grabbed from a planet's, moon's, or comet's surface to search for a homochiral property of the chemical compounds, of which the condensed matter is made of. Specific "space-suitable" polarimeters are of a most simple, light-weight, robust, and yet sensitive and reliable design. They were shown to resist mechanical, pressure, and temperature shocks usually experienced during longer space journeys and automated landing conditions in unmanned missions. Alexandra MacDermott from the University of Houston-Clear Lake and colleagues have proposed such kind of instrument (MacDermott et al. 1996; MacDermott 1997) investigating the stereochemistry of extraterrestrial chiral organic molecules under the title Search for Extraterrestrial Homochirality (SETH).

The inclusion of this kind of chiroptical instrument is foreseen in a Russian Martian Excursion, now. Moreover, one may include polarimetric measurements of the SETH or any similar type on all future missions to all solid targets, such as Mars, Mercury, and Venus, the giant planets' moons Titan, Europa, and Ganymed, and the accessible comets to be targeted in the future.[7] Had such a polarimeter instrument been included on the famous Viking Experiment to explore life on Mars already back in 1976, some of the ambiguous life detecting experiment's results might have given more straight forward answers as to the intriguing question "... does life (or extant life) exist on Mars or not?", if the knowledge about chiral compounds on Martian surface were provided at this time.

---

[7] The quest for homochirality outside Earth should be taken up of course by completely different methods independent on the above discussed purely *physical* method, because this method suffers somewhat from a number of severe limitations. There is obviously a strong need for an unequivocal *chemical* identification of chiral compounds within material recovered from extraterrestrial bodies.

9.3 The Search for Chiral Molecules on Mars

So let us turn towards Mars and ask the question whether or not it might be achievable to identify and quantify chiral organic molecules on its surface. Martian organic molecules might be relicts of stellar evolution brought to Mars via meteorites, they might be remnants of chemical evolution on Mars itself, but also remains of extinct (or present-day) life on the Red Planet. In order to distinguish between these three potential pathways of formation, the chirality of Martian organic enantiomers would be of high importance to study.

As indicated above, the current discussion on organic molecules on Mars is still deeply influenced by experiments of the Viking Landers 1 and 2 that arrived at the surface of Mars in 1976. Viking's search for organic matter in two soil samples at each landing site showed surprisingly, that no organic molecules were found at the parts per billion (ppb) level (Biemann et al. 1976). Very recently, it was proposed that Viking's experimentation might have been blind to detect (a) life (Navarro-Gonzáles et al. 2003) and in particular (b) to detect low levels of organics due to the instrument's thermal volatilization (Navarro-Gonzáles et al. 2006) suggesting that the Martian surface could have several orders of magnitude more organics than the stated Viking detection limit (see also Wu 2007). Moreover, oxidation-reactions of Martian organics by inorganic oxidants such as superoxides, peroxides, or peroxynitrates forming non-volatile compounds might have caused their non-detection (Benner et al. 2000). This accusation forced Viking scientist Klaus Biemann from the Massachusetts Institute of Technology in Cambridge USA, to republish an Antarctic soil standard used for the calibration of Viking's GC-MS (Biemann 2007). The gas chromatogram shows aliphatic, aromatic, as well as nitrogen-, and oxygen-bearing compounds and seems thus to proof design concept, operation, extensive tests, and validity of the recorded Viking GC-MS data.

So for us, it remains an open question whether or not organic molecules are present in the Martian surface or its subsurface. Upcoming space missions such as ExoMars will tackle this issue, focussing additionally on their chirality.

9.3.1 Space Invaders: Traces of Life in Martian Meteorites

Spectacular was the recent claim of detecting traces of past Martian biota in the meteorite called ALH84001 (McKay et al. 1996) supported also by isotope ratios on the meteorite EETA79001 announced by Colin Pillinger and Ian Wright from the Open University in Milton Keynes (MacDermott 1997). "If this discovery is confirmed," US-President Bill Clinton said, "it will surely be one of the most stunning insights into our universe that science has ever uncovered." Serious reinterpretations of the presented ALH84001 data today conclude that the observed results do not necessarily indicate the presence of living organism on Mars. Neither today nor to any previous times. Surprisingly, any claims on extinct or present life on Mars have so far never been supported with information on chirality and eventual enantiomeric excesses of chiral structures.

Under the extremely dry, cold, low pressure conditions of Mars the chiral properties of fossilized (micro-) organisms - if ever existed - would have easily survived over long geological time periods, as laboratory experiments by Bada and McDonald (1995) on the racemization of chiral amino acids under Martian conditions suggest. Probing solid bodies' surfaces in the Universe should include measurement on optical activity or chirality aiming to obtain results concerning the intriguing searches for the existence of life beyond Earth.

9.3.2 Chirality and Mission ExoMars

Officials from the European and North-American Space Agencies ESA and NASA understood this immense research opportunity to be of outstanding interest. In December 2005, ESA-representatives of 17 member states signed to support ESA with 8.36 billion Euros in the upcoming years. Based on this funding, ESA itself decided to design a mission to Mars ready to start in 2013. The mission is called ExoMars and includes a Mars Rover (Fig. 9.10).

The scientific objective of the ExoMars mission is to obtain information on extinct and/or present life on Mars and thus to detect carbon-containing organic molecules on the surface and subsurface of the Red Planet (Brack 2000). Amino acids, carboxylic acids, oxidized compounds, and other defined and prioritised molecular building blocks of life (see Parnell et al. 2007) should give hints to the organic and prebiotic chemistry on Mars (and Earth). The aim of the ExoMars Mission is to identify a "chemical fingerprint" of extinct and/or present life on Mars.

A large number of scientific proposals for suitable ExoMars instruments from all over the world were developed and examined by an international Science-Team at ESA headquarters in Noordwijk, the Netherlands. In 2007 the decision was taken. ExoMars will be a joint ESA-NASA mission and it will include two (sic!) instruments dedicated to the resolution of enantiomers on the surface and subsurface of Mars: MOMA and UREY.

9.3.2.1 The Enantioselective Mars Organic Molecule Analyser (MOMA)

One of these instruments is called the Mars Organic Molecule Analyser (MOMA). MOMA combines a sophisticated gas chromatograph-mass spectrometer (GC-MS) with a Laser desorption mass spectrometer (LD-MS) system (Goesmann et al. 2007a).

The idea of this integrated approach is to unite the different methods and their strengths from very high specificity at extremely low detection thresholds at the one extreme to a very wide range of detectable families of different molecules at higher detection limits at the other. The central part of MOMA is an ion trap mass

Fig. 9.10 Artist's impression of the ExoMars Rover drilling into the Martian surface. The vehicle will be able to travel a few kilometres with its 16.5 kg exobiology payload. Surface and subsurface samples of Mars will be subjected to chemical analysis in a miniaturized laboratory onboard the Rover. Here, chiral organic molecules can be separated into their enantiomers, identified, and quantified with the help of (a) enantioselective gas chromatography by MOMA and (b) enantioselective capillary electrophoresis by UREY to search for signs of past or present life. Credits of ESA – AOES Medialab

spectrometer (ITMS) for the detection and identification of molecular ions. The source to be investigated will be Martian soil samples from the surface and maximum 2 m subsurface obtained from a specifically developed drilling system. In order to transfer molecules from the soil sample to the MS two methods are employed. Firstly, Laser desorption mass spectrometry (LD-MS) will be used, where the Martian sample will be subjected to intense Laser flashes producing molecules and ions directly, even from refractory material, but in a mixture.

Secondly, in GC-MS, Martian samples will be heated (pyrolysed) or subjected to a combustion procedure. The evolving volatiles are transferred to a GC where the compounds are separated and then fed into the MS to be measured individually. The combination of methods to feed the MS in MOMA (a) via GC and (b) via LD-MS covers a wide rage of molecules from the very light (example: methanol) to medium sized (example: naphthalene) by GC-MS up to more complicated (example: peptides) by LD-MS. Rather low detection limits in the 20 ppb range for GC-MS and in the pico-mole range for LD-MS can be provided. The MOMA instrument will be designed and constructed by an international research team, the coordinators of which are Fred Goesmann at the Max Planck Institute for Solar System Research in Katlenburg-Lindau, Germany, and Luann Becker at the University of California, Santa Barbara, USA (Goesmann et al. 2007a).

If – and this is promising – organic molecules will be detected in Martian surface and subsurface samples by the MOMA-instrument, we would like to learn about the organics' chirality by measuring enantiomeric excesses. Will there be any deviation from the racemic enantiomeric distribution expected for abiotic synthesis of organic molecules? The pathways of formation of eventual chiral organic molecules on Mars, biotic versus abiotic and extra-Martian versus Martian, might be deduced from the enantiomeric ratios in different families of organic molecules.

The challenge is that no organic molecules were found on Mars by the Viking mission in 1976 but within a solid meteorite of possible Martian origin on Earth in 1996. This means that organic molecules are likely to be very rare on Mars. Furthermore, only a fraction of the conceivable organic molecules do have stereogenic centers. Hence, enantiomers will be even more rare. There is consequently a strong need for sample enrichment in order to improve detection limits for trace abundances of organic molecules. Furthermore, as the possible molecular structures are unknown, the instrument must not only be sensitive, but able to characterize the detected molecule and identify them unambiguously.

Summarizing the above, the MOMA instrument requires to be capable of performing enantioselective analyses with different chiral molecules on the surface of Mars. However, for interpretation of the results we will have to be patient until ExoMars measurements after landing which is also foreseen for 2014.

9.3.2.2 The Enantioselective UREY Instrument

The other officially selected apparatus for the ESA-NASA Mission ExoMars which focuses on the phenomenon of chirality is called UREY, referring to Nobel Prize laureate Harold Urey from the University of Chicago in whose lab Stanley Miller simulated the early Earth's atmosphere showing the synthesis of various amino acids. The UREY apparatus for the ExoMars Mission consists of a Sub-Critical Water Extractor (SCWE) to solubilize organic molecules from Martian rock/soil materials and deliver the enriched extract to the Mars Organic Detector MOD (Kminek et al. 2000). Here an analysis of amino group containing compounds such as amino acids, amines, nucleobases, amino sugars, and polycyclic aromatic hydrocarbons

(PAHs) will be performed. The detection will be done by laser-induced fluorescence, a technique capable of providing a parts-per-trillion sensitivity.

If chiral amino acids will be found, they will be further analyzed using the Micro-Capillary Electrophoresis (μCE) Unit to determine their chirality. The precise enantioselective-working mode of this instrument is not yet published. Enantioselective capillary electrophoresis is a well-known and well-proven technique for the resolution of enantiomers, however, this technique is also very vulnerable since it requires fresh buffer solutions combined with the controlled handling of various liquids. These measurements will be made at a thousand times greater sensitivity than the Viking GC-MS experiment, and will significantly advance our understanding of the organic chemistry of Martian soils.

The MOD will be fed with 1–3 mL aliquots of aqueous extract from the SCWE and start working by removing the water by freeze-drying. MOD will then slowly sublimate all volatile organic compounds including chiral amino acids onto a fluorescamine-coated target held at −10°C with a cold-finger. The chemical constituents will then be separated by controlled sublimation at increasingly higher temperatures. It is well-known that, for example, the sublimation of amino acids occurs in the temperature range of 125–350°C, whereas polyaromatic cyclic hydrocarbons sublime at approximately 450°C. Laser-induced (395 nm) fluorescence will be used to excite organic molecules such as polyaromatic cyclic hydrocarbons and amino acids that are bound to the fluorescamine on the cold-finger, in order to detect their presence. These compounds are fluorescent under ultraviolet illumination. Each analysis will be followed by a cleaning step at 900°C, in order to remove material from the sublimator and prevent contamination and memory-effects in the upcoming analysis.

What about the chirality of eventually identified amino acids or other chiral organic molecules? The Micro-Capillary Electrophoresis Unit (μCE) of the UREY-module will also be fed with liquid extracts from the SCWE. This device consists of a laminated multilayer ('lab-on-a-chip') wafer stack that performs fluidic manipulations and electrophoretic analysis. Buffers and other reagents are used to detect the chirality of any amino acids present in the liquid sample. Using this technique, amino acids and nucleotide bases of biotic and abiogenic origin can be detected and differentiated. One unknown parameter remains the long-term stability of amino acids in the Martian regolith.

Jeffrey Bada from the Scripps Institute of Oceanography in San Diego scientifically guides the UREY team. From this team, we also anxiously expect data on the chirality of extraterrestrial organic molecules that are probably not "contaminated" by the homochirality of living organisms on Earth. We'll just have to wait for the landing of the ExoMars probe until 2014.

Chapter 10
Accelerating the Carousel: Amplification Mechanisms

As we have seen in the above chapters, an enantiomeric enhancement can indeed be induced into racemic mixtures of organic molecules via internal or external asymmetric effects. We discussed random fluctuations of enantiomers, the spontaneous symmetry breaking via crystallization, the molecular parity-violating weak force, magnetochiral photochemical effects, and also the absolute asymmetric photochemical induction of an enantiomeric excess. Among all these attractive and topical models, a slight majority of scientists today favours asymmetric photoreactions to explain the origin of biomolecular asymmetry, because such reactions were successfully performed under realistic primordial conditions, as for example in simulated interstellar molecular clouds where chiral amino acid structures have been detected. We can envisage that future space missions such as the Rosetta spacecraft and others will provide us the required key information to ultimately decipher the origin of life's chirality. But attention: depending on these and other data, the scientific basis for other hypotheses mentioned above – if not entirely new – might also be strengthened. This is what makes science in general and the research field of life's molecular asymmetry in particular that attractive!

Anyhow, we have outlined that the total of the above models is capable of producing relatively small enantiomeric enhancements compared to the homochiral structures of the 20 L-amino acids and the D-ribofuranose sugar molecules embedded in biopolymers. Therefore, we have to reflect on possibly cumulative or magnifying ways to increase the enantiomeric excess from a few percent (or even much less) towards nearly homochirality. Do such models exist? This question is to be answered with a clear "yes". It seems that amplification processes able to increase small elementary enantiomer disparities to large enantiomer excesses present even a considerably lesser challenge for scientists than finding the enantiomer bias as such. A manifold of various amplification processes is abundant, at hand, and fully accepted by the scientific community (Thiemann and Bredehöft 2007).

In order to classify these amplification processes we will distinguish between (a) models that have been designed through mathematical approaches and (b) models that have been based on experimental work. I will try to treat the mathematical approaches as comprehensible as possible in order to provide a general access to the obtained highly intriguing results. Afterwards, experimental work that has been greatly stimulated by successful trials very recently will be presented. In a more

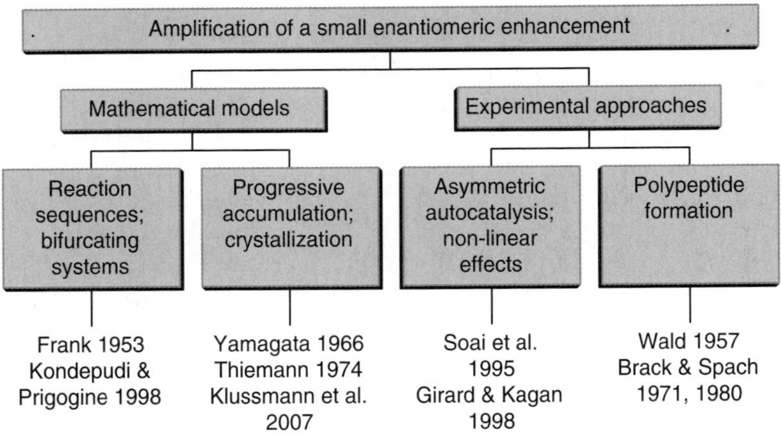

Fig. 10.1 Overview of amplification models for the generation of life's molecular homochirality. We distinguish between mathematically driven models such as bifurcating systems and the progressive accumulation as well as experimental approaches like, e.g., asymmetric autocatalysis and the enantioselective formation of polypeptides

fine-tuned classification, we will then distinguish between approaches of different research teams working internationally on the amplification of enantiomeric enhancements as it is schematically depicted in Fig. 10.1.

In the field of amplification of small enantiomeric imbalances obtained by the above-described scenarios several theoretical model mechanisms have been proposed. Starting in 1953, F. C. Frank from Bristol University suggested an autocatalytic kinetic model for spontaneous asymmetric synthesis in which the presence of one enantiomer encourages the production of itself, but inhibits the production of its optical antipode. The Frank model was later modified by Decker (1973), Kondepudi and Nelson (1985), Kondepudi (1987), and Kondepudi and Prigogine (1998) with the help of the bifurcation theory predicting that a laboratory demonstration of this work may not be impossible. In modern thermodynamics the bifurcation theory is therefore highly discussed and exciting. It will be presented in the next section.

Since recently, also experimental trials to amplify a small enantiomeric excess of 1 or 2% *e.e.* towards nearly homochirality were performed successfully at 'real' systems in the laboratory. The required asymmetric amplification can involve enantioenriched auxiliaries showing catalytic (Fenwick and Kagan 1999) or even autocatalytic (Shibata et al. 1998) function, as we will see later in this chapter.

Earlier experimental approaches stimulated by the theory of Wald (1957) predicting that the α-helical structure of a growing polypeptide chain allows the stepwise assembly of amino acid monomers in a way that small chains form with significant levels of stereoregularity (Bonner 1995a), were adequately discussed in Chap. 3.

10.1 Some Amplification Needed: The Bifurcation Theory

Traditional physico-chemical laws of thermodynamics have often been limited to the study of systems at their chemical equilibrium. Moreover, these systems involved idealized slow and reversible chemical reactions. Here, chemical kinetics were limited to first order reactions and second order reactions, easy to be treated. The term "thermostatic" might better describe the study of such systems than the grandiosely used term "thermodynamics".

Kondepudi and Prigogine (1998) outlined these limitations particularly when they tried to apply the laws of thermodynamics to biological systems (a) that are usually not at chemical equilibrium and (b) that do not involve exclusively first and second order kinetics. Kondepudi and Prigogine thus founded a new branch of physical chemistry, called "Modern Thermodynamics", allowing us now to better understand non-equilibrium phenomena such as the bifurcation and the breaking of symmetry. Moreover, advanced software programmes help us today to solve complex differential equations allowing thus the understanding of bifurcating systems. In former times, people even assumed that bifurcating systems resisted any scientific description by appropriate laws.

To introduce bifurcation theory we should visualize that bifurcation is a common phenomenon in macroscopic and microscopic scale: water streams on a flat horizontal surface at "non-equilibrium conditions" are capable of bifurcating into two or more flows. A river such as the source in the Swiss Oberengadin at Piz Lunghin bifurcates into streams serving simultaneously (a) the Rhein flowing into the Northern See, (b) the Inn flowing via the Danube towards the Black Sea, and also (c) the Italian Po flowing into the Mediterranean Ocean. To give another example: close to Detmold in Germany, where I was born, the small river Hase bifurcates into two streams. One serves the river Ems, the other serves the Weser.

Macroscopic bifurcations can be observed also in "cellular rotations" of the hydrodynamic Bénard instability (Försterling and Kuhn 1985), in fractal structures such as the branching of trees towards leafs, and also in the hexagonal branches of snow crystals. A very impressive multiple bifurcating system is the so-called zinc metal-leaf, an often used practical-training experiment appreciated by students: here, a central carbon electrode in a zinc sulphate ($ZnSO_4$) solution is surrounded by a cyclic zinc electrode of about 30 cm diameter. After applying a DC voltage of 3 to 10 V to the electrodes, beautiful metallic zinc-leafs grow in two dimensions showing a fractal structure (see e.g. Matsushita 1989). Due to their dendritic growth this is called a zinc-leaf or zinc-tree, showing bifurcations at various levels as depicted in Fig. 10.2. In a modified version of this experiment, a zinc-forest showing various bifurcations can be observed by the parallel orientation of two linear electrodes.

Having a closer look on bifurcating structures, the question arises, which general characteristics of systems are necessary to develop one or more bifurcations? Which parameters determine a bifurcation and how? And is the bifurcation a random or determinate driven ramification? In order to answer these questions we will now turn towards bifurcation theory by treating a simple mathematical system showing one bifurcation only.

Fig. 10.2 Zinc crystal "zinc-leaf" of 25 cm diameter. The purely inorganic crystal shows multiple bifurcations; it is classified as fractal structure. The fractal dimension can be determined by the "Box-Counting" method

In general, bifurcation and symmetry breaking follow a non-linear differential equation, which is called the bifurcation equation or sometimes Langevin equation (10.1). In this equation, α simply gives the amplitude of an asymmetry and λ is a parameter describing the distance of the system from chemical equilibrium. The higher α, the higher the asymmetry of the system; the higher λ, the more the system is pushed away from equilibrium conditions. As we will see later, the parameters α and λ are to be adapted to specific chemical systems.

$$\frac{d\alpha}{dt} = -\alpha^3 + \lambda\alpha \tag{10.1}$$

Equation (10.1) is a homogeneous differential equation of first order. It contains a polynomial of third order and no quadratic summands.

Now, we will consider a simple chemical system described by the Langevin equation in its steady state conditions. In general, a steady state characterizes a system that allows the transport of matter into and out of the system in a way that the concentrations of chemical species within the system remain unaltered. Therewith, in the steady state $d\alpha$ over dt in Eq. (10.1) becomes zero. With $\lambda < 0$ we obtain one real solution for the asymmetry α ($\alpha = 0$) and with $\lambda > 0$ we obtain three solutions

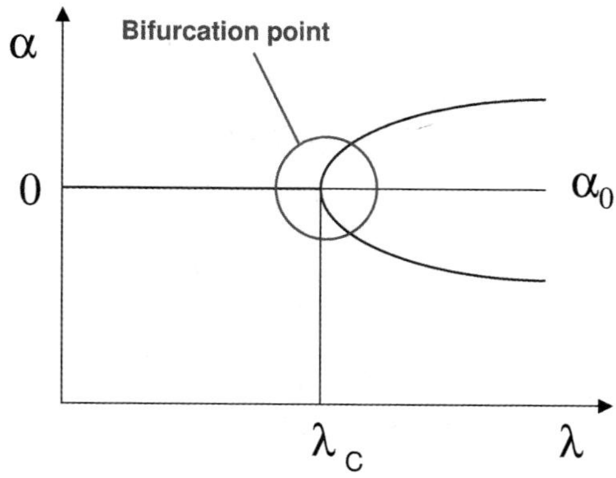

Fig. 10.3 Bifurcation diagram including the solutions of the Langevin equation which are $\alpha = 0$ for $\lambda < \lambda_C$ and $\alpha = \pm\sqrt{\lambda}$ for $\lambda > \lambda_C$. The thin line representing $\alpha = 0$ for $\lambda > \lambda_C$ corresponds to an instable solution

($\alpha = \pm\sqrt{\lambda}$ and $\alpha = 0$). One of the three solutions ($\alpha = 0$) becomes unstable, which can be demonstrated by stability analysis. The system bifurcates at $\lambda = 0$. More generally, it can be shown that the system bifurcates at a critical value for lambda which is $\lambda = \lambda_C$. If we now plot α as a function of λ, we obtain the bifurcation diagram with its bifurcation point as it is given in Fig. 10.3.

It is important to understand that if we shift the system away from the thermodynamic equilibrium by increasing values of λ (i.e., moving in the bifurcation diagram from the left to the right), the system passes the bifurcation point and is thus – at a critical value of λ_C – forced to realize either the upper or the lower branch. The repetition of an experiment under ideal symmetric conditions can be assumed to realize 50% cases the upper branch and 50% cases the lower branch. Here, the evolution of α is not deterministic. 'Natural' and biological systems often develop in sequences by realizing multiple branching, meaning that one branch follows another (see for example the zinc-leaf).

The treated phenomenon presenting bifurcating solutions is by no means an exceptional case. It is a rather common phenomenon. It is valid for non-linear equations, which might be simple algebraic equations but also more complex coupled differential equations as well as partial differential equations. If this phenomenon is indeed regarded as a general one, does any chemical system show a chiral bifurcation? Can we imagine a chemical system that amplifies a small enantioenrichment towards homochirality just by increasing its λ-value? The following kinetic reaction sequence was developed for these purposes.

10.2 Amplification by Kinetic Reaction Sequences

The amplifying model that will be presented here is a modern version of the Frank model (1953) that showed how enantioselective autocatalysis could amplify a small initial asymmetry. It was worked out by Dilip Kondepudi at the University of North Carolina in the US and Nobel Prize laureate Ilya Prigogine at the University of

Brussels (1998), modifying Frank's reaction scheme so that its non-equilibrium aspects, instability, and bifurcation of symmetry breaking states can be clearly seen.

The amplifying Kondepudi-Prigogine model assumes a chemical system, which is initially (a) close to chemical equilibrium and (b) in a symmetric state. The symmetric state can be expressed by chiral reactants furnished in a racemic mixture or still by achiral reactants. Here, we will introduce the achiral reactants **S** and **T** into the system. The chiral products $\mathbf{X_L}$ and $\mathbf{X_D}$ will be synthesized by specific reactions of the achiral reactants. Moreover, an inactive product **P** forms that will be taken out of the system. A schematic view of the Kondepudi-Prigogine model is given in Fig. 10.4.

The initial formation of the chiral enantiomers $\mathbf{X_L}$ and $\mathbf{X_D}$ by the achiral reactants **S** and **T** is described by Eq. (10.2). Please note that the rate constant k_{1f} is identical for the formation of $\mathbf{X_L}$ and for the formation of $\mathbf{X_D}$. The backwards reaction determined by rate constant k_{1r} is also identical for the two enantiomers $\mathbf{X_L}$ and $\mathbf{X_D}$. So far, no asymmetry is induced into the system. It remains highly symmetric.

$$\mathrm{S}+\mathrm{T} \underset{k_{1r}}{\overset{k_{1f}}{\leftrightarrows}} \mathrm{X_L} \quad \text{and} \quad \mathrm{S}+\mathrm{T} \underset{k_{1r}}{\overset{k_{1f}}{\leftrightarrows}} \mathrm{X_D} \tag{10.2}$$

Moreover, we will allow – and this is of crucial importance – the chemical system to synthesize the chiral products $\mathbf{X_L}$ and $\mathbf{X_D}$ by an additional reaction in an autocatalytic process. The characteristics for autocatalytic reactions are that a reactant (here $\mathbf{X_L}$ or $\mathbf{X_D}$) increases its quantity in a reaction catalyzed by itself without being consumed. This behaviour is particular but not very exotic (neither in human society nor on the molecular level). The required autocatalytic step will be introduced into the reaction sequence as it is described in Eq. (10.3) for enantiomer $\mathbf{X_L}$ and in Eq. (10.4) for enantiomer $\mathbf{X_D}$. Also here, the system remains highly symmetric since rate constants k_{2f} and k_{2r} remain again identical for the two enantiomers.

$$\mathrm{S}+\mathrm{T}+\mathrm{X_L} \underset{k_{2r}}{\overset{k_{2f}}{\leftrightarrows}} 2\mathrm{X_L} \tag{10.3}$$

$$\mathrm{S}+\mathrm{T}+\mathrm{X_D} \underset{k_{2r}}{\overset{k_{2f}}{\leftrightarrows}} 2\mathrm{X_D} \tag{10.4}$$

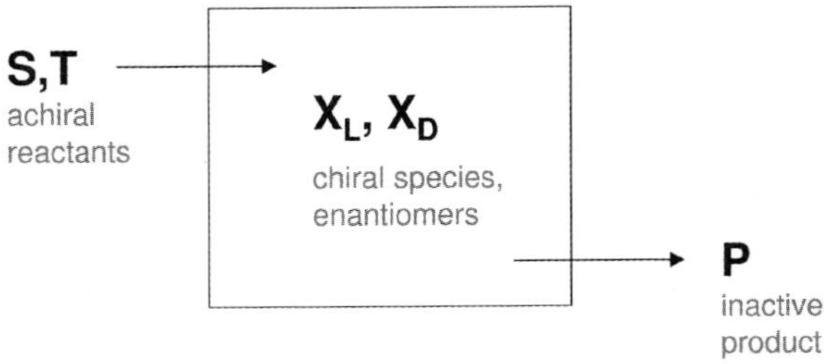

Fig. 10.4 The Kondepudi-Prigogine model assumed an open system into which the achiral reactants **S** and **T** are introduced, forming chiral species $\mathbf{X_L}$ and $\mathbf{X_D}$. The achiral product **P** is to be taken out of the system

Finally, the inactive product **P** will form irreversibly by reaction of the two enantiomers with one another Eq. (10.5). Product **P** will be removed from the system by rate constant k_3. Due to the irreversibility of this last reaction, the reverse reaction is ignored:

$$\mathbf{X}_\mathrm{L} + \mathbf{X}_\mathrm{D} \xrightarrow{k_3} \mathrm{P} \tag{10.5}$$

In this example of the Kondepudi-Prigogine reaction sequence we will also ignore a racemization reaction $\mathbf{X_L} \leftrightarrows \mathbf{X_D}$ that might be introduced into the chemical system as done by Kondepudi and Asakura (2001). This step is interesting from two viewpoints, (a) it introduces an evolutionary element into the chirally autocatalytic scheme that takes into account (long-term) racemization processes and (b) it results that any batch scenario will evolve into a racemic equilibrium state, i.e., in this case an open system becomes essential. Moreover, the scheme can be extended to accommodate unequal reaction rates for the two enantiomers, to include thermal fluctuations, and other factors such as asymmetric destruction rates of the two enantiomers by β-radiolysis or other chiral environmental influences (MacDermott and Tranter, 1989).

In order to fully predict the evolution of an eventual enantiomeric excess within this system, we have to apply chemical kinetics by developing the related kinetic equations. We are particularly interested in the evolution of the concentrations of the individual enantiomers over time, denoted as d$[X_L]$ over dt and d$[X_D]$ over dt. With this information, we will be able to compare the two concentrations and to predict conditions under which an enantiomeric excess can be amplified by any reaction sequence following the Kondepudi-Prigogine system. According to rules in chemical kinetics, the kinetic equations (Eq. 10.6 for enantiomer $\mathbf{X_L}$ and Eq. 10.7 for enantiomer $\mathbf{X_D}$) on the above system can be expressed as:

$$\begin{aligned}\frac{d[X_L]}{dt} &= k_{1f}[S][T] - k_{1r}[X_L] + k_{2f}[S][T][X_L] - k_{2r}[X_L]^2 \\ &\quad -k_3[X_L][X_D]\end{aligned} \tag{10.6}$$

$$\begin{aligned}\frac{d[X_D]}{dt} &= k_{1f}[S][T] - k_{1r}[X_D] + k_{2f}[S][T][X_D] - k_{2r}[X_D]^2 \\ &\quad -k_3[X_L][X_D]\end{aligned} \tag{10.7}$$

These kinetic differential equations, in which square brackets denote concentrations, were numerically solved with the help of the software programme *Mathematica*®. We can now follow the evolution of the concentration for the two individual enantiomers over time starting with a very small excess of $\mathbf{X_L}$ at $t = 0$ (see below) and as we surprisingly see in Fig. 10.5, the concentration of the L-enantiomer increases after about 5 min, whereas the concentration of the D-enantiomer decreases.

After temporal evolution of about 30 min, a highly significant enantiomeric excess evolved from the autocatalytic Kondepudi-Prigogine reaction sequence.

In this particular system the parameter λ, which denotes the system's distance of its thermodynamic equilibrium, is defined as the product of the concentrations of

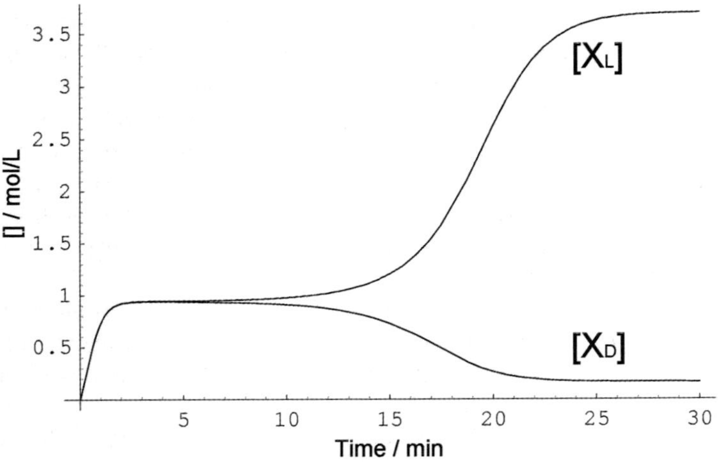

Fig. 10.5 Spot the difference: time evolution of the concentration of $\mathbf{X_L}$ and $\mathbf{X_D}$ in the autocatalytic Kondepudi-Prigogine reaction sequence. A tiny excess in the $\mathbf{X_L}$ enantiomer is increased to a high enantiomeric excess after 30 min of reaction time. Please consider that the given figure is no bifurcation diagram

the achiral reactants $\lambda = [S] \cdot [T]$. In order to push the system away from the thermodynamical equilibrium (and to increase λ), we used relatively high values for the concentrations of the achiral reactants. At thermodynamical equilibrium (i.e., with lower concentrations of **S** and **T** and thus a lower λ-value) with $\lambda < \lambda_C$ the reaction sequence evolves symmetrically, resulting in $[X_L] = [X_D]$. At thermodynamic equilibrium there can be no enantiomeric excess (Kondepudi and Asakura 2001). Here, each small asymmetry falls back to the symmetric state with $\alpha = 0$. As the input concentrations of **S** and **T** are increased, the racemic process becomes metastable and switches spontaneously into the one or the other enantiomeric homochiral reaction sequence (Mason 1984).

Furthermore, we used a relatively high value of the rate constant $\mathbf{k_3}$ for the generation of product **P**. Precise values for rate constants and initial concentrations of the reactants are given below[1] together with the *Mathematica*® commands (see also Kondepudi and Prigogine 1998). Summarizing, an inflow of **S** and **T** and an outflow of the product **P** maintained the system in non-equilibrium conditions.

The initial concentration of the L-enantiomer was selected as $\mathbf{X_L}(t=0) = 0.001$ mol L^{-1} and the initial concentration of the D-enantiomer is given by $\mathbf{X_D}(t=0) = 0$ mol L^{-1}. Therewith we introduced a tiny initial asymmetry into the system that

[1] *Mathematica*® commands:
k1f=0.5;k1r=0.01;k2f=0.5;k2r=0.2;k3=1.5;S=1.25;T=1.25;
Soln1=NDSolve[{XL'[t]==k1f×S×T-k1r×XL[t]+k2f×S×T×XL[t]-k2r×
(XL[t])^2-k3×XL[t]×XD[t],XD'[t]==k1f×S×T-k1r×XD[t]+k2f×S×T×XD[t]-
k2r×(XD[t])^2-k3×XL[t] ×XD[t],
XL[0]==0.001, XD[0]==0.0}, {XL, XD}, {t, 0, 100}, MaxSteps→500]
Plot[Evaluate[{XL[t], XD[t]}/.Soln1], {t, 0, 30}]

successfully evolved to high enantiomeric excess after 30 min. Such systems are characterized by an extreme sensitivity – Stephen Mason entitled them "hypersensitive" (Mason 1984) – for tiny asymmetries that might be amplified, triggering a selection between two alternative outputs. Kondepudi (1987) calculated that such reaction sequence models might even amplify asymmetries originated from parity non-conserving energy differences to reach a significant effect on the selection of biomolecular chirality. More generally spoken, a tiny asymmetry that arose from one (or more) of the models that we have discussed in the preceding chapters has the potential to be noticeably amplified.

Asymmetries caused by external perturbations to the system but also intrinsic asymmetries are unified in the complete Langevin equation (10.8). Intrinsic asymmetries such as parity non-conserving energy differences ΔE^{PNC} are represented in the g-term of this equation with $g = \Delta E^{PNC}/kT$, where k is the Boltzmann constant and T the temperature. Thermal fluctuations, which increase with a rise in temperature, offset the hypersensitivity of the reactions sequence at the bifurcation to chiral perturbations. An increase in temperature thus decreases g-values, being proportional to $\Delta E/kT$.

The external chiral influence of the environment such as circularly polarized light or other macroscopic electric, magnetic, and gravitational chiral fields is expressed by the term $C'\eta f_2(t)$ where $f_2(t)$ is assumed to be a normalized Gaussian white noise. Furthermore, random thermodynamic fluctuations including the normalized Gaussian white noise $f_1(t)$ given by $\sqrt{\varepsilon} f_1(t)$ were respected in Eq. 10.8. A, B, and C are kinetic constants, and ε can also be calculated from the chemical kinetics (Kondepudi and Nelson 1985; Kondepudi and Asakura 2001).

$$\frac{d\alpha}{dt} = -A\alpha^3 + B(\lambda - \lambda_C)\alpha + Cg + C'\eta f_2(t) + \sqrt{\varepsilon} f_1(t) \qquad (10.8)$$

In principle, this equation allows the calculation of an asymmetry in a bifurcating system such as the biomolecular asymmetry that had fallen towards the L-enantiomers of amino acids. The analysis of Kondepudi and Nelson (1985) and Kondepudi and Asakura (2001) of this situation resulted in the counterintuitive result that even when the magnitude of the systematic asymmetry expressed in the g-term is smaller than the root-mean-square value of the random fluctuations, the systematic asymmetry factor could tip the balance to a preferred asymmetric state with high probability. A minute but systematic chiral preference, no stronger than the preference caused by the parity non-conserving energy differences, was proposed to have been amplified by this sequence reaction over a period of about 15 000 years to determine which enantiomer will dominate (Kondepudi and Nelson 1987). Unfortunately, this time scale remained difficult to reproduce in the laboratory for any experimental verification. A common criticism of the application of the complete Langevin equation is that such a unified calculation requires the precise knowledge of each of the Langevin parameters. Today, we are still far from that precise knowledge.

Since recently, only computer simulations of this sort of stereospecific autocatalytic reaction sequences have produced promising results: Thomas Buhse, now at Morelos State University in Mexico, stimulated by the Kondepudi-Prigogine system, studied the autooxidation reaction of tetralin, a simple hydrocarbon molecule (Buhse et al. 1993a). Tetralin is achiral and well known to be oxidized under mild conditions in the liquid phase. The oxidation product tetralin hydroperoxide is chiral and partly decomposes to peroxyl radicals (ROO ·) that are chiral as well and which react with tetralin forming the chiral tetralol. A computer analysis involving the kinetic equations predicted stereospecific and autocatalytic behaviour of the system resulted in a measurable enantiomeric excess. Experimental confirmation of the reaction sequence by investigation of the tetralin-system in the laboratory, however, had remained difficult to obtain (Buhse et al. 1993b).

The situation changed dramatically some years ago when experimental approaches successfully demonstrated the amplification of small enantiomeric excesses. These experiments will be outlined in Sect. 10.4. Before, Yamagata's accumulation principle that is based on a mathematical model will be presented.

10.3 Amplification by Progressive Accumulation

Yamagata (1966) proposed an amplification model which he called "accumulation principle". This model describes the progressive amplification of a small enantiomeric enhancement starting from an achiral or racemic substrate and terminating with enantiomerically enriched or even homochiral products corresponding to enantiomeric excesses of $+1$ or -1. Here, small differences between successive enantiomeric stages are progressively accumulated in a time-extended, uniformly cumulative process. The differences are manifested in activation parameters, for example of n enantiomeric monomers evolving towards polymers. The obtained enantiomeric excess of the favoured polymer depends on the degree of polymerization. Polypeptides (Mason 1984) but also DNA (Yamagata 1966) have been discussed as such polymers.

In a general criticism of the "accumulation principle" it is argued that the polymerization needs to be enantiospecific, but with no enantiomeric antagonism (MacDermott and Tranter 1989). The monomers are furthermore required to be optically labile, i.e., they undergo rapid racemization, otherwise the polymerization of an initially racemic mixture of monomers cannot result in the production of a homochiral polymer. These conditions are usually not applicable to biopolymerization reactions often including enantiomeric antagonism and slowly racemizing chemical species such as amino acids.

Anyhow, the "accumulation principle" is ideally applicable to the resolution of racemic substances by crystallization. Here, the precipitate of racemic tartrate solution serves as example which was demonstrated to show a tiny but reproducible optical activity (Thiemann and Wagener 1970). Such kind of amplification we have moreover discussed in detail in Chap. 4 on the crystallization of quartz, sodium chlorate solutions, binaphthyl melts, and others. As we have seen, here the criterion of optical lability is often fulfilled as achiral monomer units were used for the construction of an enantiomorphous crystal. The implications of homochiral crystals for the origin of biomolecular asymmetry were sufficiently discussed in Chap. 4, including the distribution of *d*-(+)- and *l*-(−)-quartz on Earth.

10.4 Amplification by the Soai Reaction

Even if the theoretical basis for the amplification of minute chiral imbalances of enantiomers was founded more than half a century ago by Frank (1953), we had to wait for its experimental proof until 1995 and the landmark results of Soai et al. (1995). Organic chemist Kenso Soai from the University of Tokyo and coworkers studied an asymmetric autocatalytic reaction that used a catalyst with initial low enantiomeric excess. This reaction yielded the same catalyst as a product with very high enantiomeric excess. The first such system to be found was the asymmetric autocatalysis of 5pyrimidyl alkanol with 2% initial enantiomeric excess (Soai et al. 1995) proceeding without the need of any other chiral auxiliaries.

In this system the chiral initiator (5-pyrimidyl alkanol) showing low enantiomeric excess determined the absolute configuration of the product showing itself (5-pyrimidyl alkanol) overwhelming enantiomeric excesses close to enantiomeric purity of 99.5% (Soai et al. 2000, 2001). Since 1995, a wide variety of chiral 5-pyrimidyl, 3-pyrimidyl-, and 3-quinolyl alkanols have been shown to be asymmetric autocatalysts. In the enantioselective addition of diisopropylzinc (*i*-Pr_2Zn) to an achiral aldehyde these compounds automultiply the initial enantiomeric excess of a chiral alkanol as it is illustrated in Fig. 10.6. Today, this kind of reaction is subsumed under the term "Soai reaction".

Thus, a small enantiomeric asymmetry in the chiral initiator was highly amplified by the above Soai reaction. Are other chiral initiators also able to tip the balance towards the products' *S*- or *R*-configuration?

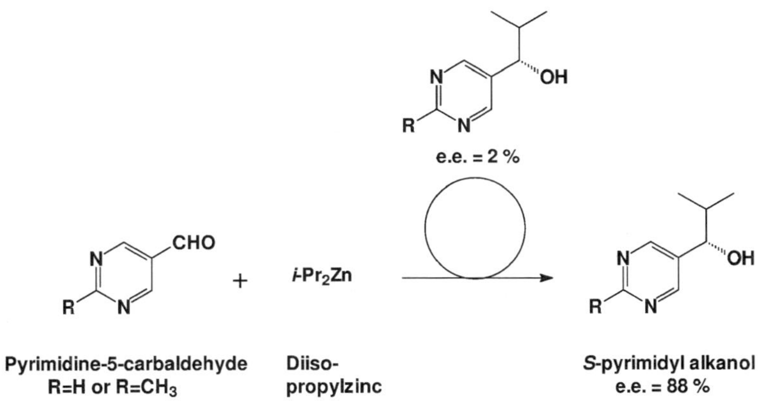

Fig. 10.6 The chiral *S*-pyrimidyl alkanol with 2% enantiomeric excess is capable of autocatalyzing its own production in the Soai reaction between *i*-Pr_2Zn and an aldehyde yielding a high enantiomeric excess of 88% and more (Soai et al. 1995)

Yes, indeed. Four years after the discovery of the Soai reaction, Soai et al. (1999) tried enantiomorphous quartz crystals as chiral initiators. The inspected chemical reaction was – similar to the above system – the asymmetric addition of *i*-Pr_2Zn to a heteroaromatic aldehyde chemically described as 2-(*tert*-butylethynyl)-pyrimidine-5-carbaldehyde in toluene. And indeed, the chiral product, a secondary alcohol, was formed in its *S*-configuration by using (+)-*d*-quartz powder as chiral initiator. On the other hand, in the presence of (−)-*l*-quartz powder as chiral promoter the product was obtained in *R*-configuration. The determined enantiomeric excesses were significantly high with 93–97%. The authors concluded that chiral enantiomorphous quartz crystals might have been involved in the origin of the biomolecular homochirality through catalytic asymmetric synthesis. In addition to the use of quartz crystals as chiral initiators, the Soai reaction was reported to produce high enantiomeric enrichments with enantiomorphous sodium chlorate $NaClO_3$ crystals (Sato et al. 2000) as well as chiral two-component crystals of tryptamine/para-chlorobenzoic acid (Kawasaki et al. 2005). This kind of reaction serves as an intriguing example of "chirality propagation" from an inorganic to an organic reaction system.

But not enough to the Soai reaction: instead of quartz crystals in high optical purity, an amino acid itself was tested in a set of experiments to serve as chiral initiator. The chosen amino acid leucine was introduced into the Soai reaction with a small *e.e.* of + 2% and − 2%. Asymmetric photochemical reactions with circularly polarized light are known to be capable of inducing an enantiomeric excess of this value into racemic mixtures of amino acids, particularly into leucine (Flores et al. 1977; Meierhenrich et al. 2005b). By adding the Soai reactants *i*-Pr_2Zn and 2-methylpyrimidine-5-carbaldehyde to L-leucine with an 2% enantiomeric excess in a toluene solution, the product (*R*-2-methyl-1-(2-methyl-5-pyrimidyl) propan-1-ol) was reproducibly obtained in its *R*-configuration showing 21% enantiomeric excess. Conversely, in the presence of D-leucine with an 2% enantiomeric excess, enantioselective alkylation under the same conditions gave the approx. opposite enantiomer-enriched *S*-pyrimidyl alkanol with 26% enantiomeric excess (Shibata et al. 1998). Similar results were obtained by use of the amino acid valine and also with methyl mandelate, 2-butanol or a chiral carboxylic acid as chiral initiator. All these chemical species serve as chiral initiators or chiral promoters in the Soai system due to which small amounts of the chiral autocatalyst, the pyrimidyl alkanol itself, form. It is important to understand that quartz, amino acids, methyl mandelate, 2-butanol, and chiral carboxylic acids cannot be considered themselves as autocatalysts in the Soai system.

The detailed chemical mechanism in particular for the autocatalytic step of the Soai reaction remained unknown, until the team of Thomas Buhse at Morelos State University in Mexico contributed to its kinetic understanding very recently. With the help of non-linear differential equations the chemical kinetics of the Soai reaction were modelled. And the results show indeed that a chiral amplification starting from an initial enantiomeric excess of $< 10^{-6}\%$ to $> 60\%$ is feasible. The kinetic model used was based on the Frank model involving stereospecific autocatalysis and mutual inhibition (Rivera Islas et al. 2005). For further insights into the autocatalytic kinetics of the Soai reaction, kinetic sequences were proposed by Lavabre et al. (2008), which were based on the Frank model, as well.

At the University of South-Paris in Orsay, France, Henri Kagan has been working on "surprising phenomena" on asymmetric amplification such as an enantiomerically impure chiral auxiliary or ligands that give a stereoselection higher than its own and even equivalent to the pure one. Kagan entitles these effects generally as "non-linear effects in asymmetric synthesis". A wide variety of different chemical systems showing this effect was summarized by Girard and Kagan (1998). These non-linear effects describe that the relation between the enantiomeric excess value of the chiral auxiliary and the enantiomeric excess value of the product deviates from linearity. This deviation might be caused by molecular aggregation or molecular organization in a particular environment reflecting molecular interactions and complexity in an eventual rather subtle reaction mechanism. Furthermore, Ryoji Noyori, Nobel laureate of Chemistry in 2001, contributed reflections on the possible reaction mechanisms of "non-linear effects" and concluded that the degree of nonlinearity is highly affected not only by the structures and purity of catalysts but also by various reaction parameters, all of them summarized in Noyori et al. (2001).

Despite of all these beautiful amplifications of small enantiomeric excesses by the Soai reaction or non-linear effects we should remain realistic, open our eyes and see that the dialkylzinc chemistry involved in the amplifying Soai reaction, however, is unlikely to have been of basic importance in an aqueous prebiotic environment.

Consequently, the area of amino acid catalysis moved recently into the center of scientific interest. Here, first experimental trials are promising, showing for example in the case of a proline-mediated reaction quite unexpectedly accelerating reaction rates and an amplified, temporally increasing enantiomeric excess of the product (Mathew et al. 2004). The authors were intrigued to find that when the reaction was carried out with non-enantiopure proline, the enantiomeric excess of the product was higher than that expected for a linear relationship and that enantiomeric excess increased during the course of the reaction.

10.5 Transfer, Memory, and Switching of Chiral Properties

In the general context of amplification of small enantiomeric enhancements it is reasonable to mention that some modern research activities focus on transfer, memory, and switching of chiral properties (see Huck et al. 1996). Chiral information can be transferred from chiral non-racemic organic molecules into liquid crystals and then amplified. Such a chirality propagation is usually based on a nematic to cholesteric phase transition of the liquid crystal which depends on the chiral information in the organic dopant (Solladie and Zimmermann 1984; Green et al. 1998; Irie 2000). It was even demonstrated that a chiral bicyclic ketone trigger can induce the reversible switching of a liquid crystal between its nematic and cholesteric form by irradiation with CPL followed by unpolarized light (Burnham and Schuster 1999). The other way round, the transfer from a structural chiral information stored in a liquid crystal can be transferred and amplified into chiral trioxalato chromate guest molecules (Teutsch 1988; Thiemann and Teutsch 1990).

In this context, Ben Feringa and his team at the University of Groningen, The Netherlands, made very recently furore by proposing an amplification of a low enantiomeric enhancement in amino acids by sublimation[2] only (Fletcher et al. 2007). The amino acid leucine with an initial enantiomeric excess of up to 10% was partially sublimated from the solid into the gaseous phase. After condensation of the gas phase leucine molecules an enantiomeric excess of 82% was detected in the sublimate. The author assumed to have used "astrophysically-relevant conditions" and to have performed relevant work in order to better understand the origin of biomolecular's asymmetry.

Nevertheless, the reported results are not surprising. In the mentioned experiment, L-amino acid crystals were added to racemic D,L-amino acid crystals before sublimation. Thus two types of crystals were mixed having different lattice structures. This mixture of crystals can be separated under specific temperature/pressure conditions, since these crystals have very different physico-chemical properties such as melting points but also sublimation temperatures. Under the chosen conditions, L-amino acids sublimed preferably, whereas racemic crystals remained in the solid state. An enantiomeric excess of 82% in the sublimate is absolutely not surprising (and depends on the absolute amount of sublimate which is not given by the authors).

However, a more profound contribution to amino acid enantiomer enrichment would be an experiment in which one mixes L-amino acid crystals with their counterpart D-amino acid crystals in order to obtain the starting 10% enantiomeric excess to be sublimated. In this case, melting points and sublimation temperatures are identical. I predict that here the authors would have obtained only the same 10% enantiomeric excess also in the sublimate. The statement of Fletcher et al. (2007) to have applied "astrophysically-relevant conditions" requires thus a sound explanation and has to be taken cautiously.

10.6 Concluding Remarks

Mechanisms for the asymmetric amplification of small enantiomeric excesses are important for designing models of biogenesis which all demand a supply of sufficient enantioenrichments in precursor molecules, thus triggering self-organization far from thermodynamic equilibrium (Krueger and Kissel 1989). Such a mechanism being provided lets us assume that the required high enantiomeric excess of one chiral species would not proceed necessarily ab initio, because reaction sequences involving only one powerful autocatalytic step would suffice to select one stereoisomer (Eigen and Schuster 1978a, 1978b). Tipping the balance to the one or the other stereoisomer might have occurred just by chance or by one of the attractive determinate mechanisms presented in the above chapters. The mirror species not selected by the first living organisms for oligomer and polymer production could then be racemized, e.g., by ultraviolet radiation, simply decomposed, or consumed as "nutrient".

---

[2] Sublimation describes the phase transition of a chemical compound from the solid state directly into the gaseous phase.

Appendix

Table 1 Trivial and IUPAC names of amino acids hitherto identified in the Murchison meteorite. Corresponding chemical structures for amino acids with up to five carbon atoms can be found in Fig. 1, structures with six and more carbon atoms are given in Fig. 2 (Bredehöft and Meierhenrich 2007)

#	#C	Amino acid	IUPAC name	Type	Reference
1	2	Glycine	2-Aminoethanoic acid	m/m	a
2	3	Sarcosine; N-Methylglycine	2-Methylaminoethanoic acid	m/m	a
3	3	Alanine	3-Aminopropanoic acid	m/m	a
4	3	β-Alanine	2,3-Diaminopropanoic acid	m/m	a
5	3	3-Aminoalanine	2-Amino-3-hydroxypropanoic acid	d/m	m
6	3	Serine	2-Methylamino-ethanoic acid	m/m	f
7	4	α-Aminobutyric acid	2-Aminobutanoic acid	m/m	b
8	4	β-Aminobutyric acid	3-Aminobutanoic acid	m/m	b
9	4	γ-Aminobutyric acid	4-Aminobutanoic acid	m/m	b
10	4	2,3-Diaminobutyric acid	2,3-Diaminobutanoic acid	d/m	m
11	4	2,4-Diaminobutyric acid	2,4-Diaminobutanoic acid	d/m	m
12	4	Threonine	2-Amino-3-hydroxybutanoic acid	m/m	f
13	4	Aspartic acid	2-Aminobutane-1,4-dibutanoic acid	m/d	b
14	4	α-Aminoisobutyric acid	2-Amino-2-methylpropanoic acid	m/m	a
15	4	β-Aminoisobutyric acid	3-Amino-2-methylpropanoic acid	m/m	b
16	4	3,3-Diaminoisobutyric acid	3,3-Diamino-2-methyl-propanoic acid	d/m	m
17	4	N-Methylalanine	2-Methylamino-propanoic acid	m/m	b
18	4	N-Methyl-β-alanine	3-Methylamino-propanoic acid	m/m	c
19	4	N-Ethylglycine	2-Ethylamino-ethanoic acid	m/m	b

Table 1 (continued)

#	#C	Amino acid	IUPAC name	Type	Reference
20	4	N,N-Dimethylglycine	2-Dimethylamino-ethanoic acid	m/m	c
21	5	Norvaline	2-Aminopentanoic acid	m/m	b
22	5	3-Ethyl-β-alanine	3-Aminopentanoic acid	m/m	i
23	5	γ-Aminovaleric acid	4-Aminopentanoic acid	m/m	c
24	5	δ-Aminovaleric acid	5-Aminopentanoic acid	m/m	d
25	5	Ornithine	2,5-Diaminopentanoic acid	d/m	m
26	5	Glutamic Acid	2-Amino-pentane-1,5-dioic acid	m/d	a
27	5	Proline	Pyrrolidine-2-carboxylic acid	m/m	a
28	5	Isovaline	2-Amino-2-methylbutanoic acid	m/m	b
29	5	Valine	2-Amino-3-methylbutanoic acid	m/m	a
30	5	Methionine	2-amino-4-(methylsulfanyl)-butanoic acid	m/m	g
31	5	2,3-Dimethyl-β-alanine	3-Amino-2-methylbutanoic acid	m/m	j
32	5	Allo-2,3-dimethyl-β-alanine	Allo-3-amino-2-methyl-butanoic acid	m/m	j
33	5	3,3-Dimethyl-β-alanine	3-Amino-3-methylbutanoic acid	m/m	j
34	5	2-Methyl-γ-aminobutyric acid	4-Amino-2-methylbutanoic acid	m/m	j
35	5	3-Methyl-γ-aminobutyric acid	4-Amino-3-methylbutanoic acid	m/m	j
36	5	4,4-Diaminoisopentanoic acid	4,4-Diamino-2-methylbutanoic acid	d/m	m
37	5	N-Methylaspartic acid	2-Methylamino-butane-1,4-dioic acid	m/d	j
38	5	2-Methylaspartic acid	2-Amino-2-methylbutane-1,4-dioic acid	m/d	j
39	5	3-Methylaspartic acid	2-Amino-3-methylbutane-1,4-dioic acid	m/d	j
40	5	Allo-3-methylaspartic acid	Allo-2-amino-3-methylbutane-1,4-dioic acid	m/d	j
41	5	2-Ethyl-β-alanine	3-Amino-2-ethylpropanoic acid	m/m	j
42	5	2,2-Dimethyl-β-alanine	3-Amino-2,2-dimethyl-propanoic acid	m/m	j
43	6	Norleucine	2-Aminohexanoic acid	m/m	d
44	6	β-Aminocaproic acid	3-Aminohexanoic acid	m/m	j*
45	6	γ-Aminocaproic acid	4-Aminohexanoic acid	m/m	d
46	6	δ-Aminocaproic acid	6-Aminohexanoic acid	m/m	d**

*,*** At least one of the pair present; ** Identified in the Mighei meteorite

Table 1 (continued)

#	#C	Amino acid	IUPAC name	Type	Reference
47	6	α-Aminoadipic acid	2-Aminohexane-1,6-dioic acid	m/d	d
48	6	β-Aminoadipic acid	3-Aminohexane-1,6-dioic acid	m/d	n
49	6	Pipecolic acid	Piperidine-2-carboxylic acid	m/m	b
50	6	Cycloleucine	1-Aminocyclopentanecarboxylic acid	m/m	j*
51	6	3-Ethyl-γ-aminobutyric acid	3-Methylamine-pentanoic acid	m/m	j***
52	6	2-Methylnorvaline	2-Amino-2-methylpentanoic acid	m/m	h
53	6	Isoleucine	2-Amino-3-methylpentanoic acid	m/m	e
54	6	Alloisoleucine	Allo-2-amino-3-methyl-pentanoic acid	m/m	d
55	6	Leucine	2-Amino-4-methylpentanoic acid	m/m	e
56	6	3-Ethyl-3-methyl-β-alanine	3-Amino-3-methylpentanoic acid	m/m	j
57	6	2,4-Dimethyl-γ-amino-butyric acid	4-Amino-2-methylpentanoic acid	m/m	j
58	6	3,4-Dimethyl-γ-amino-butyric acid	4-Amino-3-methylpentanoic acid	m/m	j
59	6	4,4-Dimethyl-γ-amino-butyric acid	4-Amino-4-methylpentanoic acid	m/m	j*
60	6	2-Methylglutamic acid	2-Amino-2-methylpentane-1,5-dioic acid	m/d	h
61	6	3-Methylglutamic acid	2-Amino-3-methylpentane-1,5-dioic acid	m/d	n
62	6	4-Methylglutamic acid	2-Amino-4-methylpentane-1,5-dioic acid	m/d	n
63	6	2-Ethyl-α-amino-butyric acid	2-Amino-2-ethylbutanoic acid	m/m	h
64	6	2-Ethyl-3-methyl-β-alanine	3-Amino-2-ethylbutanoic acid	m/m	j
65	6	2-Methylvaline	2-Amino-2,3-dimethylbutanoic acid	m/m	h
66	6	Pseudoleucine	2-Amino-3,3-dimethylbutanoic acid	m/m	h
67	6	2,3,3-Trimethyl-β-alanine	3-Amino-2,3-dimethylbutanoic acid	m/m	j
68	6	3,3-Dimethyl-γ-amino-butyric acid	4-Amino-3,3-dimethylbutanoic acid	m/m	j***
69	7	α-Aminoenanthic acid	2-Aminoheptanoic acid	m/m	k
70	7	α-Pimelic acid	2-Aminoheptane-1,7-dioic acid	m/d	j
71	7	2-Methylnorleucine	2-Amino-2-methylhexanoic acid	m/m	j
72	7	3-Methylnorleucine	2-Amino-3-methylhexanoic acid	m/m	j

Table 1 (continued)

#	#C	Amino acid	IUPAC name	Type	Reference
73	7	Allo-3-methylnorleucine	Allo-2-amino-3-methyl-hexanoic acid	m/m	j
74	7	4-Methylnorleucine	2-Amino-4-methyl-hexanoic acid	m/m	j
75	7	Allo-4-methylnorleucine	Allo-2-amino-4-methyl-hexanoic acid	m/m	j
76	7	5-Methylnorleucine	2-Amino-5-methyl-hexanoic acid	m/m	j
77	7	2-Ethylnorvaline	2-Amino-2-ethylpentanoic acid	m/m	j
78	7	3-Ethylnorvaline	2-Amino-3-ethylpentanoic acid	m/m	j
79	7	2,3-Dimethylnorvaline	2-Amino-2,3-dimethyl-pentanoic acid	m/m	j
80	7	Allo-2,3-dimethylnorvaline	Allo-2-amino-2,3-dimethyl-pentanoic acid	m/m	j
81	7	2,4-Dimethylnorvaline	2-Amino-2,4-dimethyl-pentanoic acid	m/m	j
82	7	3,3-Dimethylnorvaline	2-Amino-3,3-dimethyl-pentanoic acid	m/m	k
83	7	3,4-Dimethylnorvaline	2-Amino-3,4-dimethyl-pentanoic acid	m/m	j
84	7	Allo-3,4-dimethylnorvaline	Allo-2-amino-3,4-dimethyl-pentanoic acid	m/m	j
85	7	4,4-Dimethylnorvaline	2-Amino-4,4-dimethyl-pentanoic acid	m/m	j
86	7	2-Ethylvaline	2-Amino-2-ethyl-3-methyl-butanoic acid	m/m	j
87	7	2,3,3-Trimethyl-γ-aminobutyric acid	2-Amino-2,3,3-trimethyl-butanoic acid	m/m	k
88	9	Phenylalanine	2-Amino-3-phenyl-propanoic acid	m/m	g
89	9	Tyrosine	2-Amino-3-(4-hydroxyphenyl)-propanoic acid	m/m	g

Type: m/m Monoamino monocarboxylic acid; m/d Monoamino dicarboxylic acid; d/m Diamino monocarboxylic acid. References: (a) Kvenvolden et al. 1970; (b) Kvenvolden et al. 1971; (c) Lawless 1973; (d) Buhl 1975; (e) Pereira et al. 1975; (f) Cronin and Moore 1976; (g) Kotra et al. 1979; (h) Cronin et al. 1981; (j) Cronin and Pizzarello 1983; (k) Cronin and Pizzarello 1986; (m) Meierhenrich et al. 2004; (n) Pizzarello et al. 2004

*,*** At least one of the pair present; ** Identified in the Mighei meteorite

Table 2 Resolution R_S of 3 stationary phases selected for the COSAC experiment onboard the ROSETTA Lander Philae

Family of compounds	Racemic analyte	Chirasil-L-Val R_S	Chiraldex G-TA R_S	Chirasil-Dex CB R_S
Alkanes	3-Methylhexane	n.d.	n.s.	2.05
	3-Methylheptane	n.s.	n.s.	3.92
	3-Methyloctane	n.s.	n.s.	3.79
	3-Methylnonane			2.83
	3-Methyldecane			2.66
	2,4-Dimethylhexane			0.78
	2,2,4-Trimethylhexane			1.44
	2,3-Dimethylheptane			2.75
	4-Methyloctane			3.04
	4-Methylnonane			2.69
	4-Methyldecane			1.52
	2,2,4,4,6,8,8-Hepta-methylnonane			14.27
Amines	2-Amino-*n*-butane	n.d.	n.d.	
	2-Amino-*n*-pentane	n.s.	n.s.	
	2-Amino-*n*-heptane	n.s.	n.s.	
	2-Amino-5-methylhexane	n.s.	n.s.	
	2-Ethyl-1-hexylamine	n.s.	n.s.	
	1-Phenylethylamine	n.s.	0.59	
Alcohols	2-Butanol	n.s.	0.40	
	2-Pentanol	n.s.	0.83	
	2-Hexanol	n.s.	0.96	
	Menthol	0.32	0.85	
Diols	1,3-Butanediol	n.s.	1.04	
	1,2-Pentanediol	n.s.	1.66	
Amines N-TFA	2-Methylpiperidine	n.s.	3.78	
	2-Amino-n-heptane	0.94	2.17	
	1-Aminoindane	3.86	1.01	
Amino acids	Phenylalanine	3.75 [gd]		
Hydroxy-carboxylic acids	Lactic acid		0.48[gd]	
	Mandelic acid		1.18[gd]	
	Malic acid		1.27[gd]	
	Tartaric acid		3.68[gd]	

R_S resolution, *n.s.* no separation of the enantiomers possible, *n.d.* no detection possible, *N-TFA* N-trifluoracetyl, *gd* gas-phase derivatisation with dimethylformamide dimethylacetal DMF-DMA.

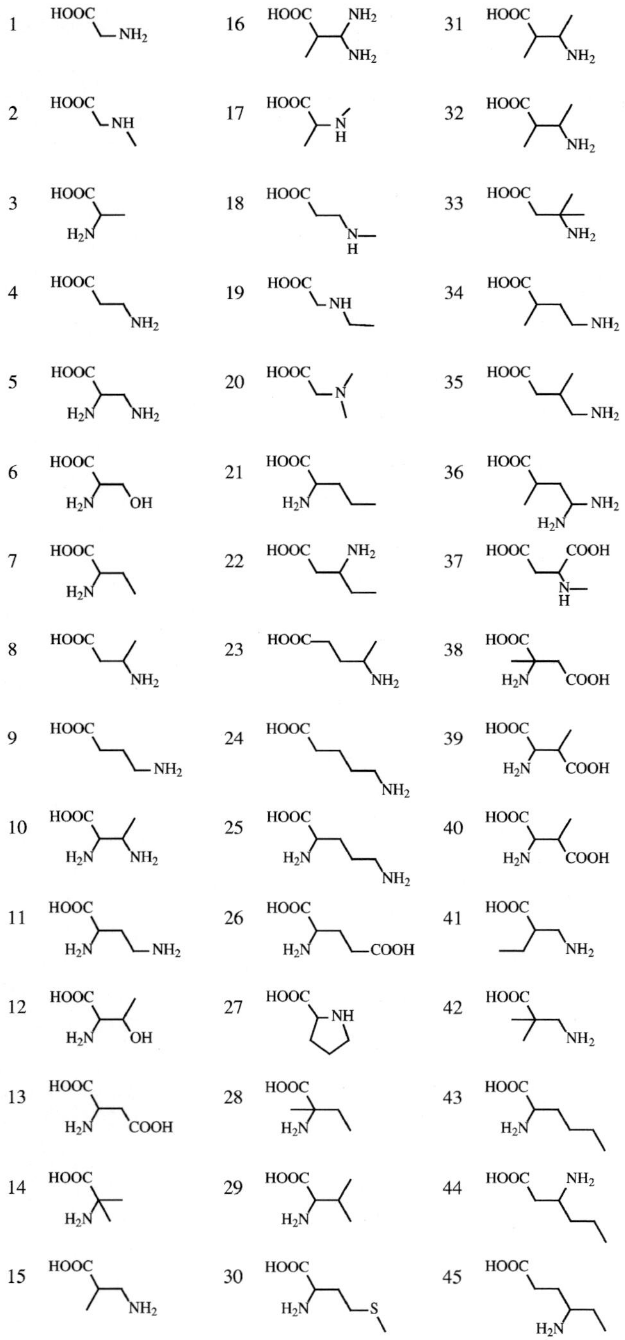

Fig. 1 Structures of amino acids with 2 to 6 carbon atoms hitherto identified in the Murchison meteorite (Bredehöft and Meierhenrich 2007)

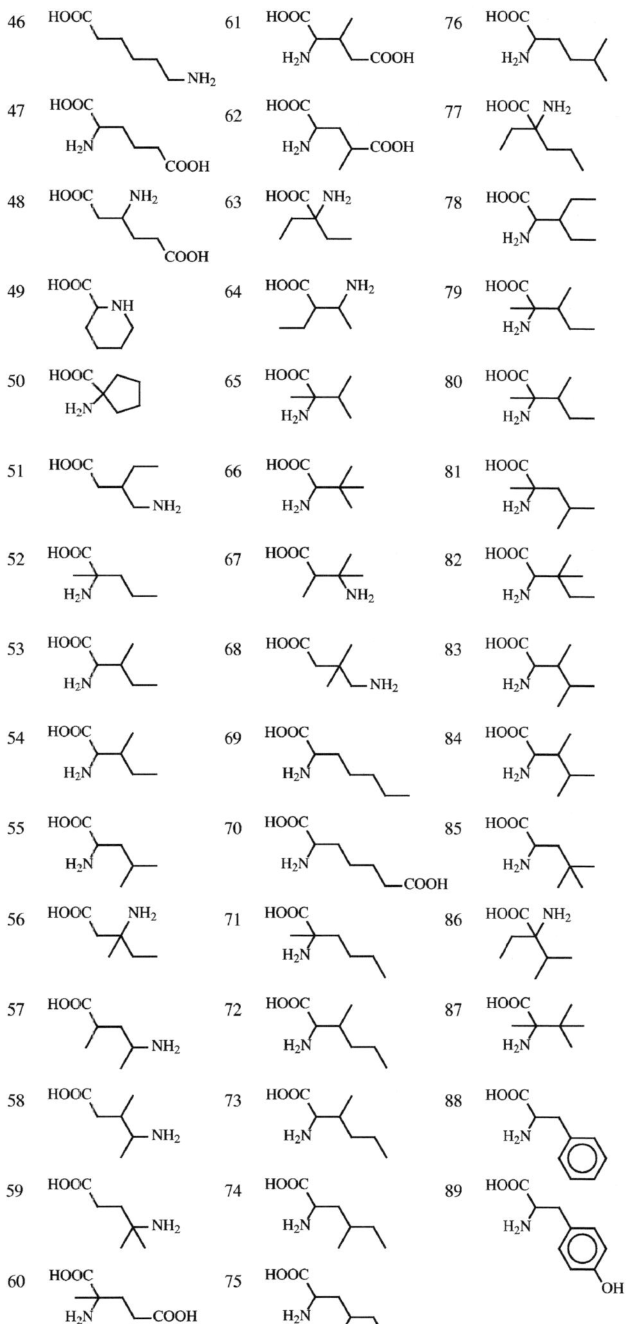

Fig. 2 Structures of amino acids hitherto identified in the Murchison meteorite with 6, 7, and 9-carbon atoms. There is a report of one "8-carbon alkanoic acid" in the literature, but it is not known, which isomer was observed (Bredehöft and Meierhenrich 2007)

Bibliography

Abe I, Fujimoto N, Nishiyama T, Terada K, Nakahara T (1996) Rapid analysis of amino acid enantiomers by chiral-phase capillary gas chromatography. J Chromatogr A 722:221–227

Adams F (2002) Origins of existence. The Free Press, New York

Agarwal VK, Schutte W, Greenberg JM, Ferris JP, Briggs R, Connor S, van de Bult CPEM, Baas F (1985) Photochemical reactions in interstellar grains photolysis of CO, NH_3, and H_2O. Orig Life 16:21–40

A'Hearn MF, Festou MC (1990) Composition of the coma. In: Huebner WF (ed) Physics and chemistry of comets. Springer, Berlin Heidelberg New York, pp 102–109

Akaboshi M, Noda M, Kawai K, Maki H, Kawamoto K (1979) Asymmetrical radical formation in D- and L-alanines irradiated with Yttrium-90 β-rays. Orig Life 9:181–186

Akaboshi M, Noda M, Kawai K, Maki H, Kawamoto K (1982) Asymmetrical radical formation in D- and L-alanines irradiated with tritium-β-rays. Orig Life 12:395–399

Alcaraz C, Thissen R, Compin M, Jolly A, Drescher M, Nahon L (1999) First polarization measurements of Ophelie: a versatile polarization VUV undulator at Super-Aco. Spie 3773:250–261

Ali H, Pätzold R, Brückner H (2006) Determination of L- and D-amino acids in smokeless tobacco products and tobacco. Food Chem 99:803–812

Anders E (1989) Pre-biotic organic matter from comets and asteroids. Nature 342:255–257

Armstrong DW, Jin HL (1990) Acylation effects on chiral recognition of racemic amines and alcohols by new polar and non-polar cyclodextrin derivative gas chromatographic phases. J Chromatogr 502:154–159

Armstrong DW, Li W, Chang C-D (1990) Polar-liquid, derivatized cyclodextrin stationary phases for the capillary gas chromatography separation of enantiomers. Anal Chem 62:914–923

Asakura K, Ikumo A, Kurihara K, Osanai S, Kondepudi DK (2000) Random chiral asymmetry generation by chiral autocatalysis in a far-from-equilibrium reaction system. J Phys Chem 104:2689–2694

Asthagiri A, Hazen RM (2007) An ab initio study of adsorption of alanine on the chiral calcite(2131) surface. Mol Simul 33:343–351

Atkins PW, Barron LD (1969) Rayleigh scattering of polarized photons by molecules. Mol Phys 16:453–466

Ávalos M, Babiano R, Cintas P, Jiménez JL, Palacios JC (2004) Symmetry breaking: an epistemological note. Tetrahedron Asym 15:3171–3175

Bada JL, McDonald GD (1995) Amino acid racemization on Mars: Implications for the preservation of biomolecules from on extinct Martian biota. Icarus 114:139–143

Bada JL, Herrmann B, Payan IL, Man EH (1989) Amino acid racemization in bone and the boiling of the German Emperor Lothar I. Appl Geochem 4:325–327

Bailey J (2000) Astronomical sources of circularly polarized light and the origin of homochirality. Orig Life Evol Biosphere 31:167–183

Bailey J (2003) Extraterrestrial chirality. In: Celnikier LM, Trân Thanh Vân J (eds) Frontiers of life. Thê Giói Publishers, Vietnam, pp 161–165

Bailey J, Chrysostomou A, Hough JH, Gledhill TM, McCall A, Clark S, Ménard F, Tamura M (1998) Circular polarization in star-formation region: Implications for biomolecular homochirality. Science 281:672–674

Bakasov A, Ha T-K, Quack M (1996) Ab initio calculation of molecular energies including parity violating interactions. In: Chela-Flores J, Raulin F (eds) Chemical evolution: Physics of the origin and evolution of life. Kluwer, Dordrecht, pp 287–296

Balavoine G, Moradpour A, Kagan HB (1974) Preparation of chiral compounds with high optical purity by irradiation with circularly polarized light, a model reaction for the prebiotic generation of optical activity. J Am Chem Soc 96:5152–5158

Bandermann LW, Kemp JC, Wolstencroft RD (1972) Circular polarization of light scattered from rough surfaces. Mon Not R Astr Soc 158:291–304

Bar-Nun A, Chang S (1983) Photochemical reactions of water and carbon monoxide in Earth's primitive atmosphere. J Geophys Res 88:6662–6672

Barron LD (1987) Fundamental symmetry aspects of chirality. BioSystems 20:7–14

Barron LD (1994a) CP violation and molecular physics. Chem Phys Letters 221:311–316

Barron LD (1994b) Can a magnetic field induce absolute asymmetric synthesis? Science 266:1491–1492

Barron LD (2000a) Chemistry: Chirality, magnetism and light. Nature 405:895–896

Barron LD (2000b) Magnetochiral influence. Personal communication, April 19

Barron LD (2004) Molecular light scattering and optical activity. Cambridge University Press, Cambridge

Barron LD (2007) Compliments from Lord Kelvin. Nature 446:505–506

Barron LD, Zhu F, Hecht L, Tranter GE, Isaacs NW (2007) Raman optical activity: An incisiver probe of molecular chirality and biomolecular structure. J Mol Structure 834–836:7–16

Bayer E (1983) Chirale Erkennung von Naturstoffen an optisch aktiven Polysiloxanen. Z Naturforsch 38b:1281–1291

Beier M, Reck F, Wagner T, Krishnamurthy R, Eschenmoser A (1999) Chemical etiology of nucleic acid structure: Comparing pentopyranosyl-($2' \rightarrow 4'$) oligonucleotides with RNA. Science 283:699–703

Benner SA, Devine KG, Matveeva LN, Powell DH (2000) The missing organic molecules on Mars. Proc Natl Acad Sci USA 97:2425–2430

Berger R, Quack M (2000) Multiconfiguration linear response approach to the calculation of parity violating potentials in polyatomic molecules. J Chem Phys 112:3148–3158

Berger R, Gottselig M, Quack M, Willeke M (2001) Parity violation dominates the dynamics of chirality in dichlorodisulfane. Angew Chem Int Ed 40:4195–4198; Paritätsverletzung dominiert die Dynamik der Chiralität in Dischwefeldichlorid. Angew Chem 113:4342–4345

Bernstein MP, Sandford SA, Allamandola LJ, Chang S, Scharberg MA (1995) Organic compounds produced by photolysis of realistic interstellar and cometary ice analogs containing methanol. Astrophys J 454:327–344

Bernstein MP, Dworkin JP, Sandford SA, Cooper GW, Allamandola LJ (2002) Racemic amino acids from the ultraviolet photolysis of interstellar ice analogues. Nature 416:401–403

Bernstein WJ, Calvin M, Burchardt O (1972) Absolute asymmetric synthesis. I. On the mechanism of the photochemical synthesis of nonracemic helicenes with circularly polarized light. Wavelength dependance of the optical yield of octahelicene. J Am Chem Soc 94:494–497

Bernstein WJ, Calvin M, Burchardt O (1973) Absolute asymmetric synthesis. III. Hindered rotation about aryl-ethylene bonds in the excited states of diaryl ethylenes. Structural effects on the asymmetric synthesis of 2- and 4-substituted hexahelicenes. J Am Chem Soc 95:527–532

Berova N, Nakanishi K, Woody RW (2000) Circular dichroims. Principles and applications. Wiley-VCH, New York

Bibring J-P, Rosenbauer H, Boehnhardt H, Ulamec S, Biele J, Espinasse S, Feuerbacher B, Gaudon P, Hemmerich P, Kletzkine P, Moura D, Mugnuolo R, Nietner G, Pätz B, Roll R,

Scheuerle H, Szegö K, Wittmann K (2007) The Rosetta Lander ("Philae") investigations. Space Sci Rev 128:205–220

Biele J, Ulamec S (2008) Capabilities of Philae, the Rosetta Lander. Space Sci Rev, in print

Biemann K (2007) On the ability of the Viking gas chromatograph-mass spectrometer to detect organic matter. Proc Natl Acad Sci USA 104:10310–10313

Biemann K, Oro J, Toulmin P, Orgel LE, Nier AO, Anderson DN, Simmonds PG, Flory D, Diaz AV, Rushneck DR (1976) Search for organic and volatile inorganic compounds in two surface samples from the Chryse Planitia region of Mars. Science 194:72–76

Blair NE, Bonner WA (1980) The radiolysis of tryptophan and leucine with ^{32}P β-radiation. J Mol Evol 15:21–28

Blocher M, Hitz T, Luisi PL (2001) Stereoselectivity in the oligomerization of racemic tryptophan *N*-carboxyanhydride (NCA-Trp) as determined by isotope labeling and mass spectrometry. Helv Chim Acta 84:842–848

Bolik S, Rübhausen M, Binder S, Schulz B, Perbandt M, Genov N, Erdmann V, Klussmann S, Betzel C (2007) First experimental evidence for the preferential stabilization of the natural D- over the nonnatural L-configuration in nucleic acids. RNA 13:1–4

Bolli M, Micura R, Eschenmoser A (1997) Pyranosyl-RNA: chiroselective self-assembly of base sequences by ligative oligomerization of tetranucleotide-2',3'-cyclophosphates (with a commentary concerning the origin of biomolecular homochirality). Chem Biol 4:309–320

Bonner WA (1974) Experiments on the origin of molecular chirality by parity non-conservation during β-decay. J Mol Evol 4:23–39

Bonner WA (1984) Experimental evidence for β-decay as a source of chirality by enantiomer analysis. Orig Life 14:383–390

Bonner WA (1991) The origin and amplification of biomolecular chirality. Orig Life Evol Biosphere 21:59–111

Bonner WA (1992) Terrestrial and extraterrestrial sources of molecular homochirality. Orig Life Evol Biosphere 21:407–420

Bonner WA (1995a) Chirality and life. Orig Life Evol Biosphere 25:175–190

Bonner WA (1995b) The quest of chirality. In: Cline DB (ed) Physical origin of homochirality in life, American Inst. Phys. Conf. Proc. AIP Press, Woodbury New York, pp 17–49

Bonner WA, Bean BD (2000) Asymmetric photolysis with elliptically polarized light. Orig Life Evol Biosphere 30:513–517

Bonner WA, Flores JF (1975) Experiments on the origins of optical activity. Orig Live 6:187–194

Bonner WA, Kavasmaneck PR (1976) Asymmetric adsorption of DL-alanine hydrochloride by quartz. J Org Chem 41:2225–2226

Bonner WA, Rubenstein E (1987) Supernovae, neutron stars and biomolecular chirality. BioSystems 20:99–111

Bonner WA, Kavasmaneck PR, Martin FS, Flores JJ (1974) Asymmetric adsorption of alanine by quartz. Science 186:143–144

Bonner WA, Kavasmanek PR, Martin FS, Flores JJ (1975a) Asymmetric adsorption by quartz: A model for the prebiotic origin of optical activity. Orig Life 6:367–376

Bonner WA, van Dort MA, Yearian MR (1975b) Asymmetric degradation of DL-leucine with longitudinally polarised electrons. Nature 258:419–421

Bonner WA, van Dort MA, Yearian MR, Zeman HD, Gloria CL (1976) Polarized electrons and the origin of optical activity. Isr J Chem 15:89–95.

Bonner WA, Lemmon RM, Noyes HP (1978) β Radiolysis of crystalline ^{14}C-labeled amino acids. J Org Chem 43:522–524

Bonner WA, Blair NE, Flores JJ (1979) Attempted asymmetric radiolysis of D,L-tryptophan with ^{32}P β radiation. Nature 281:150–151

Bonner WA, Blair NE, Dirbas FM (1981) Experiments on the abiotic amplification of optical activity. Orig Life 11:119–134

Bonner WA, Lemmon RM, Conzett HE (1982) The radiolysis and racemization of leucine on proton irradiation. Orig Life 12:51–54

Bonner WA, Greenberg JM, Rubenstein E (1999a) The extraterrestrial origin of the homochirality of biomolecules – rebuttal to a critique. Orig Life Evol Biosphere 29:215–219

Bonner WA, Rubenstein E, Brown GS (1999b) Extraterrestrial handedness: A reply. Orig Life Evol Biosphere 29:329–332

Borchers AT, Davis PA, Gershwin ME (2004) The asymmetry of existence: do we owe our existence to cold dark matter and the weak force? Exp Biol Med 229:21–32

Botta O, Glavin DP, Kminek G, Bada JL (2002) Relative amino acid concentrations as a signature for parent body processes of carbonaceous chondrites. Orig Life Evol Biosphere 32:143–163

Brack A (1987) Selective emergence and survival of early polypeptides in water. Orig Life Evol Biosphere 17:367–379

Brack A (1993) From amino acids to prebiotic active peptides: A chemical restitution. Pure Appl Chem 65:1143–1151

Brack A (1998) The molecular origins of life – assembling the pieces of the puzzle. Cambridge University Press, Cambridge

Brack A (2000) The exobiology exploration of Mars: a survey of the European approaches. Planet Space Sci 48:1023–1026

Brack A, Ferris JP (2002) From the building blocks of life. In: Contrafatto G, Minelli A (eds) Biological systematics. Encyclopedia of Life Support Systems (EOLSS), Developed under the Auspices of the Unesco. Eolss Publishers, Oxford

Brack A, Orgel LE (1975) β-Structures of alternating polypeptides and their possible prebiotic significance. Nature 256:383–387

Brack A, Spach G (1971) New example of helical polymerization. Nature 229:124–125

Brack A, Spach G (1980) β-Structures of polypeptides with L- and D-residues. Part III. Experimental evidences for enrichment in enantiomer. J Mol Evol 15:231–238

Brack A, Spach G (1986) Search for chiral molecules and optical activity in extraterrestrial systems. Council of Europe, Research group on chemical evolution, early biological evolution and exobiology, Parliamentary assembly, Strasbourg, p 17

Brack A, Baglioni P, Borruat G, Brandstätter F, Demets R, Edwards HGM, Genge M, Kurat G, Miller MF, Newton EM, Pillinger CT, Roten C-A, Wäsch E (2002) Do meteorits of sedimentary origin survive terrestrial atmospheric entry? The ESA artificial meteorite experiment STONE. Planet Space Sci 50:763–772

Bredehöft JH (2007) Study on the origin of homochirality in the case of amino acids. Ph.D. thesis, University of Bremen

Bredehöft JH, Meierhenrich UJ (2007) Amino acid structures from UV irradiation of simulated interstellar ices. In: Takenaka N (ed) Recent development of (photo) chemistry in ice. Research Signpost, Kerala, India, in print

Bredehöft JH, Breme K, Meierhenrich UJ, Hoffmann SV, Thiemann WHP (2007) Chiroptical properties of diamino carboxylic acids. Chirality 19:570–573

Bregman JD, Campins H, Witteborn FC, Wooden DH, Rank DM, Allamandola LJ, Cohen M, Tielens AGGM (1987) Airborne and groundbased spectrophotometry of comet P/Halley from 5–13 micrometers. Astron Astrophys 187:616–620

Breuer M, Ditrich K, Habicher T, Hauer B, Keßeler M, Stürmer R, Zelinski T (2002) Industrial methods for the production of optically active intermediates. Angew Chem Int Ed 43:788–824; Angew Chem 116:806–843

Briggs R, Ertem G, Ferris JP, Greenberg JM, McCain PJ, Mendoza-Gomez CX, Schutte W (1992) Comet Halley as an aggregate of interstellar dust and further evidence for the photochemical formation of organics in the interstellar medium. Orig Life Evol Biosphere 22:287–307

Brinton KLF, Engrand C, Glavin DP, Bada JL, Maurette M (1998) A search for extraterrestrial amino acids in carbonaceous antarctic micrometeorites. Orig Life Evol Biosphere 28:413–424

Brownlee D, Tsou P, Aléon J, et al. (2006) Comet 81P/Wild under a microscope Science 314:1711–1716

Brückner H, Westhauser T (2003) Chromatographic determination of L- and D-amino acids in plants. Amino Acids 24:43–55

Brunner H (1999) Rechts oder links. In der Natur und anderswo. Wiley-VCH, Weinheim

Buhl PH (1975) An investigation of organic compounds in the Mighei meteorite. Ph.D. thesis, Maryland University

Buhse T, Lavabre D, Micheau J-C, Thiemann W (1993a) Chiral symmetry breaking: Experimental results and computer analysis of a liquid-phase autooxidation. Chirality 5:341–345

Buhse T, Lavabre D, Micheau J-C, Thiemann W (1993b) Practicabilities and limits of stereospecific autocatalysis: An experimental approach. In: Ponnamperuma C, Chela-Flores J (eds) Proceedings of the Trieste Conference on Chemical Evolution and the Origin of Life, 26–30 October 1992. A Deepak Publishing, Hampton, Virginia, pp 205–218

Buhse T, Kondepudi DK, Hoskins B (1999) Kinetics and chiral resolution in stirred crystallizations of D/L-glutamic acid. Chirality 11:343–348

Buhse T, Durand D, Kondepudi D, Laudadio J, Spilker S (2000) Chiral symmetry breaking in crystallization: the role of convection. Phys Rev Lett 84:4405–4408

Burnham KS, Schuster GB (1999) Transfer of chirality from circularly polarized light to a bulk material property: Propagation of photoresolution by a liquid crystal transition. J Am Chem Soc 121:10245–10246

Buschermöhle M, Whittet DCB, Chrysostomou A, Hough JH, Lucas PW, Adamson AJ, Whitney BA, Wolff MJ (2005) An extended search for circularly polarized infrared radiation from the OMC-1 region of orion. Astrophys J 624:821–826

Cairns-Smith AG (1982) Genetic takeover and the mineral origin of life. Cambridge University Press, Cambridge

Cairns-Smith AG (1985) Seven clues to the origin of life. Cambridge University Press, Cambridge

Calvin M (1969) Chemical evolution. Oxford University Press, Oxford

Campbell DM, Farago PS (1987) Electron optic dichroism in camphor. J Phys B At Mol Phys 20:5133–5143

Cattani M, Bassalo JMF (1998) Weak interactions and the tunneling racemization. Chirality 10:519–521

Cerf C, Jorissen A (2000) Is amino-acid homochirality due to asymmetric photolysis in space? Space Sci Rev 92:603–612

Chaput JC, Switzer C (2000) Nonenzymatic oligomerization on templates containing phosphoester-linked acyclic glycerol nucleic acid analogues. J Mol Evol 51:464–470

Chyba CF (1990) Impact delivery and erosion of planetary oceans in the early inner Solar System. Nature 343:129–133

Chyba CF (1997) A left-handed Solar System? Nature 389:234–235

Chyba CF, Sagan C (1992) Endogenous production, exogenous delivery and impact-shock synthesis of organic molecules: an inventory for the origins of life. Nature 355:125–132

Cintas P (2001) Elemantary asymmetry and biochirality: no longer twinned. ChemPhysChem 2:409–410

Cintas P (2007) Tracing the origins and evolution of chirality and handedness in chemical language. Angew Chem Int Ed 46:4016–4024; Ursprünge und Entwicklung der Begriffe Chiralität und Händigkeit in der chemischen Sprache. Angew Chem 119:4090–4099

Clayden J, Greeves N, Warren S, Wothers P (2006) Organic Chemistry. Oxford University Press, Oxford

Condon EU (1937) Theories of optical rotatory power. Rev Mod Phys 9:432–457

Constante J, Hecht L, Polavarapu PL, Collet A, Barron LD (1997) Absolute configuration of bromochlorofluoromethane from experimental and ab initio theoretical vibrational Raman optical activity. Angew Chem Int Ed 36:885–887; Angew Chem 109:917–919

Conte E (1985) Investigation on the chirality of electrons from ^{90}Sr–^{90}Y beta-decay and their asymmetrical interactions with D- and L-alanines. Lett Nuovo Cimento 44:641–647

Cooper G, Kimmich N, Belisle W, Sarinana J, Brabham K, Garrel L (2001) Carbonaceous meteorites as a source of sugar-related organic compounds for the early Earth. Nature 414:879–883

Cribbs DH, Pike CJ, Weinstein SL, Velazquez P, Cotman CW (1997) All-D-enantiomers of β-amyloid exhibit similar biological properties to all-L-β-amyloid. J Biol Chem 272:7431–7436

Cronin JR (1976) Acid-labile amino acid precursors in the Murchison meteorite. Orig Life 7:337–342

Cronin JR (1998) Clues from the origin of the solar system: Meteorites. In: Brack A (ed) The Molecular Origins of Life – Assembling Pieces of the Puzzle. Cambridge University Press, Cambridge, pp 119–146

Cronin JR, Chang S (1993) Organic matter in meteorites: Molecular and isotopic analyses of the Murchison meteorite. In: Greenberg JM (ed) The chemistry of life's origins. Kluwer Academic Publishers, Dordrecht, pp 209–258

Cronin JR, Moore CB (1976) Amino acids of the Nogoya and Mokoia carbonaceous chondrites. Geochim Cosmochim Acta 40:853–857

Cronin JR, Pizzarello S (1983) Amino acids in meteorites. Adv Space Res 3:5–18

Cronin JR, Pizzarello S (1986) Amino acids of the Murchison meteorite. III – Seven carbon acyclic primary alpha-amino alkanoic acids. Geochim Cosmochim Acta 50:2419–2427

Cronin JR, Pizzarello S (1997) Enantiomeric excesses in meteoritic amino acids. Science 275:951–955

Cronin JR, Pizzarello S (1999) Amino acid enantiomer excesses in meteorites: Origin and significance. Adv Space Res 23:293–299

Cronin JR, Reisse J (2005) Chirality and the origin of homochirality. In: Gargaud M, Barbier B, Martin H, Reisse J (eds) Lectures in Astrobiology, vol 1. Springer, Berlin Heidelberg New York, pp 473–515

Cronin JR, Gandy WE, Pizzarello S (1981) Amino acids of the Murchison meteorite: I. Six carbon acyclic primary alpha-amino alkanoic acids. J Mol Evol 17:265–272

Darge W, Laczko I, Thiemann W (1976) Stereoselectivity of β irradiation of D,L-tryptophan in aqueous solution. Nature 261:522–524

Darge W, Laczko I, Thiemann W (1979) Reply to the attempted asymmetric radiolysis of D,L-tryptophan with ^{32}P β radiation. Nature 281:151

Davies P (2003) The origin of life. The Penguin Press, London

Deamer D, Dick R, Thiemann W, Shinitzky M (2007) Intrinsic asymmetries of amino acid enantiomers and their peptides: a possible role in the origin of biochirality. Chirality 19:751–763

Decker P (1973) Possible resolution of racemic mixtures by bistability in "bioids", open systems which can exist in several steady states. J Mol Evol 2:137–143

de Duve C (2003) A research proposal on the origin of life. Orig Life Evol Biosphere 33:559–574

de Duve C (2006) Singularities: Landmarks on the pathways of life. Cambridge University Press, Cambridge

Dello Russo N, Vervack RJ, Weaver HA, Biver N, Bockelée-Morvan D, Crovisier J, Lisse CM (2007) Compositional homogeneity in the fragmented comet 73P/Schwassmann-Wachmann 3. Nature 448:172–175

Denmark SE (2006) Catalysts break symmetry. Nature 443:40–41.

de Vries A (1958) Determination of the absolute configuration of α-quartz. Nature 181:1193

Duprat J, Engrand C, Maurette M, Kurat G, Gounelle M, Hammer C (2007) Micrometeorites from central Antartic snow: the Concordia collection. Adv Space Res 39:605–611

Egholm M, Burchardt O, Nielsen PE, Berg RH (1992) Peptide Nucleic Acids (PNA). Oligonucleotide analogues with an achiral peptide backbone. J Am Chem Soc 114:1895–1897

Egholm M, Behrens C, Christensen L, Berg RH, Nielsen P, Burchardt O (1993a) Peptide nucleic acids containing adenine or guanine recognize thymine and cytosine in complementary DNA sequences. J Chem Soc Chem Commun 9:800–801

Egholm M, Burchardt O, Christensen L, Behrens C, Freier SM, Driver DA, Berg RH, Kim SK, Norden B, Nielsen PE (1993b) PNA hybridizes to complementary oligonucleotides obeying the Watson-Crick hydrogen-bonding rules. Nature 365:566–568

Ehrenfreund P (1999) Molecules on a space odyssey. Science 283:1123–1124

Ehrenfreund P, Kerkhof O, Schutte WA, Boogert ACA, Gerakines PA, Dartois E, d'Hendecourt L, Tielens AGGM, van Dishoeck EF, Whittet DCB (1999) Laboratory studies of thermally processed H_2O:CH_3OH:CO_2 ice mixtures and their astrophysical implications. Astron Astrophys 350:240–253

Ehrenfreund P, Bernstein MP, Dworkin JP, Sandford SA, Allamandola LJ (2001a) The photostability of amino acids in space. Astrophys J 550:L95–L99

Ehrenfreund P, Glavin DP, Botta O, Cooper G, Bada JL (2001b) Extraterrestrial amino acids in Orgueil and Ivuna: Tracing the parent body of CI type carbonaceous chondrites. Proc Natl Acad Sci USA 98:2138–2141

Eigen M, Schuster P (1978a) The hypercycle – a principle of natural self-organization Part B: The abstract hypercycle. Naturwissenschaften 65:7–41

Eigen M, Schuster P (1978b) The hypercycle – a principle of natural self-organization Part C: The realistic hypercycle. Naturwissenschaften 65:341–369

Ellis J (2003) Antimatter matters. Nature 424:631–634

Elsila JE, Dworkin JP, Bernstein MP, Martin MP, Sandford SA (2007) Mechanisms of amino acid formation in interstellar ice analogs. Astrophys J 660:911–918

Elster H, Gil-Av E, Weiner S (1991) Amino acid racemization of fossil bone. J Archaeol Sci 18:605–617

Emmons TP, Reeves JM, Fortson EN (1983) Parity-nonconserving optical rotation in atomic lead. Phys Rev Lett 51:2089–2092

Engel MH, Macko SA (1997) Isotopic evidence for extraterrestrial non-racemic amino acids in the Murchison meteorite. Nature 389:265–268

Engel MH, Macko SA (2001) The stereochemistry of amino acids in the Murchison meteorite. Precambr Res 106:35–45

Engel MH, Nagy B (1982) Distribution of an enantiomeric composition of amino acids in the Murchison meteorite. Nature 296:837–840

Engel MH, Macko SA, Qian Y, Silfer JA (1995) Stable isotope analysis at the molecular level: a new approach for determining the origins of amino acids in the Murchison meteorite. Adv Space Res 15:99–106

Ercoli Finzi A, Bernelli Zazzera F, Dainese C, Malnati F, Magnani PG, Re E, Bologna P, Espinasse S, Olivieri A (2007) SD2 – How to sample a comet. Space Sci Rev 128:281–299

Eschenmoser A (1997) Towards a chemical etiology of nucleic acid structure. Orig Life Evol Biosphere 27:535–553

Fasel R, Parschau M, Ernst K-H (2006) Amplification of chirality in two-dimensional enantiomorphous lattices. Nature 439:449–452

Fenwick DR, Kagan HB (1999) Asymmetric amplification. In: Denmark SE (ed) Topics in stereochemistry, Vol. 22. Wiley, New York, pp 257–296

Feringa G, Kellog RM, Hulst R, Zondervan C, Kruizinga WH (1994) Attempts to carry out enantioselective reactions in a static magnetic field. Angew Chem Int Ed 33:1458–1459; Angew Chem 106:1526–1527

Ferris J (1993) Catalysis and prebiotic RNA synthesis. Orig Life Evol Biosphere 23:307–315

Feynman RP (1997) Surely you're joking Mr. Feynman! Norton, New York London

Fletcher SP, Jagt RBC, Feringa BL (2007) An astrophysically-relevant mechanism for amino acid enantiomer enrichment. Chem Commun 2578–2580

Flores JJ, Bonner WA, Massey GA (1977) Asymmetric photolysis of (RS)-leucine with circularly polarized ultraviolet light. J Am Chem Soc 99:3622–3625

Försterling H-D, Kuhn H (1985) Praxis der Physikalischen Chemie, Zweite Auflage. VCH Verlagsgesellschaft Weinheim, Weinheim

Frank FC (1953) On spontaneous asymmetric synthesis. Biochim Biophys Acta 11:459–463

Frank H, Nicholson GJ, Bayer E (1978) Chirale Polysiloxane zur Trennung von optischen Antipoden. Angew Chem 90:396–397

Frondel C (1978) Characters of quartz fibers. American Mineralogist 63:17–27

Fujii N (2002) D-Amino acids in living higher organisms. Orig Life Evol Biosphere 32:103–127

Garay AS (1968) Origin and role of optical isomery in life. Nature 219:338–340

Garay AS, Ahlgren-Beckendorf JA (1990) Differential interaction of chiral β-particles with enantiomers. Nature 346:451–453

Gardner M (2005) The new ambidextrous universe. Dover Publications, Mineola New York

Gerakines PA, Schutte WA, Greenberg JM, van Dishoeck EF (1995) The infrared band strengths of H_2O, CO and CO_2 in laboratory simulations of astrophysical ice mixtures. Astron Astrophys 296:810–818

Gerakines PA, Whittet DCB, Ehrenfreund P (1999) Observations of solid carbon dioxide in molecular clouds with the infrared space observatory. Astrophys J 522:357–377

Gesteland RF, Cech TR, Atkins JF (1999) The RNA world. Second edn. Cold Spring Habor Laboratory Press, Cold Spring Habor New York

Gibb EL, Whittet DC, Schutte WA, Boogert ACA, Chiar JE, Ehrenfreund P, Gerakines PA, Keane JV, Tielens AGGM, van Dishoeck EF, Kerkhof O (2000) An inventory of interstellar ices toward the embedded protostar W33A. Astrophys J 536:347–356

Gibb EL, Whittet DC, Chiar JE (2001) Searching for ammonia in grain mantles toward massive young stellar objects. Astrophys J 558:702–716

Gibb EL, Whittet DCB, Boogert ACA, Tielens AGGM (2004) Interstellar ice. The infrared space observatory legacy. Astrophys J 151:35–73

Gilbert W (1986) Origin of life: The RNA world. Nature 319:618

Girard C, Kagan HB (1998) Nonlinear effects in asymmetric synthesis and stereoselective reactions: Ten years of investigation. Angew Chem Int Ed 37:2922–2959; Angew Chem 110:3088–3127

Glassmeier K-H, Boehnhardt H, Koschny D, Kührt E, Richter I (2007) The Rosetta mission: Flying towards the origin of the solar system. Space Sci Rev 128:1–21

Goesmann F, Raulin F, Becker L, Ehrenfreund P, Hilchenbach M (2007a) MOMA, the Martian Organic Molecule Analyser; current developments and capabilities of a combined GC/MS and LD-MS instrument. Geophys Res Abs 9:5953

Goesmann F, Rosenbauer H, Roll R, Szopa C, Raulin F, Sternberg R, Israel G, Meierhenrich U, Thiemann W, Muñoz Caro GM (2007b) COSAC, the cometary sampling and composition experiment on Philae. Space Sci Rev 128:257–280

Gölitz P (1994) Enantioselective reactions in a static magnetic field – a false alarm! Angew Chem Int Ed 33:1457; Enantioselektive Reationen im statischen Magnetfeld? – Falscher Alarm! Angew Chem 106:1525

Goldanskii VI, Khrapov VV (1963) Comparison of the effect of electron irradiation on the optical activity of racemates and optical antipodes. Soviet Physics JETP 16:582–585

Goldanskii VI, Kuz'min VV (1988) Spontaneous mirror symmetry breaking in nature and the origin of life. Z phys Chemie 269:216–274

Goldanskii VI, Kuz'min VV (1991) Chirality and cold origin of life. Nature 352:114

Goldberg SI (2008) Experimental evidence leading to an alternative explanation of why D-tyrosine sometimes crystallizes faster than its L-enantiomer. Orig Life Evol Biosphere 38:149–153

Goldberg SI, Crosby JM, Iusem ND, Younes UE (1987) Racemic origins of the stereochemically homogeneous biosphere. Biased stereoselectivities in the foremation of oligomeric peptides. J Am Chem Soc 109:823–830

Goldhaber M, Grodzins L, Sunyar AW (1957) Evidence for circular polarization of Bremsstrahlung produced by beta rays. Phys Rev 106:826–828

Goossens H, de Leeuw JW, Schenck PA, Brassell SC (1984) Tocopherols as likely precursers of pristane in ancient sediments and crude oils. Nature 312:440–442

Goossens H, Due A, de Leeuw JW, van de Graaf B, Schenck PA (1988) The pristane formation index, a new molecular maturity parameter. A simple method to assess maturity by pyrolysis/evaporation-gas chromatography of unextracted samples. Geochim Cosmochim Acta 52:1189–1193

Gottselig M, Luckhaus D, Quack M, Stohner J, Willeke M (2001) Mode selective stereomutation and parity violation in disulfane isotopomers H_2S_2, D_2S_2, T_2S_2. Helv Chim Acta 84:1846–1861

Green MM, Selinger JV (1998) Cosmic chirality. Science 282:880–881

Green MM, Zanella S, Gu H, Sato T, Gottarelli G, Jha SK, Spada GP, Schoevaars AM, Feringa B, Teramoto A (1998) Mechanism of the transformation of a stiff polymer lyotropic nematic liquid crystal to the cholesteric state by dopant-mediated chiral information transfer. J Am Chem Soc 120:9810–9817

Green MM, Park J-W, Sato T, Teramoto A, Lifson S, Selinger RLB, Selinger JV (1999) The macromolecular route to chiral amplification. Angew Chem Int Ed 38:3138–3154; Die makromolekulare Route zur Chiralitätsverstärkung. Angew Chem 111:3328–3345

Greenberg JM (1982) What are comets made of? A model based on interstellar dust. In: Wilkening LL, Matthews MS (eds) Comets. University Arizona Press, Tucson, pp 131–163

Greenberg JM (1986) Predicting that comet Halley is dark. Nature 321:385

Greenberg JM (1993) Physical and chemical composition of comets – from interstellar space to the Earth. In: Greenberg JM (ed) The chemistry of life's origins. Kluwer, Dordrecht, pp 195–207

Greenberg JM (1996) Chirality in Interstellar Dust and in Comets: Life from Dead Stars. In: Greenberg JM, Mendoza-Gomez CX (eds) Physical origin of homochirality in life. American Inst. of Physics, New York, pp 185–205

Greenberg JM, Kouchi A, Niessen W, Irth H, van Paradijs J, de Groot M, Hermsen W (1994) Interstellar dust, chirality, comets and the origins of life: Life from dead stars? J Biol Phys 20:61–70

Griesbeck AG, Meierhenrich UJ (2002) Asymmetric photochemistry and photochirogenesis. Angew Chem Int Ed 41:3147–3154; Asymmetrische Photochemie und Photochirogenese. Angew Chem 114:3279–3286

Gübitz G, Schmid MG (2004) Chiral separation principles: An introduction. In: Gübitz G, Schmid MG (eds) Chiral separations. Methods and protocols. Methods in molecular biology. Humana Press, Totowa, pp 1–28

Haesler J, Schindelholz I, Riguet E, Bochet CG, Hug W (2007) Absolute configuration of chirally deuterated neopentane. Nature 446:526–529

Hanner MS (1999) The silicate material in comets. Space Sci Rev 90:99–108

Hashimoto A, Kumashiro S, Nishikawa T, Oka T, Takahashi K, Mito T, Takashima S, Doi N, MizutaniY, Yamazaki T, Kaneko T, Ootomo E (1993) Embryonic development and postnatal changes in free D-aspartate and D-serine in the human prefrontal cortex. J Neuchem 61:348–351

Hazen RM (2001) Life's rocky start. Sci Am 4:76–85; Les minéraux et la naissance de la vie. Pour la Science 6:38–44

Hazen RM, Filley TR, Goodfriend GA (2001) Selective adsorption of L- and D-amino acids on calcite: implications for biochemical homochirality. Proc Natl Acad Sci USA 98:5487–5490

Hegstrom RA (1984) Parity nonconservation and the origin of biological chirality: theoretical calculations. Orig Life 14:405–411

Hellwich K-H, Siebert CD (2006) Stereochemistry workbook. 191 Problems and solutions. Springer, Berlin Heidelberg New York

Hennet RJ-C, Holm NG, Engel MH (1992) Abiotic synthesis of amino acids under hydrothermal conditions and the origin of life: a perpetual phenomenon? Naturwissenschaften 79:361–365

Herrero S, Usón MA (1995) A straightforward method for assigning stereochemical Λ/Δ descriptors to octahedral coordination compounds. J Chem Educ 72:1065–1066

Herschel JWF (1822) On the rotation impressed by plates of rock crystal on the planes of polarization of the rays of light as connected with certain peculiarities in its crystallization. Trans Cambridge Philos Soc 1:43–50

Holm NG, Andersson EM (1998) Hydrothermal systems. In: Brack A (ed) The molecular origins of life – assembling pieces of the puzzle. Cambridge University Press, Cambridge, pp 86–99

Holm NG, Charlou JL (2001) Initial indications of abiotic formation of hydrocarbons in the Rainbow ultramafic hydrothermal system, Mid-Atlantic Ridge. Earth and Planetary Science Letters 191:1–8

Huang Z-H, Wang J, Gage DA, Watson JT, Sweeley CC (1993) Characterization of N-ethoxycarbonyl ethyl esters of amino acids by mass spectrometry. J Chromatogr 635:271–281

Huber C, Wächtershäuser G (1997) Activated acetic acid by carbon fixation on (Fe,Ni)S under primordial conditions. Sience 276:245–247

Huck NPM, Jager WF, de Lange B, Feringa BL (1996) Dynamic control and amplification of molecular chirality by circular polarized light. Science 273:1686–1688

Huebner WF, Boice DC (1992) Comets as a possible source of prebiotic molecules. Orig Life Evol Biosphere 21:299–315

Hug W (2003) Virtual enantiomers as the solution of optical activity's deterministic offset problem. Appl Spectrosc 57:1–13

Hughes WB, Holba AG, Dzou LIP (1995) The ratios of dibenzothiophene to phenanthrene and pristane to phytane as indicators of depositional environment and lithology of petroleum source rocks. Geochim Cosmochim Acta 59:3581–3598

Idelson M, Blout ER (1958) Polypeptides. XVIII. A kinetic study of the polymerization of amino acid N-carboxyanhydrides initiated by strong bases. J Am Chem Soc 80:2387–2393

Inoue Y (2004) Enantiodifferentiating photosensitized reactions. In: Inoue Y, Ramamurthy V (eds) Chiral photochemistry. Marcel Dekker, New York, pp 129–178

Inoue Y, Ramamurthy V (2004) Chiral photochemistry. Marcel Dekker, New York

Inoue Y, Tsuneishi H, Hakushi T, Yagi K, Awazu K, Onuki H (1996) First absolute asymmetric synthesis with circularly polarized synchrotron radiation in the vacuum ultraviolet region: Direct photoderacemization of (E)-cyclooctene. Chem Commun 23:2627–2628

Irie M (2000) Diarylethenes for memories and switches. Chem Rev 100:1685–1716

Irvine WM, Schloerb FP, Crovisier J, Fegley B Jr, Mumma MJ (2000) Comets: A link between interstellar and nebular chemistry. In: Mannings V, Boss AP, Russell SS (eds) Protostars and planets IV. University Arizona Press, Tucson, pp 1159–1200

Jakubetz H, Czesla H, Schurig V (1997) On the feasibility of miniaturized enantiomeric separation by liquid chromatography (OTLC) and open turbular electrochromatography (OTEC). J Micro Sep 9:421–431

Jessberger EK (1999) Rocky cometary particulates: their elemental, isotopic and mineralogical ingredients. Space Sci Rev 90:91–97

Johns RB, Seuret MG (1970) Photochemistry of biological molecules. III. Mechanism of photodamage of alanine peptides in the solid state. Photochem Photobiol 12:405–417

Jorda L, Rickman H (1995) Comet P/Wirtanen, summary of observational data. Planet Space Sci 43:575–579

Jordan IK, Kondrashov FA, Adzhubei IA, Wolf YI, Koonin EV, Kondrashov AS, Sunyaev S (2005) A universal trend of amino acid gain and loss in protein evolution. Nature 433:633–638

Joyce GF (2002) The antiquity of RNA-based evolution. Nature 418:214–221

Joyce GF, Visser GM, van Boeckel CAA, van Boom JH, Orgel LE, van Westrenen J (1984) Chiral selection in poly(C)-directed synthesis of oligo(G). Nature 310:602–604

Joyce GF, Schwartz AW, Miller SL, Orgel LE (1987) The case for an ancestral genetic system involving simple analogues of the nucleotides. Proc Natl Acad Sci USA 84:4398–4402

Karagunis G, Coumoulos G (1938) A new method of resolving a racemic compound. Nature 142:162–163

Kasting JF (1992) Proterozoic climates: the effect of changing atmospheric carbon dioxide concentrations. In: Schopf JW, Klein C (eds) The proterozoic biosphere: A Multidisciplinary Study. Cambridge University Press, Cambridge, pp 165–168

Kasting JF, Brown LL (1998) The early atmosphere as a source of biogenic compounds. In: Brack A (ed) The molecular origins of life – assembling pieces of the puzzle. Cambridge University Press, Cambridge, pp 35–56

Kawasaki T, Jo K, Igarashi H, Sato I, Nagano M, Koshima H, Soai K (2005) Asymmetric amplification using chiral cocrystals formed from achiral organic molecules by asymmetric autocatalysis. Angew Chem Int Ed 44:2774–2777; Angew Chem 117:2834–2837

Kemp JC, Wolstencroft RD (1971) Elliptical polarization by surface-layer scattering. Nature 231:170–171

Kemp JC, Swedlund JB, Murphy RE, Wolstencroft RD (1971a) Circularly polarized visible light from Jupiter. Nature 231:169–170

Kemp JC, Wolstencroft RD, Swedlund JB (1971b) Circular polarization: Jupiter and other planets. Nature 232:165–168

Kerridge JF (1999) Formation and processing of organics in the early Solar System. Space Sci Rev 90:275–288

Keszthelyi L (1984) Parity violation as a source of chirality in nature. Orig Life 14:375–382
Keszthelyi L (2001) Homochirality of biomolecules: counter-arguments against critical notes. Orig Life Evol Biosphere 31:249–256
Khare BN, Sagan C (1971) Synthesis of cystine in simulated primitive conditions. Nature 232:577–579
Kipping FS, Pope WJ (1898) Enantiomorphism. J Chem Soc Trans 73:606–617
Kissel J, Krueger FR (1987) The organic component in dust from comet Halley as measured by the PUMA mass spectrometer on board Vega1. Nature 326:755–760
Kissel J, Sagdeev RZ, Bertaux JL, Angarov VN, Audouze J, Blamont JE, Büchler K, Evlanov EN, Fechtig H, Fomenkova MN, von Hoerner H, Inogamov NA, Khromov VN, Knabe W, Krueger FR, Langevin Y, Leonas VB, Levasseur-Regourd AC, Managadze GG, Podkolzin SN, Shapiro VD, Tabaldyev SR, Zubkov BV (1986a) Composition of comet Halley dust particles from Vega observations. Nature 321:280–282
Kissel J, Sagdeev RZ, Bertaux JL, Angarov VN, Audouze J, Blamont JE, Büchler K, Evlanov EN, Fechtig H, Fomenkova MN, von Hoerner H, Inogamov NA, Khromov VN, Knabe W, Krueger FR, Langevin Y, Leonas VB, Levasseur-Regourd AC, Managadze GG, Podkolzin SN, Shapiro VD, Tabaldyev SR, Zubkov BV (1986b) Composition of comet Halley dust particles from Giotto observations. Nature 321:336–337
Klabunovskii EI (1963) Asymmetrische Synthese. VEB, DVW, Berlin, 213 pp
Klabunovskii EI (2001) Can enantiomorphic crystals like quartz play a role in the origin of homochirality on Earth? Astrobiol 1:127–131
Klabunovskii E, Thiemann W (2000) The role of quartz in the origin of optical activity on Earth. Orig Life Evol Biosphere 30:431–434
Klánová J, Klán P, Heger D, Holoubek I (2003) Comparison of the effects of UV, H_2O_2/UV and γ-irradiation processes on frozen interstellar ices. Photoch Photobio Sci 2:1023–1031
Klussmann M, Iwamura H, Mathew SP, Wells Jr DH, Pandya U, Armstrong A, Blackmond DG (2006) Thermodynamic control of asymmetric amplification of amino acid catalysis. Nature 441:621–623
Klussmann M, Izumi T, White AJP, Armstrong A, Blackmond DG (2007) Emergence of solution-phase homochirality via crystal engineering of amino acids. J Am Chem Soc 129:7657–7660
Kminek G, Bada JL, Botta O, Glavin DP, Grunthaner F (2000) MOD: an organic detector for the future robotic exploration of Mars. Planet Space Sci 48:1087–1091
Kobayashi K, Kaneko T, Hashimoto H, Kouchi A, Saito T, Yamashita M (1998) Abiotic formation of bioorganic compounds in space. Biol Sci Space 12:102–105
Kobayashi K, Kaneko T, Kouchi A, Hashimoto H, Saito T, Yamashita M (1999a) Synthesis of amino acids in Earth orbit: Proposal. Adv Space Res 23:401–404
Kobayashi K, Kaneko T, Saito T (1999b) Characterization of complex organic compounds formed in simulated planetary atmospheres by the action of high energy particles. Adv Space Res 24:461–464
König WA, Icheln D, Runge T, Pforr I, Krebs A (1990) Cyclodextrins as chiral stationary phases in capillary gas chromatography. Part VII: Cyclodextrins with an inverse substitution pattern – synthesis and enantioselectivity. J High Res Chrom 13:702–707
Kondepudi DK (1987) Selection of molecular chirality by extremely weak chiral interactions under far-from-equilibrium conditions. BioSystems 20:75–83
Kondepudi DK, Asakura K (2001) Chiral autocatalysis, spontaneous symmetry breaking, and stochastic behaviour. Acc Chem Res 34:946–954
Kondepudi DK, Nelson GW (1985) Weak neutral currents and the origin of biomolecular chirality. Nature 314:438–441
Kondepudi DK, Prigogine I (1998) Modern Thermodynamics. From heat engines to dissipative structures. Wiley, New York
Kondepudi DK, Kaufman RJ, Singh N (1990) Chiral symmetry breaking in sodium chlorate crystallization. Science 250:975–976
Koppitz M, Nielsen PE, Orgel LE (1998) Formation of oligonucleotide-PNA-chimeras by template-directed ligation. J Am Chem Soc 120:4563–4569

Koshima H (2004) Chiral solid-state photochemistry including supramolecular approaches. In: Inoue Y, Ramamurthy V (eds) Chiral photochemistry. Marcel Dekker, New York, pp 485–532

Kotra RK, Shimoyama A, Ponnamperuma C, Hare PE (1979) Amino acids in a carbonaceous chondrite from Antarctica. J Mol Evol 13:179–184

Kovacs KL, Keszthelyi L (1981) Unconsidered sources of chirality in nature. Orig Life 11:93–103

Krueger FR, Kissel J (1989) Biogenesis by cometary grains – thermodynamic aspects of self-organization. Orig Life Evol Biosphere 19:87–93

Kuan YJ, Charnley SB, Huang H-C, Tseng W-L, Kisiel Z (2003) Interstellar glycine. Astrophys J 593:848–867

Kuhn W (1930) The physical significance of optical rotatory power. Trans Faraday Soc 26:293–310

Kuhn W, Braun E (1929) Photochemische Erzeugung optisch aktiver Stoffe. Die Naturwissenschaften 17:227–228

Kuhn W, Knopf E (1930) Photochemische Erzeugung optisch aktiver Stoffe. Die Naturwissenschaften 18:183

Kullman JP, Chen X, Armstrong DW (1999) Evaluation of the enantiomeric composition of amino acids in tobacco. Chirality 11:669–673

Kunnas AV, Jauhiainen T-P (1993) Separation and identification of free amino acid enantiomers in peat by capillary gas chromatography. J Chrom 628:269–273

Kuroda R (2004) Circular dichroism in the solid state. In: Inoue Y, Ramamurthy V (eds) Chiral photochemistry. Marcel Dekker, New York, pp 385–414

Kvenvolden KA, Lawless J, Pering K, Peterson E, Flores J, Ponnamperuma C, Kaplan IR, Moore C (1970) Evidence for extraterrestrial amino acids and hydrocarbons in the Murchison meteorite. Nature 228:923–926

Kvenvolden KA, Lawless JG, Ponnamperuma C (1971) Nonprotein amino acids in the Murchison meteorite. Proc Natl Acad Sci USA 68:486–490

Lahav M, Weissbuch I, Shavit E, Reiner C, Nicholson GJ, Schurig V (2006) Parity violating energy difference and enantiomorphous crystals-caveats; reinvestigation of tyrosine crystallization. Orig Life Evol Biosphere 36:151–170

Lamy PL, Toth I, Weaver H, Jorda L, Kaasalainen M (2003) The nucleus of comet 67P/Churyumov-Gerasimenko, the new target of the Rosetta Mission. Bull Am Astron Soc 35:970

Lamy PL, Toth I, Davidsson BJR, Groussin O, Gutiérrez P, Jorda L, Kaasalainen M, Lowry SC (2007) A portrait of the nucleus of comet 67P/Churyumov-Gerasimenko. Space Sci Rev 128:23–66

Lang KR, Willson RF (1983) The circularly polarized Sun at 12.6 cm wavelength. Astron Astrophys 127:135–139

Large DJ, Gize AP (1996) Pristane/phytane ratios in the mineralized Kupferschiefer of the Fore-Sudetic Monocline, southwest Poland. Ore Geol Rev 11:89–103

Larter SR, Solli H, Douglas AG, de Lange F, de Leeuw JW (1979) Occurrence and significance of prist-1-ene in kerogene pryolysates. Nature 279:405–408

Lavabre D, Micheau J-C, Rivera Islas J, Buhse T (2008) Kinetic insight into specific features of the autocatalytic Soai reaction. Top Curr Chem 284:in print

Lawless JG (1973) Amino acids in the Murchison meteorite. Geochim Cosmochim Acta 37:2201–2212

Lee TD, Yang CN (1956) Question of parity conservation in weak interactions. Phys Rev 104:254–258

Lemmlein GG (1973) Morphology and genesis of crystals. Nauka Publisher, Moskow

Li M, Larter SR, Taylor P, Jones M, Bowler B, Bjoroy M (1995) Biomarkers or not biomarkers? A new hypothesis for the origin of pristane involving derivation from methyltrimethyltridecylchromans (MTTCs) formed during diagenesis from chlorophyll and alkylphenols. Org Geochem 23:159–167

Lisse CM, VanCleve J, Adams AC, A'Hearn MF, Fernández YR, Farnham TL, Armus L, Grillmair CJ, Ingalls J, Belton MJS, Groussin O, McFadden LA, Meech KJ, Schultz PH, Clark

BC, Feaga LM, Sunshine JM (2006) Spitzer spectral observations of the deep impact ejecta. Sience 313:635–640
Lucas PW, Hough JH, Bailey J, Chrysostomou A, Gledhill TM, McCall A (2005) UV circular polarization in star formation regions: the origin of homochirality? Orig Life Evol Biosphere 35:29–60
Luisi PL (2006) The emergence of life: From chemical origins to synthetic biology. Cambridge University Press, Cambridge
Lundberg RD, Doty P (1957) Polypeptides XVII. A study of the kinetics of the primary amine-initiated polymerization of N-carboxy-anhydrides with special reference to configurational and stereochemical effects. J Am Chem Soc 79:3961–3972
Lutz S (1988) Gaschromatographische Enantiomerentrennung an neuen chiralen Phasen. Ph.D. thesis, University of Hamburg
MacDermott AJ (1993) The weak force and the origin of life. In: Ponnamperuma C, Chela-Flores J, (eds) Proceedings of the Trieste Conference on Chemical Evolution and the Origin of Life, 26–30 October 1992. A. Deepark Publishing, Hampton, Virginia, pp 85–99
MacDermott AJ (1995) Electroweak enantioselection and the origin of life. Orig Life Evol Biosphere 25:191–199
MacDermott AJ (1997) Distinguishing the chiral signature of life in the solar system and beyond. ProcSPIE 3111:272–279
MacDermott AJ, Tranter GE (1989) Electroweak bioenantioselection. Croatica Chem Acta 62:165–187
MacDermott AJ, Barron LD, Brack A, Buhse T, Drake AF, Emery R, Gottarelli G, Greenberg JM, Haberle R, Hegstrom RA, Hobbs K, Kondepudi DK, McKay C, Moorbath S, Raulin F, Sandford M, Schwartzmann DW, Thiemann WH-P, Tranter GE, Zarnecki JC (1996) Homochirality as the signature of life: The SETH cigar. Planet Space Sci 44:1441–1446
Mason SF (1984) Origins of biomolecular handedness. Nature 311:19–23
Mason SF (1997) Extraterrestrial handedness. Nature 389:804
Mason SF (2000) Extraterrestrial handedness revisited. Orig Life Evol Biosphere 30:435–437
Mason SF, Tranter GE (1983a) Energy inequivalence of peptide enantiomers from parity non-conservation. J Chem Soc Chem Commun 117–119
Mason SF, Tranter GE (1983b) The parity-violating energy difference between enantiomeric molecules. Chem Phys Lett 94:34–37
Mathew SP, Iwamura H, Blackmond DG (2004) Amplification of enantiomeric excess in a proline-mediated reaction. Angew Chem Int Ed 43:3317–3321; Angew Chem 116:3379–3383
Matsushita M (1989) Experimental observations of aggregations. In: Avnir D (ed) The fractal approach to heterogeneous chemistry. Wiley, New York, pp 161–179
Matthews CN (1992) Dark matter in the solar system: hydrogen cyanide polymers. Orig Life Evol Biosphere 21:421–434
Matthews CN (2004) The HCN world: Establishing protein-nucleic acid life via hydrogen cyanide polymers. In: Seckbach J (ed) Origins. Kluwer, Dordrecht, pp 121–135
Matthews CN, Minard RD (2006) Hydrogan cyanide polymers, comets and the origin of life. Faraday Discuss 133:393–401
Maurette M (2006) Micrometeorites and the mysteries of our origins. Springer, Berlin Heidelberg New York
McKay DS, Gibson Jr EK, Thomas-Kaprta KL, Vali H, Romanek CS, Clemett SJ, Chillier XDF, Maechling CR, Zare RN (1996) Search for past life on Mars: Possible relic biogenic activity in Martian meteorite ALH84001. Science 273:924–929
McKinney DE, Carson DM, Clifford DJ, Minard RD, Hatcher PG (1995) Off-line thermochemolysis versus flash pyrolysis for the in situ methylation of lignin: is pyrolysis necessary? J Anal Appl Pyrol 34:41–46
Mennella V, Muñoz Caro GM, Ruiterkamp R, Schutte WA, Greenberg JM, Brucato JR, Colangeli L (2001) UV photodestruction of CH bonds and the evolution of the 3.4 μm feature carrier. II. The case of hydrogenated carbon grains. Astron Astrophys 367:355–361

Meierhenrich U (2002) Präbiotische Chemie – Kometen und irdisches Leben. Nachr Chem 50:1338–1341

Meierhenrich UJ (2004) Review of Chiral Separations – Methods and Protocols. Angew Chem Int Ed 43:4251–4252; Angew Chem 116:4347–4350

Meierhenrich UJ, Thiemann WH-P (2004) Photochemical concepts on the origin of biomolecular asymmetry. Orig Life Evol Biosphere 34:111–121

Meierhenrich UJ, Thiemann WH-P, Rosenbauer H (1999) Molecular parity violation by comets? Chirality 11:575–582

Meierhenrich U, Barbier B, Jacquet R, Chabin A, Alcaraz A, Nahon L, Brack A (2001a) Photochemical origin of biomolecular asymmetry. In: Ehrenfreund P, Angerer O, Battrick B (eds) Exo-/Astro-Biology. European Space Agency, ESA SP-496, Noordwijk, The Netherlands, pp 167–170

Meierhenrich UJ, Thiemann WH-P, Goesmann F. Roll R, Rosenbauer H (2001b) Enantiomer separation of hydrocarbons in preparation of Rosetta's 'chirality-experiment'. Chirality 13:454–457

Meierhenrich U, Thiemann WH-P, Rosenbauer H (2001c) Pyrolytic methylation assisted enantioseparation of chiral hydroxycarboxylic acids. J Anal Appl Pyrolysis 60:13–26

Meierhenrich, UJ, Thiemann, WH-P, Rosenbauer H (2001d) Stereochemical investigations of cometary matter onboard the Rosetta Lander. Enantiomer 6:97–99

Meierhenrich U, Thiemann W, Schubert C, Barbier B, Brack A (2001e) Isoprenoid enantiomers as molecular biomarkers in ancient sediments. In: Nakashima S, Maruyama S, Brack A, Windley BF (eds) Geochemistry and the origin of life. Universal Academy Press, Tokyo, pp 269–283

Meierhenrich UJ, Thiemann WH-P, Muñoz Caro G M, Schutte WA, Greenberg JM (2001f) Simulated cometary matter as a test for enantiomer separating chromatography for use on comet 46P/Wirtanen. Adv Space Res 27:329–334

Meierhenrich UJ, Muñoz Caro GM, Schutte WA, Barbier B, Arcones Segovia A, Rosenbauer H, Thiemann WH-P, Brack A (2002a) The prebiotic synthesis of amino acids – Interstellar vs. atmospheric mechanisms. In: Sawaya-Lacoste H, (ed) Exo-/Astro-Biology. European Space Agency, ESA SP-518, Noord-wijk, The Netherlands, pp 25–30

Meierhenrich UJ, Thiemann WH-P, Barbier B, Brack A, Alcaraz C, Nahon L, Wolstencroft R (2002b) Circular polarization of light by planet Mercury and enantiomorphism of ist surface minerals. Orig Life Evol Biosphere 32:181–190

Meierhenrich UJ, Nguyen M-J, Barbier B, Brack A, Thiemann WH-P (2003a) Gas chromatographic separation of saturated aliphatic hydrocarbon enantiomers on permethylated β-cyclodextrin. Chirality 15:S13–S16

Meierhenrich U, Thiemann WH-P, Rosenbauer H (2003b) The search for enantiomeric excesses in cometary matter. In: Celnikier LM, Trân Thanh Vân J (eds) Frontiers of Life. Thê Gió́i Publishers, Vietnam, pp 23–26

Meierhenrich UJ, Muñoz Caro GM, Bredehöft JH, Jessberger EK, Thiemann WH-P (2004) Identification of diamino acids in the Murchison meteorite. Proc Natl Acad Sci USA 101:9182–9186

Meierhenrich UJ, Muñoz Caro GM, Schutte WA, Thiemann WHP, Barbier B, Brack A (2005a) Precursors of biological cofactors from ultraviolet irradiation of circumstellar/interstellar ice analogues. Chem Eur J 11:4895–4900

Meierhenrich UJ, Nahon L, Alcaraz C, Bredehöft JH, Hoffmann SV, Barbier B, Brack A (2005b) Asymmetric vacuum UV photolysis of the amino acid leucine in the solid state. Angew Chem Int Ed 44:5630–5634; Asymmetrische Vakuum-UV-Photolyse der Aminosäure Leucin in fester Phase. Angew Chem 117:5774–5779

Meiring WJ (1987) Nuclear β-decay and the origin of biomolecular chirality. Nature 329:712–714

Merk R, Prialnik D (2006) Combined modeling of thermal evolution and accretion of trans-neptunian objects – occurrence of high temperatures and liquid water. Icarus 183:283–295

Merwitz O (1976) Selective radiolysis of enantiomers. Rad and Environm Biophys 13:63–69

Merwitz O, Schleser GH, Koppenhoefer B, Trettin U (1991) The origin of differences in ^{13}C-content of amino acid enantiomers. Amino Acids 1:144

Merwitz O, Nitzsche H-M, Hochhäuser E (1998) Isotope fractionation during radiation-induced decarboxylation and deamination of L-leucine. Isotopenpraxis 34:319–324

Mie G (1908) Beiträge zur Optik trüber Medien, speziell kolloidaler Metallösungen. Ann Phys 330:377–445

Miller SL (1953) A production of amino acids under possible primitive Earth conditions. Science 117:528

Miller SL (1998) The endogenous synthesis of organic compounds. In: Brack A (ed) The molecular origins of life – assembling pieces of the puzzle. Cambridge University Press, Cambridge, pp 59–85

Miller SL, Lazcano A (2002) Formation of the building blocks of life. In: Schopf JW (ed) Life's origin. University of California Press, Berkeley, pp 78–112

Milton RCL, Milton SCF, Kent SBH (1992) Total chemical synthesis of a D-enzyme: the enantiomers of HIV-1 protease show demonstration of reciprocal chiral substrate specificity. Science 256:1445–1448

Minard RD, Hatcher PG, Gourley RC, Matthews CN (1998) Structural investigations of hydrogen cyanide polymers: new insights using TMAH thermochemolysis/GC-MS. Orig Life Evol Biosphere 28:461–473

Momose Y, Fujii N, Kodama T, Yamagaki T, Nakanishi H, Kodama M (1998) Specific racemization of the aspartyl residue of αA-crystallin in aged mouse lenses. Viva Origino 26:329–340

Moradpour A, Nicoud JF, Balavoine G, Kagan H, Tsoucaris G (1971) Photochemistry with circularly polarized light. The synthesis of optically active hexahelicene. J Am Chem Soc 93:2353–2354

Muñoz Caro GM (2003) From photoprocessing of interstellar ice to amino acids and other organics. Ph.D. thesis, University of Leiden

Muñoz Caro GM, Schutte WA (2003) UV-photoprocessing of interstellar ice analogs: New infrared spectroscopic results. Astron Astrophys 412:121–132

Muñoz Caro GM, Ruiterkamp R, Schutte WA, Greenberg JM, Mennella V (2001) UV photodestruction of CH bonds and the evolution of the 3.4 μm feature carrier. I. The case of aliphatic and aromatic molecular species. Astron Astrophys 367:347–354

Muñoz Caro GM, Meierhenrich UJ, Schutte WA, Barbier B, Arcones Segovia A, Rosenbauer H, Thiemann WH-P, Brack A, Greenberg JM (2002) Amino acids from ultraviolet irradiation of interstellar ice analogues. Nature 416:403–406

Muñoz Caro GM, Meierhenrich UJ, Schutte WA, Greenberg JM, Thiemann WH-P (2004) UV-photoprocessing of interstellar ice analogs: Detection of hexamethylenetetramine-based species. Astron Astrophys 413:209–216

Musigmann M, Busalla A, Blum K, Thompson DG (1999) Asymmetries in collisions between electrons and oriented chiral molecules. J Phys B: At Mol Opt Phys 32:4117–4128

Nagata Y, Tagashira J (1998) Rate of racemization and degradation of amino acids during acid-hydrolysis of synthetic peptides. Viva Origino 26:321–328

Nagata Y, Masui R, Akino T (1992a) The presence of free D-serine, D-alanine and D-proline in human plasma. Birkhäuser Verlag, Basel, Experientia 48:986–988

Nagata Y, Yamamoto K, Shimojo T (1992b) Determination of D- and L-amino acids in mouse kidney by high-performance liquid chromatography. J Chromatogr Biomed Appl 575:147–152

Nagata Y, Horiike K, Maeda T (1994) Distribution of free D-serine in vertebrate brains. Brain Research 634:291–295

Nagata Y, Fujiwara T, Kawaguchi-Nagata K, Fukumori Y, Yamanaka T (1998) Occurence of peptidyl D-amino acids in soluble fractions of several eubacteria, archaea and eukaryotes. Biochim Biophys Acta 1379:76–82

Nahon L, Alcaraz C (2004) SU5: a calibrated variable-polarization synchrotron radiation beam line in the vacuum-ultraviolet range. Appl Optics 43:1024–1037

Nahon L, Corlier M, Peaupardin P, Marteau F, Marcouillé O, Brunelle P, Alcaraz C, Thiry P (1997) A versatile electromagnetic planar/helical crossed undulator optimized for the SU5 low energy/high resolution beamline at Super-ACO. Nucl Instr and Meth in Phys Res A 396:237–250

Nahon L, Alcaraz C, Marlats J-L, Lagarde B, Polack F, Thissen R, Lepère D, Ito K (2001) Very high spectral resolution obtained with SU5: A vacuum ultraviolet undulator-based beamline at Super-ACO. Rev Sci Instrum 72:1320–1329

Navarro-Gonzáles R, Rainey FA, Molina P, Bagaley DR, Hollen BJ, de la Rosa J, Small AM, Quinn RC, Grunthaner FJ, Cáceres L, Gomez-Silva B, McKay CP (2003) Mars-like soils in the Atacama desert, Chile, and the dry limit of microbial life. Sience 302:1018–1021

Navarro-Gonzáles R, Navarro KF, de la Rosa J, Iñiguez E, Molina P, Miranda LD, Morales P, Cienfuegos E, Coll P, Raulin F, Amils R, McKay CP (2006) The limitations on organic detection in Mars-like soils by thermal volatilization-gas chromatography-MS and their implications for the Viking results. Proc Natl Acad Sci USA 103:16089–16094

Nelson KE, Levy M, Miller SL (2000) Peptide nucleic acids rather than RNA may have been the first genetic molecule. Proc Natl Acad Sci 97:3868–3871

Nielsen PE (1993) Peptide nucleic acid (PNA): A model structure for the primordial genetic material? Orig Life Evol Biosphere 23:323–327

Nielsen PE, Egholm M, Berg RH, Buchardt O (1991) Sequence-selective recognition of DNA by strand displacement with a thymine-substituted polyamide. Science 254:1497–1500

Nip M (1987) Chemical characterization of coals, coal macerals and their precursors. A study by analytical pyrolysis. Ph.D. thesis, Technische Universiteit Delft

Nishino H, Kosaka A, Hembury GA, Shitomi H, Onuki H, Inoue Y (2001) Mechanims of pH-dependent photolysis of aliphatic amino acids and enantiomeric enrichment of racemic leucine by circularly polarized light. Org Lett 3:921–924

Nishino H, Kosaka A, Hembury GA, Aoki F, Miyauchi K, Shitomi H, Onuki H, Inoue Y (2002) Absolute asymmetric photoreactions of aliphatic amino acids by circularly polarized synchrotron radiation: Critically pH-dependent photobehavior. J Am Chem Soc 124:11618–11627

Nordén B (1977) Was photoresolution of amino acids the origin of optical activity in life? Nature 266:567–568

Nordén B, Liljenzin J-O, Tokay RK (1985) Stereoselective decarboxylation of amino acids in the solid state, with special reference to chiral discrimination in prebiotic evolution. J Mol Evol 21:364–370

Noyori R, Suga S, Oka H, Kitamura M (2001) Self and nonself recognition of chiral catalysts: the origin of nonlinear effects in the amino-alcohol catalysed asymmetric addition of diorganozincs to aldehydes. Chem Rec 1:85–100

Nuevo M (2005) Photolysis of interstellar ices and production of organic molecules: Laboratory simulations (in French). Ph.D. thesis, Université de Paris-Sud

Nuevo M, Meierhenrich UJ, Muñoz Caro GM, Dartois E, d'Hendecourt L, Deboffle D, Auger G, Blanot D, Bredehöft JH, Nahon L (2006) The effects of circularly polarized light on amino acid enantiomers produced by the UV irradiation of interstellar ice analogs. Astron Astrophys 457:741–751

Nuevo M, Meierhenrich UJ, d'Hendecourt L, Muñoz Caro GM, Dartois E, Deboffle D, Thiemann WH-P, Bredehöft JH, Nahon L (2007) Enantiomeric separation of complex organic molecules produced from irradiation of interstellar/circumstellar ice analogs. Adv Space Res 39:400–404

Oberbeck VR, Aggarwal H (1992) Comet impacts and chemical evolution on the bombarded Earth. Orig Life Evol Biosphere 21:317–338

Oró J (1961) Comets and the formation of biochemical compounds on the primitive Earth. Nature 190:389–390

Oró J, Updegrove WS, Gibert J, McReynolds J, Gil-Av E, Ibanez J, Zlatkis A, Flory DA, Levy RL, Wolf C (1970) Organogenic elements and compounds in surface samples from the sea of tranquillity. Science 167:765–767

Oró J, Nakaparksin S, Lichtenstein H, Gil-Av E (1971) Configuration of amino-acids in carbonaceous chondrites and a pre-cambrian chert. Nature 230:107–108

Pätzold R, Brückner H (2006) Gas chromatographic determination and mechanism of formation of D-amino acids occurring in fermented and roasted cocoa beans, cocoa powder, chocolate and cocoa shell. Amino Acids 31:63–72

Palache C, Berman GB, Frondel C (1962) Relative frequencies of left and right quartz. In: Frondel C (ed) The system of mineralogy. Yale University, Wiley, New York, p 17

Parnell J, Cullen D, Sims MR, Bowden S, Cockell CS, Court R, Ehrenfreund P, Gaubert F, Grant W, Parro V, Rohmer M, Sephton M, Stan-Lotter H, Steele A, Toporski J, Vago J (2007) Searching for life on Mars: Selection of molecular targets for ESA's Aurora ExoMars Mission. Astrobiol 7:578–604

Penny D (2006) Defining moments. What are the major principles that controlled the origin of life? Nature 442:745–746

Pereira WE, Summons RE, Rindfleisch TC, Duffield AM, Zeitman B, Lawless JG (1975) Stable isotope mass fragmentography: identification and hydrogen-deuterium exchange studies of eight Murchison meteoritic amino acids. Geochim Cosmochim Acta 39:163–172

Petsko GA (1992) On the other hand. Science 256:1403–1404

Pillinger CT (1984) Light element stable isotopes in meteorites – from grams to picograms. Geochim Cosmochim Acta 48:2739–2766

Pizzarello S, Cronin JR (2000) Non-racemic amino acids in the Murray and Murchison meteorites. Geochim Cosmochim Acta 64:329–338

Pizzarello S, Weber AL (2004) Prebiotic amino acids as asymmetric catalysts. Science 303:1151

Pizzarello S, Zolensky M, Turk KA (2003) Nonracemic isovaline in the Murchison meteorite: Chiral distribution and mineral association. Geochim Cosmochim Acta 67:1589–1595

Pizzarello S, Huang Y, Fuller M (2004) The carbon isotopic distribution of Murchison amino acids. Geochim Cosmochim Acta 68:4963–4969

Pollock GE, Cronin SE, Kvenvolden KA, Cheng CN (1975) Stereoisomers of isovaline in the Murchison meteorite. Geochim Cosmochim Acta 39:1571–1573

Proskurowski G, Lilley MD, Seewald JS, Früh-Green GL, Olson EJ, Lupton JE, Sylva SP, Kelley DS (2008) Abiogenic hydrocarbon production at Lost City hydrothermal field. Science 319:604–607

Quack M (1989) Structure and dynamics of chiral molecules. Angew Chem Int Ed 28:571–586; Angew Chem 101:588–604

Quack M (1993) The symmetries of time and space and their violations in chiral molecules and molecular processes. In: Costa G, Calucci G, Giorgi M (eds) Conceptual tools for understanding nature. World Scientific, Singapore, pp 172–208

Quack M (2002) How important is parity violation for molecular and biomolecular chirality. Angew Chem Int Ed 41:4618–4630; Wie wichtig ist Paritätsverletzung für die molekulare und biomolekulare Chiralität. Angew Chem 114:4812–4825

Quack M, Stohner J (2001) Molecular chirality and the fundamental symmetries of physics: Influence of parity violation on rovibrational frequencies and thermodynamic properties. Chirality 13:745–753

Rau H (2004) Direct asymmetric photochemistry with circularly polarized light. In: Inoue Y, Ramamurthy V (eds) Chiral photochemistry. Marcel Dekker, New York, pp 1–44

Rauchfuß H (2005) Chemische Evolution und der Ursprung des Lebens. Springer, Berlin Heidelberg New York

Rein D (1992) Die wunderbare Händigkeit der Moleküle: vom Ursprung des Lebens aus der Asymmetrie der Natur. Birkhäuser Verlag, Basel

Rein DW, Hegstrom RA, Sandars PGH (1979) Parity non-conserving energy difference between mirror image molecules. Phys Lett 71A:499–502

Rhodes W, Dougherty RC (1978) Effects of electric and magnetic fields on prochiral chemical reactions: macroscopic electric and magnetic fields can cause asymmetric synthesis. J Am Chem Soc 100:6247–6248

Ribó JM, Crusats J, Sagués F, Claret J, Rubires R (2001) Chiral sign induction by vortices during the formation of mesophases in stirred solutions. Science 292:2063–2066

Rikken GLJA (2004) Magnetochiral anisotropy in asymmetric photochemistry. In: Inoue Y, Ramamurthy V (eds) Chiral photochemistry. Marcel Dekker, New York, pp 107–128

Rikken GLJA, Raupach E (1998) Pure and cascaded magnetochiral anisotropy in optical absorption. Phys Rev E 58:5081–5084

Rikken GLJA, Raupach E (2000) Enantioselective magnetochiral photochemistry. Nature 405:932–935

Rivera Islas JR, Lavabre D, Grevy J-M, Lamoneda RH, Cabrera HR, Micheau J-C, Buhse T (2005) Mirror-symmetry breaking in the Soai reaction: a kinetic understanding. Proc Natl Acad Sci USA 102:13743–13748

Rodger A, Nordén B (1997) Circular dichroism and linear dichroism. Oxford University Press, Oxford

Rosenbauer H, Fuselier SA, Ghielmetti A, Greenberg JM, Goesmann F, Ulamec S, Israel G, Livi S, MacDermott AJ, Matsuo T, Pillinger CT, Raulin F, Roll R, and Thiemann W (1999) Exobiologically-oriented space methodologies – the COSAC experiment on the lander of the ROSETTA mission. Adv Space Res 23:333–340

Rubenstein E, Bonner WA, Noyes HP, Brown GS (1983) Supernovae and life. Nature 306:118

Sagan C, Khare BN (1971) Long-wavelength ultraviolet photoproduction of amino acids on the primitive Earth. Science 173:417–420

Saghatelian A, Yokobayashi Y, Soltani K, Ghadiri MR (2001) A chiroselective peptide replicator. Nature 409:797–801

Sakamoto M (2004) Absolute asymmetric photochemistry using spontaneous chiral crystallization. In: Inoue Y, Ramamurthy V (eds) Chiral photochemistry. Marcel Dekker, New York, pp 415–462

Salam A (1980) Gauge unification of fundamental forces. Rev Mod Phys 52:525–538

Sato I, Kadowaki K, Soai K (2000) Asymmetric synthesis of an organic compound with high enantiomeric excess induced by inorganic ionic sodium chlorate. Angew Chem Int Ed 39:1510–1512; Angew Chem 112:1570–1572

Schleser GH, Merwitz O, Tokay RK (1991) Carbon isotope fractionation during radiation-induced decarboxylation of solid phenylalanine-1-^{13}C. Amino Acids 1:143–144

Schmidt JG, Christensen L, Nielsen PE, Orgel LE (1997a) Information transfer from DNA to peptide nucleic acids by template-directed synthesis. Nucl Acid Res 25:4792–4796

Schmidt JG, Nielsen PE, Orgel LE (1997b) Information transfer from peptide nucleic acids to RNA by template-directed synthesis. Nucl Acid Res 25:4797–4802

Schöning K-U, Scholz P, Guntha S, Wu X, Krishnamurthy R, Eschenmoser A (2000) Chemical etiology of nucleic acid structure: the α-threofuranosyl-($3' \rightarrow 2'$) oligonucleotide system. Science 290:1347–1351

Schopf JW (2002) Life's origin. University of California Press, Berkeley Los Angelos London

Schurig V (1986) Current methods for determination of enantiomeric compositions (Part 3): Gas chromatography on chiral stationary phases. Kontakte 1:3–22

Schurig V, Jakubetz H (1997) Enantiomer separation by OTLC and by OTEC with a capillary column coated with Chirasil-Dex. GIT Special Edn "Prof. Bayer" 20–24

Schurig V, Nowotny H-P (1988) Separation of enantiomers on diluted permethylated β-cyclodextrin by high-resolution gas chromatography. J Chromatogr 441:155–163

Schurig V, Nowotny H-P, Schmalzing D (1989) Gas chromatographic enantiomer separation of unfunctionalized cycloalkanes on permethylated β-cyclodextrin. Angew Chem 101:785–786; Angew Chem Int Ed 28:736–737

Schurig V, Jung M, Mayer S, Negura S, Fluck M, Jakubetz H (1994) Toward unified enantioselective chromatography with a single capillary column coated with Chirasil-Dex. Angew Chem Int Ed 33:2222–2223; Enantioselektive Trennung von Hexobarbital durch GC, SFC, LC und CEC an einer mit Chirasil-Dex belegten Kapillarsäule. Angew Chem 106:2265–2267

Schwartz AW (2002) Commentary: Chiral symmetry-breaking by crystallization. Orig Life Evol Biosphere 32:283

Schwartz AW, Chang S (2002) From Big Bang to primordial planet. In: Schopf JW (ed) Life's origin. University of California Press, Berkeley, pp 46–77

Schwell M (2007) Photochemistry of molecules of exobiology-interest in the far ultraviolet (in French). Habilitation, Université Paris 7 Denis Diderot

Schwell M, Jochims H-W, Baumgärtel H, Dulieu F, Leach S (2006) VUV photochemistry of small biomolecules. Planet Space Sci 54:1073–1085

Scolnik Y, Portnaya I, Cogan U, Tal S, Haimovitz R, Fridkin M, Elitzur AC, Deamer D, Shinitzky M (2006) Subtle differences in structural transitions between poly-L- and poly-D-amino acids of equal length in water. Phys Chem Chem Phys 8:333–339

Shibata T, Yamamoto J, Matsumoto N, Yonekubo S, Osanai S, Soai K (1998) Amplification of a slight enantiomeric imbalance in molecules based on asymmetric autocatalysis: the first correlation between high enantiomeric enrichment in a chiral molecule and circularly polarized light. J Am Chem Soc 120:12157–12158

Shinitzky M, Nudelman F, Barda Y, Haimovitz R, Chen E, Deamer D (2002) Unexpected differences between D- and L-tyrosine lead to chiral enhancement in racemic mixtures. Orig Life Evol Biosphere 32:285–297

Shock E (2002) Seeds of life? Nature 416:380–381

Sicoli G, Jiang Z, Jicinsky L, Schurig V (2005) Modified linear dextrins ("acyclodextrins") as new chiral selectors for the gas-chromatographic separation of enantiomers. Angew Chem Int Ed 44:4092–4095; Angew Chem 117:4161–4164

Siegel JS (1998) Homochiral imperative of molecular evolution. Chirality 10:24–27

Siegel JS (2001) Single-handed cooperation. Nature 409:777–778.

Sinninghe Damsté JS, de Leeuw JW (1995) Comments on "Biomarkers or not bio-markers? A new hypothesis for the origin of pristane involving derivation from methyltrimethyltridecylchromans (MTTCs) formed during diagenesis from chlorophyll and alkylphenols" from M Li, SR Larter, P Taylor, DM Jones, B Bowler and M Bjoroy. Org Geochem 23:1085–1087

Smith JM, Szathmáry E (1999) The origins of life. Oxford University Press, Oxford

Smith JV (1998) Biochemical evolution: I. Polymerization on internal, organophilic silica surfaces of dealuminated zeolithes and feldspars. Proc Natl Acad Sci USA 95:3370–3375

Soai K, Shibata T, Morioka H, Choji K (1995) Asymmetric autocatalysis and amplification of enantiomeric excess of a chiral molecule. Nature 378:767–768

Soai K, Osanai S, Kadowaki K, Yonekubo S, Shibata T, Sato I (1999) *d*- and *l*-quartz-promoted highly enantioselective synthesis of a chiral organic compound. J Am Chem Soc 121:11235–11236

Soai K, Shibata T, Sato I (2000) Enantioselective automultiplication of chiral molecules by asymmetric autocatalysis. Acc Chem Res 33:382–390

Soai K, Sato I, Shibata T (2001) Asymmetric autocatalysis and the origin of chiral homogeneity in organic compounds. Chem Record 1:321–332

Solladie G, Zimmermann RG (1984) Liquid crystals: a tool for chirality studies. Angew Chem Int Ed 23:348–362; Angew Chem 96:335–49

Spach G (1984) Chiral versus chemical evolutions and the appearance of life. Orig Life Evol Biosphere 14:433–437

Spach G, Brack A (1983) Reflections on molecular asymmetry and appearance of life. In: Hélène C (ed) Structure, dynamics, interactions and evolution of biological macromolecules. D. Reidel Publishing, Dordrecht Boston, pp 383–394

Stöbel A-O (1985) Magnetic field induced enantioselective synthesis on the example of trans-stilbenoxide (in German). Diplomarbeit, University of Bremen

Stoks PG, Schwartz AW (1979) Uracil in carbonaceous meteorites. Nature 282:709–710

Strasdeit H (2005) New studies on the Murchison meteorite shed light on the pre-RNA world. ChemBioChem 6:801–803

Sutherland JD (2007) Looking beyond the RNA structural neighborhood for potentially primordial genetic systems. Angew Chem Int Ed 46:2354–2356

Szabó-Nagy A, Keszthelyi L (1999) Demonstration of the parity-violating energy difference between enantiomers. Proc Natl Acad Sci USA 96:4252–4255

Szopa C, Sternberg R, Coscia D, Cottin H, Raulin F, Goesmann F, Rosenbauer H (1999) Gas chromatography for in situ analysis of a cometary nucleus. J Chromatogr A 863:157–169

Szopa C, Sternberg R, Coscia D, Raulin F, Vidal-Madjar C (2000) Gas chromatography for in situ analysis of a cometary nucleus II. J Chromatogr A 904:73–85

Szopa C, Meierhenrich U, Coscia D, Janin L, Goesmann F, Sternberg R, Brun J-F, Israel G, Cabane M, Roll R, Raulin F, Thiemann W, Vidal-Madjar C, Rosenbauer H (2002) Gas chromatography for in situ analysis of a cometary nucleus IV. Study of capillary column robustness for space application. J Chrom A 982:303–312

Szopa C, Goesmann F, Rosenbauer H, Sternberg R, the COSAC team (2007) The COSAC experiment of the Rosetta mission: Performance under representative conditions and expected scientific return. Adv Space Res 40:180–186

Takahashi J, Masuda H, Kaneko T, Kobayashi K, Saito T, Hosokawa T (2005) Photochemical abiotic synthesis of amino-acid precursors from simulated planetary atmospheres by vacuum ultraviolet light. J Appl Phys 98:024907

Takano Y, Horiuchi T, Kobayashi K, Marumo K, Urabe T (2003a) Large enantiomeric excesses of L-form amino acids in deep-sea hydrothermal sub-vent of 156 °C fluids at the Suiyo Seamount, Izu-Bonin Arc, Pacific Ocean. Chem Lett 32:970–971

Takano Y, Ushio K, Kaneko T, Kobayashi K (2003b) Amino acid precursors from carbon monoxide in simulated interstellar dust ice mantle by UV irradiation at 10 K. Chem Lett 32:612–613

Takano Y, Sato R, Kaneko T, Kobayashi K, Marumo K (2003c) Biological origin for amino acids in a deep subterranean hydrothermal vent, Toyoha mine, Hokkaido, Japan. Org Geochem 34:1491–1496

Takano Y, Kobayashi K, Yamanaka T, Marumo K, Urabe T (2004a) Amino acids in the 308 °C deep-sea hydrothermal system of the Suiyo Seamount, Izu-Bonin Arc, Pacific Ocean. Earth Planet Sci Lett 219:147–153

Takano Y, Ohashi A, Kaneko T, Kobayashi K (2004b) Abiotic synthesis of high-molecular-weight organics from an inorganic gas mixture of carbon monoxide, ammonia, and water by 3 MeV proton irradiation. Appl Phys Lett 84:1410–1412

Takano Y, Takahashi J, Kaneko T, Marumo K, Kobayashi K (2007) Asymmetric synthesis of amino acid precursors in interstellar complex organics by circularly polarized light. Earth Planet Sci Lett 254:106–114

Tang YC, Stauffer M (1995) Formation of pristene, pristane and phytane: kinetic study by laboratory pyrolysis of Monterey source rock. Org Geochem 23:451–460

ten Haven HL, de Leeuw JW, Rullkotter J, Sinninghe Damsté JS (1987) Restricted utility of the pristane/phytane ratio as a palaeoenvironmental indicator. Nature 330:641–643

Teutsch H. (1988) The origin of optical activity in the biosphere: selected photoreactions for asymmetric syntheses (in German). Ph.D. thesis, University of Bremen

Teutsch H, Thiemann W (1986) Asymmetric photoreactions as a model for evolution of chirality. Orig Life 16:420

Teutsch H, Thiemann W (1990) Possible amplification of enantiomer excess through structural properties of liquid crystals – a model for origin of optical activity in the biosphere? Orig Life 20:121–126

Thiemann W (1974) Disproportionation of enantiomers by precipitation. J Mol Evol 4:85–97

Thiemann W (1975) Is the detection of optical activity in extraterrestrial samples a safe indicator for life? In: Sneath PHA (ed) Life Sciences and Space Research XIII. Proceedings of the Open Meeting of the Working Group on Space Biology of the 7th Plenary Meeting of COSPAR, Sao Paulo, Brazil, June 1974. Akademie-Verlag, Berlin, pp 63–69

Thiemann W (1984) Speculations and facts on the possible inductions of chirality through Earth magnetic field. Orig Life 14:421–426

Thiemann WH-P (1998) Homochirality of the evolution of biospheres. Biol Sci Space 12:73–77

Thiemann W, Bredehöft JH (2007) The origin of homochirality. In: Navarro-Gonzales R, Basiuk V (eds) Astrobiology. American Scientific Publishers, Los Angelos, in print

Thiemann W, Darge W (1974) Experimental attempts for the study of the origin of optical activity on Earth. Orig Life 5:263–283

Thiemann W, Jarzak U (1981) A new idea and experiment related to the possible interaction between magnetic field and stereoselectivity. Orig Life Evol Biosphere 11:85–92

Thiemann WH-P, Meierhenrich U (2000) Future space experiment will concern the origin of homochirality. Orig Life Evol Biosphere 30:220

Thiemann WH-P, Meierhenrich U (2001) ESA mission ROSETTA will probe for chirality of cometary amino acids. Orig Life Evol Biosphere 31:199–210

Thiemann W, Teutsch H (1990) Possible amplification of enantiomer excess through structural properties of liquid crystals – a model for origin of optical activity in the biosphere? Orig Life Evol Biosphere 20:121–126

Thiemann W, Wagener K (1970) Is there an energy difference between enantiomorphic structures? Angew Chem Int Ed 9:740–741; Angew Chem 82:776–777

Tisse S, Peulon-Agasse V, Cardinaël P, Bouillon J-P, Combret J-C (2006) Capillary gas chromatographic properties of three new mono-ester permethylated β-cyclodextrin derivatives. Anal Chim Acta 560:207–217

Tranter GE (1985a) Parity-violating energy differences of chiral minerals and the origin of biomolecular homochirality. Nature 318:172–173

Tranter GE (1985b) The parity-violating energy difference between enantiomeric reactions. Chem Phys Lett 115:286–290

Ulamec S, Block J, Fenzi M, Feuerbacher B, Haerendel G, Hemmerich P, Maibaum M, Rosenbauer H, Schiewe B, Schmidt HP, Schütze R, Wittmann K (1997) RoLand: A long-term lander for the Rosetta mission. Space Technology 17:59–64

Ulbricht TLV, Vester F (1962) Attempts to induce optical activity with polarized β-radiation. Tetrahedron 18:629–637

van de Meent D, Brown SC, Philp RP, Simoneit BRT (1980) Pyrolysis-high resolution gas chromatography and pyrolysis gas chromatography-mass spectrometry of kerogens and kerogen precursors. Geochim Cosmochim Acta 44:999–1013

van der Velden W, Schwartz AW (1977) Search for purines and pyrimidines in the Murchison meteorite. Geochim Cosmochim Acta 41:961–968

van Graas G, de Leeuw JW, Schenck PA, Haverkamp J (1981) Kerogen of Toarcian shales of the Paris Basin. A study of its maturation by flash pyrolysis techniques. Geochim Cosmochim Acta 45:2465–2474

van House J, Rich A, Zitzewitz PW (1984) Beta decay and the origin of biological chirality: new experimental results. Orig Life 14:413–420

Vester F (1974) The (hi)story of the induction of molecular asymmetry by the intrinsic asymmetry in β-decay. J Mol Evol 4:1–13

Viedma C (2001) Enantiomeric crystallization from DL-aspartic and DL-glutamic acids: Implications for biomolecular chirality in the origin of life. Orig Life Evol Biosphere 31:501–509

Viedma C (2005) Chiral symmetry breaking during crystallization: complete chiral purity induced by nonlinear autocatalysis and recycling. Phys Rev Lett 94:065504/1–065504/4

Wächtershäuser G (1991) Biomolecules: The origin of the optical activity. Med Hypothesis 36:307–311

Wächtershäuser G (1992) Groundworks for an evolutionary biochemistry: the iron-sulphur world. Prog Biophys Molec Biol 58:85–201

Wada T, Inoue Y (2004) Supramolecular asymmetric photoreactions. In: Inoue Y, Ramamurthy V (eds) Chiral photochemistry. Marcel Dekker, New York, pp 341–384

Wald G (1957) The origin of optical activity. Ann NY Acad Sci 69:352–368

Walker JCG, Hays PB, Kasting JF (1981) A negative feedback mechanism for the long-term stabilization of Earth's surface temperature. J Geophys Res 86:9776–9782

Wang W, Yi F, Ni Y, Zhao Z, Jin X, Tang Y (2000) Parity violation of electroweak force in phase transitions of single crystals of D- and L-alanine and valine. J Biol Phys 26:51–65

Wang W, Min W, Liang Z, Wang L-Y, Chen L, Deng F (2003) NMR and parity violation: low-temperature dependance in ^{1}H CRAMPS and ^{13}C CP/MAS ssNMR spectra of alanine enantiomer. Biophys Chem 103:289–298

Ward PD, Brownlee D (2000) Rare Earth – why complex life is uncommon in the universe. Springer, Berlin Heidelberg New York

Weber P, Greenberg JM (1985) Can spores survive in interstellar space? Nature 316:403–407

Weinberg S (1980) Conceptual foundations of the unified theory of weak and electromagnetic interactions. Rev Mod Phys 52:515–523

Weissbuch I, Addadi L, Berkovitch-Yellin Z, Gati E, Lahav M, Leiserowitz L (1984) Spontaneous generation and amplification of optical activity in α-amino acids by enantioselective occlusion into centrosymmetric crystals of glycine. Nature 310:161–164

Wesendrup R, Laerdahl JK, Compton RN, Schwerdtfeger P (2003) Biomolecular homochirality and electroweak interactions. The Yamagata hypothesis. J Phys Chem A 107:6668–6673

Woese C (1998) The universal ancestor. Proc Natl Acad Sci USA 95:6854–6859

Woon DE (2002) Pathways to glycine and other amino acids in ultraviolet-irradiated astrophysical ices determined via quantum chemical modeling. Astrophys J 571:L177–L180

Wu C (2007) Secrets of the Martian soil. Nature 448:742–744

Wu CS, Ambler E, Hayward RW, Hoppes DD, Hudson RP (1957) Experimental test of parity conservation in beta decay. Phys Rev 105:1413–1415

Yamagata Y (1966) A hypothesis for the asymmetric appearance of biomolecules on Earth. J Theoret Biol 11:495–498

Yokoyama Y, Saito M (2004) Chirality in photochromism. In: Inoue Y, Ramamurthy V (eds) Chiral photochemistry. Marcel Dekker, New York, pp 235–260

Zadel G, Eisenbraun C, Wolff G-J, Breitmaier E (1994) Enantioselective reactions in a static magnetic field. Angew Chem Int Ed 33:454–456; Enantioselektive Reaktionen im statischen Magnetfeld. Angew Chem 106:460–463

Zepik H, Shavit E, Tang M, Jensen TR, Kjaer K, Bolbach G, Leiserowitz L, Weissbuch I, Lahav M (2002) Chiral amplification of oligopepdides in two-dimensional crystalline self-assemblies on water. Science 295:1266–1269

Index

Printing: Krips bv, Meppel, The Netherlands
Binding: Stürtz, Würzburg, Germany